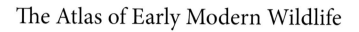

The Atlas of Early Modern Wildlife

THE ATLAS oF
EARLY MoDERN
WILDLIFE

BRITAIN AND IRELAND BETWEEN THE MIDDLE AGES
AND THE INDUSTRIAL REVOLUTION

LEE RAYE

PELAGIC PUBLISHING

Published in 2023 by
Pelagic Publishing
20–22 Wenlock Road
London N1 7GU, UK

www.pelagicpublishing.com

*The Atlas of Early Modern Wildlife: Britain and Ireland between
the Middle Ages and the Industrial Revolution*

https://doi.org/10.53061/DDBT1693

A CIP record for this book is available from the British Library

ISBN 978-1-78427-407-8 Hbk
ISBN 978-1-78427-408-5 ePub
ISBN 978-1-78427-409-2 PDF

Typeset by BBR Design, Sheffield

Front cover image: A Wolf from *The History of Four-Footed Animals
and Serpents* by Edward Topsell, E. Cotes, 1658.
Spine image: A Golden Eagle from *The Ornithology of
Francis Willughby* edited by John Ray, Royal Society, 1678.
Back cover images: A Rabbit from *The History of Four-Footed Animals and
Serpents* by Edward Topsell, E. Cotes, 1658; a Burbot from *Historia Piscium*
by Francis Willughby and John Ray, Royal Society, 1686.
Cover background photo: The Highlands, Scotland by K. Mitch Hodge on Unsplash.

Printed in the UK by Short Run Press Ltd

To Laurie, who will be pleased to find that this book is now pogge-free.

In memory of Pauline Allen (1940–2022).

CONTENTS

Acknowledgements

Considering that this book is a historical atlas, it is ironic how difficult it is to trace back when and where it was inspired. The seed of the idea might have been planted in 2015, as I walked around the newly created Natterjack Toad ponds at the RSPB Lodge Nature Reserve in Bedfordshire, and chatted with the insightful Reserves Archaeologist about the need for an authoritative single resource on species history. The idea was certainly actively growing when the RSPB's *State of Nature* report came out the next year and compared the present and past distribution of species in different parts of the UK. The project was planned close to its present form by 2017, when a learned professor at the University of Bristol helped me develop the idea into a research proposal. That research proposal was ultimately unsuccessful, but the work has been enjoyable regardless.

However, I think the germ of the idea has deeper roots then that. Ultimately, this atlas just maps old records of wildlife from historical sources. The tools used here to compile and analyse and present the data are new, but the data itself has been waiting to be collated for centuries, as we shall see.

The creation of the atlas itself has been supported by many individuals over the last few years. Thanks are due to the Society for the History of Natural History for providing two grants from their Small Research Fund, once in 2020 to obtain a copy of a manuscript by John Hooker from the South West Heritage Trust in Devon Archives, and again in 2021 to visit the unpublished archives of William Henry at the Public Record Office of Northern Ireland. I also especially need to thank Trinity College Dublin Archives for providing copies of Samuel Molyneux's notes on natural history (TCD MS 883.1). Access to historical sources provided by Early English Books Online, Google Books, the Internet Archive, the Biodiversity Heritage Library, Eighteenth Century Collections Online, the National Library of Scotland, the National Library of Wales, The Open University Library, Cardiff University Library and Swansea University Library has also been absolutely essential for the success of this project.

I am very fortunate to have been accepted into some local, national and international communities, and this atlas has also benefitted from day-to-day conversations within these communities. I am grateful for inspiring discussions with the Animal History Group and the Society for the History of Natural History, focused work with the Open University Writing Club and excited conversations with Cardiff University biodiversity gang, and with other wildlife and history enthusiasts on Twitter. I have also very much benefitted from reading the work of scholars in the NiCHE group, especially those focused on HGIS.

Whilst writing the atlas I worked on two papers for *Mammal Communications* and *Bird Study*, disclosing some of my early results. I am especially grateful to the editors and reviewers there for their kind and thoughtful comments. I would also like to especially acknowledge the help and support from Nigel, David and Sarah at Pelagic Publishing.

Finally, I need to thank a few individuals not included above for providing feedback and help with the atlas as I was working on it: Pauline Allen, Kieran Buckley, Ian Carter, Frazer Coomber, Charlie Cooper, Peter Cooper, Peter Evans, Julie Ewald, Derek Gow, Lauren Hartny-Mills, Alen Knox, Charlie Le Marquand, Jack Elliot Marley, Jim Martin, Alec Moore, Stephen Newton, Laurie Raye, Simon Sanghera, Richard Seaby, Dhruti Shah, Robin Somes, Harvey Tweats, Rene Winkler.

With special apologies to those I have inevitably forgotten – your work was not in vain!

All mistakes are my own.

COPYRIGHT PERMISSIONS

The maps presented in this atlas are based on data taken directly from historical primary sources.

The county borders on the maps are based on freely available data by wikishire.co.uk (2021). These files contain border data provided by the Historic County Borders Project © Historic Counties Trust. May contain Ordnance Survey data © Crown credit and database right 2014–18. May contain public sector information licensed under the Northern Ireland Open Government Licence. May contain public sector information licensed under the Isle of Man Open Government Licence. May contain Ordnance Survey Ireland data licensed under the Creative Commons 4.0 Licence.

The inset maps that show the modern distribution of each species are very simplified versions of copyrighted distribution data provided in other sources. All are used with permission from the copyright holders:

- *Mammals of Europe, North Africa and the Middle East* (Aulagnier et al. 2018).
- *The Atlas of the Mammals of Great Britain and Northern Ireland* (Crawley et al. 2020).
- The cetacean fact sheets by the Sea Watch Foundation (Sea Watch Foundation 2020).
- *The RSPB Handbook of British Birds* (Holden and Gregory 2021).
- *Identification Guide to the Inshore Fish of the British Isles* (Henderson 2014).
- *Freshwater Fishes in Britain* (Davies et al. 2004).
- *Amphibians and Reptiles* (Beebee 2013).
- *The Diver's Guide to Marine Life of Britain and Ireland* (Wood 2018).

The species pictures are all from the early modern period. All are in the public domain. The sources are:

- *Ornithology* (Ray 1678) for all of the birds.
- *Historia Piscium* (Willughby & Ray 1686) for most of the fishes.
- *History of Four-Footed Animals and Serpents* (Topsell 1658) for the frogs and most of the mammals.
- *The Natural History of Cornwall* (Borlase 1758) for the Dolphin, Porpoise, Adder, Slow Worm, Leatherback Turtle, Limpet, Scallop and sea urchins.
- *The Natural History of Lancashire, Cheshire and the Peak in Derbyshire* (Leigh 1700) for the cockle, the Cuttlefish and the seal.
- *Historiae sive Synopsis Methodicae Conchyliorum* (Lister 1752) for the Common Mussel.
- *Phalainologia nova* (Sibbald 1692) for the Sperm Whale.

INTRODUCTION

This book maps and discusses where mammals, birds, fishes, amphibians, reptiles, molluscs, crustaceans and echinoderms were recorded in Britain and Ireland before the Industrial Revolution. By piecing together very old records of wildlife provided by naturalists, geographers, travellers and historians it is possible to build up a detailed idea of the state of nature between 250 and 500 years ago.

There are two common traps that we can walk into when we think about wildlife in the past. The first is to fall prey to shifting baseline syndrome. This is to assume that the species that we see and remember around us are the same species that have always been present in the area. Studying historical sources can help with this. As we shall find, in Britain and Ireland, the early modern period was the last great age of the Wolf, the Sea Eagle, the Red Kite and the Bustard. Brown Rats and Grey Squirrels had yet to become established, but Wildcats and Pine Martens wandered through the woods. Rabbits and Herring Gulls had not finished moving inland but Sturgeon and Burbot still swam in the rivers. All of this has changed over the last 250 years.

But there is a second common trap. This is to assume that before the modern period the natural world existed in a pristine, harmonious state. Studying historical sources can help with this too. Similar to today, life in early modern Britain and Ireland was defined by a climate anomaly ('The Little Ice Age') which was at its height between 1550 and 1700 CE. This resulted in regular crop failures and famines. Between 1695 and 1704 the Little Ice Age lowered the temperature of the North Sea to the extent that Cod and Herring fisheries failed. Even aside from the climate, some species around these islands were doing poorly anyway. Lynxes, Roe Deer, Beavers, Red Squirrels, Cranes, Capercaillies and Great Auks were already declining due to hunting. Brown Bears, Wild Boar, Right Whales and others already seem to have been extinct, probably for the same reason. It is important not to undermine the almost unprecedented seriousness of our modern biodiversity and climate crises, but coming to a more sophisticated understanding of the history of nature can provide us with the context to properly understand our modern challenges.

SCOPE

The geographical scope of this atlas is Britain and Ireland, as shown in Figure 1. That covers the island of Great Britain (including mainland Scotland, Wales and England), the island of Ireland (including both what is now the Republic of Ireland and Northern Ireland) and other neighbouring British islands such as the Isle of Man, the Northern Isles (Shetland and Orkney), the Hebrides (including St Kilda), the Isles of Scilly, the Channel Islands (including Jersey and the Bailiwick of Guernsey), as well as all the islands closer to the British and Irish mainlands.

In this atlas, Britain and Ireland are further subdivided into smaller regions, to facilitate regional analysis of the data. These include: (i) Lowland Scotland, (ii) Highland Scotland, (iii) Ulster, (iv) Connacht, (v) Leinster, (vi) Munster, (vii) Wales, (viii) South West England, (ix) South England, (x) the Midlands of England and (xi) North England. When it comes to the regional analysis, records from the Scilly Isles and Channel Islands are included with South West England, records from the Isle of Man are included with Lowland Scotland, and records from the Hebrides and the Northern Isles are included with Highland Scotland.

This scope is historically uncomfortable. Britain and Ireland are not and perhaps never have been a politically harmonious territory. The inhabitants have different histories, cultures and languages. The future constitutional status of parts of these islands is uncertain. These tensions, of course, play

Figure 1 Map showing the geographical range of this atlas.

out in the sources. To write a description of an area is to claim to be an authority there, so writing natural history is a political act. To pick one example, the *Natural History of Ireland* (O'Sullivan 2009) was written by the Irish Catholic Philip O'Sullivan Beare from exile in Spain following the Tudor Conquest of Ireland and was finished around 1626.[1] It was expressly written to refute the poorly informed natural history written by Gerald of Wales following the Anglo-Norman Invasion of Ireland centuries earlier.

The reason the Atlantic Archipelago is studied together here is that Britain and Ireland share a biodiversity shaped by thousands of years of separation from mainland Europe and similar fashions for introducing and persecuting animals. Studying these islands together also makes the contrast between them more obvious, as this atlas will show. The success of volumes like the *New Atlas of the British and Irish Flora* (Preston et al. 2002), *The Bird Atlas, 2007–11* (Balmer et al. 2013) and the *Mammals*

---

1   Note that the dates given in citations of primary texts in this volume are the dates when the text was edited for publication. These are often centuries after the texts themselves were written.

of the British Isles Handbook (ed. Harris and Yalden 2008) attest to this approach. If this atlas must be political, it is at least apologetically so.

The temporal scope of the atlas is the early modern period. This is defined differently by different historians. For this volume, the start in Britain and Ireland is considered to be with the renaissance naturalists contemporary with Conrad Gessner (1516–1565), such as John Caius (1510–1573) and William Turner (1509/10–1568). It must especially include the work of the Royal Society naturalists like John Ray (1627–1705) and Martin Lister (d. 1712). The end of the early modern period and the beginning of the late modern period can probably be associated with the scholars who followed Carl Linnaeus (1707–1778) and adopted his system of nomenclature, such as Thomas Pennant (1726–1798), Patrick Browne (*c.*1720–1790) and Gilbert White (1720–1793) (Allen 1976: 40–3; see Fisher 1966: 61–5). There are, however, a few peripheral texts which have been accepted into the database (Raye 2023). These include the accounts published by and in the pattern of the Physico-Historical Society in Dublin (ed. Timoney 2013; ed. Boyd 1974; ed. King 1892; Rutty 1772; Smith 1746, 1750a,b, 1756; Harris and Smith 1744) as described by Magennis (2002), and the natural history of Northumberland by Wallis (1769). I have included them as early modern sources for two main reasons: (i) these accounts are unlikely to be considered by modern naturalists, and (ii) there are few reliable accounts of those areas written earlier than this. The primary sources I have included in the database therefore span the period 1519–1772, although very few texts are included from after 1760.

EARLY MODERN NATURAL HISTORY

In the second half of the seventeenth century, Robert Boyle, in co-ordination with other fellows of the Royal Society of London, started a citizen science recording project. The project's method was the distribution of a questionnaire: 'General heads for the natural history of a country'. The questionnaire was intended to be so thorough that any 'Gentlemen, Seamen and others' visiting an unfamiliar area could use it to obtain enough data so that even naturalists who had never been there could write a local natural history about it (Boyle 1666, 1692). The use of questionnaires or 'general heads' for natural history research was first popularised by Francis Bacon, so the genre of work produced using this method now sometimes called 'Baconian' natural history (Yale 2016; Fox 2010; Cooper 2007: 114–40; Hunter 2007) – although surveys had been used to gather information about natural resources before this (Viana et al. 2022; Yalden 1987). The data gathered on this project can still be used today to help modern historians and naturalists reconstruct natural history as it existed in the early modern period.

Many of the sources used for this atlas can be considered Baconian natural histories. This is not to say their authors lacked any field skills – on the contrary, from Aubrey to Morton the authors generally continued to write about their local areas and drew principally on their own experience (Emery 1958). However, the accounts were also sometimes based on crowd-sourced data. Other versions of the natural history questionnaire were circulated in Britain by various authors through the rest of the seventeenth and eighteenth centuries (Fox 2010). Most of them were either unsuccessful in gathering data or the results were lost, but most of the local data on wildlife that was published, and some that survived in manuscript form, has been drawn into this atlas.

This means that many of our early modern texts share an unusual feature: the recorders who compiled them knew that they were creating primary sources. It was part of the original design of these texts that the natural history data produced would eventually be synthesised and furnish evidence for speculations about the natural world (Yale 2016: 4–5; Boyle 1692). A first attempt was even made to synthesise the English regional sources (along with the county histories and genealogies etc.) into the six volumes of *Magna Britannia Antiqua & Nova* (1720–1738). Unfortunately, from a natural history perspective this was unsatisfactory, since only around 150 of our 10,000+ records were included in this text and (surprisingly for a book entitled *Magna Britannia*) the text omitted Scotland, and of course did not include Ireland.

The synthesis of the local natural history data into a single atlas might, therefore, have been seen by the early modern naturalists positively, as a partial accomplishment of the project of Baconian local natural history. Smith describes the ultimate aim as this:

> humbler minds must still be contented to assist only in collecting materials, and as it were to dig in the quarries of nature, until more such spirits as they [Francis Bacon and Isaac Newton] shall arise to make some use of their drudgery, in erecting such a system of natural inquiries.
> (Smith 1756: 233)

But there is a limit to how far this atlas fulfils the dreams of the Baconian naturalists. The book is about exclusively wild animals, rather than domesticated animals, plants, minerals, fuels, technologies, recipes, medicines and other natural resources which would have seemed more obvious objects of study to the early modern naturalists. Many readers of this atlas are also likely to be people who love nature for its own sake, and who have an interest in its conservation. This would have seemed a peculiar view of nature to the early modern 'naturalists', who often only valued nature to the extent that it could be exploited for human use.

Discussion of Baconian local natural history also must contend with the genre's problematic legacy. When benefiting from these sources we must acknowledge that the early modern data-gathering that enabled the creation of this atlas also set the stage for acts of colonialism across the world which have had a long-lasting legacy (Kumar 2017). This problematic legacy is shared even by documents written within and about Britain and Ireland. Robert Payne's *A brief description of Ireland* (ed. 1841) provides important records of the Grey Partridge and Black Grouse in Ireland in the sixteenth century, but was also written to encourage other English people to settle in the plantation of Munster. This plantation prefixed the replacement of Irish landowners with English adventurers during the transplantation to Connacht a century later (Cunningham 2011). In the New World in particular, natural history collecting was also funded by the slave trade and facilitated using networks set up by slave traders and plantation owners (Das and Lowe 2018; Roos 2019: 77–9). Between 1665 and 1700, there was at least one report on the natural history of the New World in every issue of *Philosophical Transactions*, the journal of the Royal Society of London (Irving-Stonebraker 2019). The slave trade was less relevant to the collection of data from early modern Britain and Ireland, but this is to draw a modern distinction. When the most important list of fishes from early modern Cornwall (George Iago's *Catalogus quorandam piscium rariorum*) was published in John Ray's *Historia Piscium* (1713a), it appeared immediately after descriptions of the fishes of Jamaica and the fishes of the Antillean islands. There is also a considerable amount of xenophobia and English exceptionalism within the early modern sources considered for this atlas, to the extent that descriptions of Scotland, Ireland and Wales written by English writers are sometimes deeply prejudiced and uncomfortable to read.

KEY SOURCES

The *Atlas of Early Modern Wildlife* is based on data from more than 200 sources published between 1519 and 1772 CE. These sources describe 900+ species, of which 151 are mapped in this volume and provide 10,000+ records, of which around 3,000 are the top-quality records used for statistical analysis (as explained in Figure 2). The raw data consulted for the atlas has been deposited online and can be consulted for free (Raye 2023).

The most important of the early modern sources are the local natural histories. These key sources describe the natural resources of a specific area – usually a single historical county or small group of islands – and they usually provide the most reliable and specific records.

Figure 2 Graph visualising the database (Raye 2023) used for this atlas.

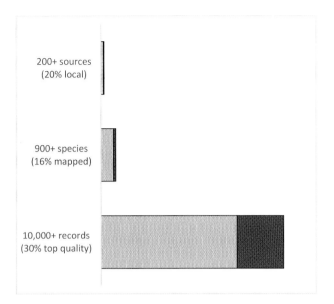

Although the rest of this atlas will be focused on wildlife, it is important to start by considering the human authors and the sources they produced, as many of the conclusions rest on their evidence. The following is a list of the key sources, sorted by the region and by the number of species mentioned:

LOWLAND SCOTLAND AND THE ISLE OF MAN

1. *The History, Ancient and Modern, of the Sheriffdoms of Fife and Kinross* (1710). The author Robert Sibbald (1641–1722) was a physician based in Edinburgh. Edition consulted: Sibbald 1803. Refers to 177 local species.
2. *An Account of the Fishes and other Aquatic Animals taken in the Firth of Forth* (1701). Manuscript consulted: Sibbald 1701. Refers to 114 local species. This is the same Robert Sibbald that wrote *The history, ancient and modern, of the Sheriffdoms of Fife and Kinross* (Key Source #1).
3. *Large Description of Galloway* (1684). The author Andrew Symson (*c.*1638–1712) was a member of the clergy based between Wigtownshire and Perthshire. Edition consulted: Symson 1823. Refers to 35 local species.
4. *An Account of the Isle of Man* (1702). The author William Sacheverell (*c.*1664–1715) was a civic officer based on the Isle of Man. Edition consulted: Sacheverell 1859. Refers to 30 local species.

HIGHLAND SCOTLAND AND THE NORTHERN ISLES AND HEBRIDES

5. *Description of the Western Isles* (1703). The author Martin Martin (d. 1718) was a teacher based between Skye and London. Editions consulted: Martin and Monro 2018; Martin 1703. Refers to 85 local species.
6. *An Account of the Islands of Orkney* (1700). The author James Wallace (born 1684) grew up on Orkney, edited this text from the *Description of the Isles of Orkney* written by his father (#9). Edition consulted: Wallace 1700. Refers to 61 local species.
7. *Historical Description of the Zetland Islands in the Year 1733* (1733). The author Thomas Gifford (d. 1760) was a civic officer based in Shetland. Edition consulted: Gifford 1879. Refers to 54 local species.
8. *The Description of the Islands of Orknay and Zetland* (1633). The author Robert Monteith was based in Orkney. Edition consulted: Monteith 1711. Refers to 54 local species.

9. *A Description of the Isles of Orkney* (1693). The author James Wallace (1642–1688) was a member of the clergy based in Orkney. Edition consulted: Wallace 1693. Refers to 53 local species. This is the original account of James Wallace senior, which was amended in 1700 for the *Account of the Islands of Orkney* (#6).

10. *A Brief Description of Orkney, Zetland, Pightland-Firth & Caithness* (1701). The author, John Brand (1669–1738), was a member of the clergy based in West Lothian. Edition consulted: Brand 1701. Refers to 52 local species.

11. *Genealogical History of the Earldom of Sutherland* (1630–51). The author Robert Gordon (1580–1656) was an aristocrat based between Sutherland and Wiltshire. Edition consulted: Gordon 1813. Refers to 50 local species.

12. *A Late Voyage to St Kilda* (1698). Editions consulted: Martin and Monro 2018; Martin 1698. Refers to 32 local species. The author was the same Martin Martin that wrote the *Description of the Western Isles* (#5).

ULSTER

13. *The Antient and Present State of the county of Down* (1744). The authors were Walter Harris (1686–1761), a pensioned historian based in Dublin, and Charles Smith, who wrote *The antient and present state of the county and city of Cork* (#20, also #21 and #23). Edition consulted: Harris and Smith 1744. Refers to 67 local species.

14. *The Description of Ardes Barony* (1683). The author William Montgomery (1633–1707) was a gentleman based in Co. Down. Edition consulted: Montgomery 1896. Refers to 46 local species.

15. *Hints towards a Natural and Typographical History of the Counties Sligoe, Donegal, Fermanagh and Lough Erne* (1739). The author William Henry (d. 1768) was a member of the clergy based in Co. Tyrone. The editions are incomplete (Timoney 2013; Simms 1960; King 1892), the full version is unpublished (Henry 1739a). Refers to 41 local species.

16. *Topographical Description of the Coast of County Antrim and North Down* (1739). Edition: Boyd 1974 is incomplete, most of the records are only contained in the unpublished extended version of the text (Henry 1739b). Refers to 39 species. The author was the same William Henry that published *Hints towards a natural and typographical history of the counties Sligoe, Donegal, Fermanagh and Lough Erne* (#15).

17. *Description of the County of Antrim* (1683). The author Richard Dobbs (1634–1701) was a civic officer based in Co. Antrim. Edition consulted: Hill 1873: 376–389. Refers to 31 local species.

CONNACHT

18. *Chorographical Description of West or h-Iar Connaught* (1684). The author Roderic O'Flaherty (*c.*1630–*c.*1717) was from an aristocratic family and based in Iar Connacht. Edition consulted: O'Flaherty 1846. Refers to 59 local species.

LEINSTER

19. *An Essay towards a Natural History of the County of Dublin* (1772). The author John Rutty (1698–1775) was a physician based in Dublin. Edition consulted: Rutty 1772. Refers to 223 local species.

MUNSTER

20. *The Antient and Present State of the County and City of Cork* (1750). The author Charles Smith (*c.*1715–1762) originally worked as an apothecary in Co. Waterford and by this point was based in Dublin. Edition consulted: Smith 1750a,b. Refers to 187 local species.

21. *The Ancient and Present State of the County and City of Waterford* (1746). Edition consulted: Smith 1746. Refers to 116 local species. This is the same Charles Smith that wrote *The antient and present state of the county and city of Cork* (Key Source #20, also #13 and #23).

22. *Propugnaculum Catholicae Veritatis* (1669). The author Anthony Bruodin (d. 1680) was a friar based in Prague, who grew up in Co. Clare. Edition consulted: MacBrody 1669: 825–7, 955–9 esp. There is a partial translation (O'Dalaigh 1998). Refers to 42 local species, but with no site-level records.

23. *The Antient and Present State of the County of Kerry* (1756). Edition consulted: Smith 1756. Refers to 35 local species. The author was the same Charles Smith that wrote *The antient and present state of the county and city of Cork* (#20, also #13 and #21).

WALES

24. *Description of Pembrokeshire* (1603). The author George Owen of Henllys (1552–1613) was a civic officer based in Pembrokeshire. Edition consulted: Owen 1994. Refers to 100 local species.

25. *A History of the Island of Anglesey* (1763). The author John Thomas (d. 1769) was a member of the clergy based in Caernarvonshire and Anglesey. Edition consulted: Dodsley 1775. See Ramage (1987: 263–4) for author and date. Refers to 42 species.

SOUTH WEST ENGLAND AND THE SCILLY ISLES AND CHANNEL ISLANDS

26. *The Natural History of Cornwall* (1758). The author William Borlase (1696–1772) was a member of the clergy based in Cornwall. Edition consulted: Borlase 1758. Refers to 144 local species.

27. *Survey of Cornwall* (1602). The author Richard Carew (1555–1620) was a gentleman based in Cornwall. Edition consulted: Chynoweth et al. 2004. Refers to 134 local species.

28. *A View of Devonshire in MDCXXX* (1630). The author Thomas Westcote (bap. 1567–*c*.1637) was a gentleman based in Devon. Edition consulted: Westcote 1845. Refers to 74 local species.

29. *The Description of the Hundred of Berkeley* (1605). The author John Smyth (1567–1641) was a lawyer and steward based between Berkeley and London. Edition consulted: Smyth 1885: 319 esp. Refers to 49 local species.

30. *The Natural History of Wiltshire* (1690–91). The author John Aubrey (1626–1697) was from a gentry family and based in Wiltshire. Edition consulted: Aubrey 1847. Refers to 40 local species.

31. *The History and Antiquities of the County of Dorset* (data gathered around 1739). The author John Hutchins (1698–1773) was a member of the clergy based in Dorset. Edition consulted: most records are Hutchins 1774: lxxvii. Refers to 34 local species.

SOUTH ENGLAND

32. *Notes and Letters on the Natural History of Norfolk* (a retrospective collection from *c*.1662–1668). The author Thomas Browne (1605–1682) was a physician based in Norfolk. Edition consulted: Browne 1902. Refers to 219 local species.

33. *History and Antiquities of Harwich and Dovercourt* (1730). The authors, Silas Taylor (1624–1678) an army officer, and Samuel Dale (bap. 1659–1739) an apothecary and physician, were based in Essex. Edition consulted: Taylor and Dale 1730. Refers to 148 local species.

34. *The Breviary of Suffolk* (1618). The author Robert Reyce (1555–1638) was a gentleman based in Suffolk. Edition consulted: Reyce 1902. Refers to 100 local species, but with county-level records only.

35. *A Description of the River Thames* (1758). The authors Roger Griffiths & Robert Binnell may have been based in London. Edition consulted: Griffiths and Binnell 1758. Refers to 69 local species.

MIDLANDS OF ENGLAND

36. *The Natural History of Northampton-shire* (1712). The author John Morton (1671–1726) was a member of the clergy based in Northamptonshire. Edition consulted: Morton 1712. Refers to 133 local species.

37. *The Natural History of Stafford-shire* (1686). The author Robert Plot (baptised 1640–1696) was a teacher based in Oxford. Edition consulted: Plot 1686. Refers to 44 local species.

NORTH ENGLAND

38. *The Natural History and Antiquities of Northumberland* (1769). The author John Wallis (1714–1793) was a member of the clergy based in Northumberland. Edition consulted: Wallis 1769. Refers to 145 local species.
39. *Natural History of Lancashire, Cheshire and the Peak in Derbyshire* (1700). The author Charles Leigh (1662–1701) was a physician based in Lancashire. Edition consulted: Leigh 1700. Refers to 88 local species.
40. A retrospective collection of journal entries and notes (1692–98). The author Thomas Machell (bap. 1647–1698) was a member of the clergy based in Westmoreland. Edition consulted: Ewbank 1963. Refers to 45 local species.
41. *A Perambulation of Cumberland* (1688). The author Thomas Denton (1637–1698) was a lawyer based in Cumberland. Edition consulted: Denton 2003. Refers to 37 local species.
42. *The Vale-Royall of England or the County Palatine of Chester* (1656). The authors were William Smith (*c.*1550–1618) and William Webb. Edition consulted: Smith and Webb 1656. Refers to 34 local species. An additional four species are mentioned in the (1656) *Treatise on the Isle of Man*, printed in the same edition.

Other early modern local natural histories, which include reliable references to fewer than 30 different, identifiable species of wildlife have also been included in the database (Raye 2023) but are not mentioned in the above list of key sources.

We can make a few observations about the key sources, just from the information given in this list. First, geographically, most of the key sources were written by locals, and they have a decent geographical spread across Britain and Ireland. They are mapped in Figure 4. But at the same time, some of the sources are far more comprehensive than others. In fact, the 11 sources with the least records mention only 171 unique species between them (with many more attestations to the same common species), which is fewer than each than the number of species mentioned in each of the four top key sources individually.

Second, although this atlas deals with the whole of the early modern period, the most reliable evidence is biased towards the late seventeenth and early eighteenth centuries, as shown on Figure 3. In fact, there are no key sources from before 1600, and three quarters of the key sources were written in the last century – between 1675 and 1772. That means that wildlife that declined earlier in the early modern period is only attested in less reliable sources and may be less visible in the atlas. County histories were common in the sixteenth and early seventeenth century too, but these early sources usually omitted wildlife (Emery 1958).

Some of the recorders provide more than one key source: there are two from Robert Sibbald, two from Martin Martin, two from William Henry and – most importantly – Charles Smith contributes to four key sources from Ireland. Even if this had not been the case, the sources lack diversity and are likely to share some biases. Although very little is known about some of the authors on the list, they all published under masculine-sounding names. In fact, if we removed texts authored by people with the first names Robert, John and William, only half of these key sources would be left. All of the authors whose portraits have been preserved appear to be white.

These writers to some extent can be seen as part of the educated elite rather than being representative of society as a whole. In terms of occupation, about a quarter of the authors were physicians and apothecaries; they learned to identify species as part of their training in pharmacology. A second

Figure 3 Graph showing when the key sources were produced.

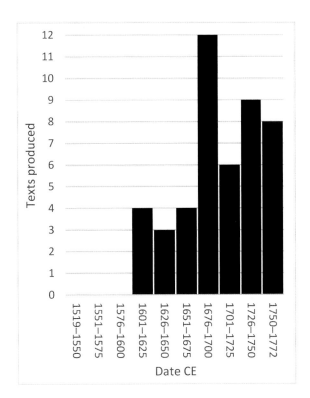

quarter were members of the clergy; these men also held advanced degrees and their contributions were often solicited by members of the Royal Society as representatives of their local areas. Studying nature could also be a way for them to understand God through natural theology. A third quarter were professionals in other fields: teachers, lawyers, army officers and so forth. These people had varying reasons for recording. The final quarter were members of the idle gentry or aristocracy, who lived off inherited wealth by extorting rent from tenancies, or by investing in industry, plantations and the slave trade. These authors had more time to devote to intellectual pursuits and could also show pride in their estates by describing what made them special, alongside accounts of genealogy and the deeds of their ancestors.

This is not to say that people who did not fit this description did not contribute to early modern natural history, it is just that their contributions were not as visible and were quickly forgotten. The Baconian local natural history methodology required writers to draw on ordinary people as informants (see 'Early modern natural history', p. 3). In practice, this increasingly included those living in rural areas, such as many anonymous fishers, farmers and fowlers (Grafton 2018; Ogilvie 2008, chap. 2). Researchers on larger projects sometimes hired amanuenses to help with research and writing. European natural history writers travelling away from home hired local guides and established working relationships with local experts. When they left Europe this regularly involved partnerships with experts of Indigenous, mixed-race and African descent, including enslaved people, as well as with people of European descent (Das and Lowe 2018; Yoo 2018). It was also not only men that contributed to early modern natural history. Well-connected, educated women were sometimes acknowledged as providing data and descriptions of areas to men compiling natural history texts. For example, Anne, Countess of Erroll (1656–1719) contributed ornithological descriptions, plates and two local accounts to Robert Sibbald's (1684) *Scotia Illustrata* (Raye 2019). Martin Lister's daughters Susanna Lister (*c.*1670–1738) and Anna Lister (1671–1700) were prolific natural history illustrators (they drew the picture of the Common Mussel, included on p. 328) and were proficient in the use of a microscope to observe and distinguish specimens (Roos 2018: 91–117). Celia Fiennes (1662–1741), a lady, provided records of 24 species observed on her *Journeys* (ed. 1888). Data recorded by other less well-connected women is likely to be hidden in natural history sources published under men's names.

OTHER SOURCES

In addition to the key sources, the atlas also draws on data from many early modern national natural history texts (Sibbald 2020 [1684]; O'Sullivan 2009 [1626]; Lhuyd [*c.*1696] 1909, 1910, 1911; Morris 1747; Merrett 1666). These texts often used the same Baconian method, but had as their scope the natural history of the whole of England, Scotland, Ireland, Wales, or Britain, and are therefore generally of less use for mapping the regional and local distribution of species. The national texts did, however, draw on correspondence from multiple local naturalists who did not write their own local natural history, and so they do occasionally offer unique site-level records, which can be included on our maps.

Perhaps the most ambitious of the published national natural histories, Robert Sibbald's (1684) *Scotia Illustrata*, alone drew on data from 77 principal local respondents and synthesised these accounts to describe the whole of Scotland (Withers 2001: 256–62). The manuscripts of Sibbald's local source material still survive and have been edited (together with an additional set of responses to MacFarlane in the eighteenth century) by Mitchell in the *Geographical Collections Relating to Scotland* series (1907, 1908).

Some of those that were not finished were even more ambitious. In 1696 Edward Lhuyd is known to have circulated four thousand copies of his questionnaire for *The Natural History of Wales*, but the data was never written up. The surviving returns have been edited as the *Parochialia*, with some later additions (Fox 2010: 603; Emery 1974; Lhuyd 1909, 1910, 1911).

Comparable recording projects were also started elsewhere. From 1682 to 1685 William Molyneux, in association with Roderic O'Flaherty, collected more than 20 county natural histories from at least 14 contributors for his *Natural History of Ireland*. This was originally intended to form part of Moses Pitt's proposed *English Atlas*, a grand atlas of the world, with every part of it described. The project ultimately failed due to financial difficulties, but Irish contributions survive in manuscript form (Molyneux 1685) and many have since been edited (Ó Muraíle 2002; Gillespie and Moran 1991: 207–11; Logan 1971; O'Sullivan 1971; Montgomery 1896; Hill 1873; Hore 1859, 1862a,b; Piers 1786). Records from both the edited and unedited texts have been included individually in our database (Raye 2023).

As well as records from the local and national recorders, the database for this atlas includes early modern travelogues (Defoe 1971 [1724–27]; Fiennes 1888 [1702]; Pococke 1887 [1747–60]; Ray and Derham 1846 [1658–62]; Taylor 1630). It also draws from what we would today call handbooks of natural history (natural history texts describing all the species in a particular group, such as fish or birds (e.g. Ray 1678, 1713a; Willughby & Ray 1686)). The purpose of these texts is to offer an encyclopaedic description of all known species, but several of them provide locations where British and Irish species can be found. Where they include records of wildlife, early modern chronicles and histories (Sutton 2010 [1575]; Holinshed Project 2008a [1577–87]) have also been included in the database. These need to be treated with special care because they frequently rely on previous sources rather than recent field observations, well into the eighteenth century.

The standard of evidence for inclusion in this atlas is high: the database does not include references to species from international handbooks without location information. It also does not include records of dead animals (i.e. meat listed in kitchen accounts, furs listed in export duties), except where it is absolutely clear that the animal has been killed in a particular place (e.g. in the case of the Lestrange household records which occasionally specify when birds were killed locally with gun, crossbow, Sparrowhawk, or hunting hound (Gurney 1832)). This is because animals were being shipped internationally in the early modern period for fur and food, medicine and sport and so a dead animal recorded in a particular place may not have come from that local area (Greenlee 2020, chap. 8; Shrubb 2013: 24; Hoffmann 1994). Evidence not meeting this standard has been discussed in the account for each species, but not included on the maps.

Some early modern natural history texts collected much older records of stranded cetaceans in their local areas. These have not been mapped for this volume, except where the author of the account in question implies that the stranded species still occurred there regularly.

INTERPRETING THE SOURCES

Before we can interpret the data, we first have to decipher the locations and species identities provided in the early modern sources. This is the biggest potential source of errors in creating an atlas. It is often difficult to identify the species intended in the records, in part because they were made before Carl Linnaeus systematised the binomial nomenclature used for species names, and because vernacular names were often generic and could vary from region to region. The age of the records also means it is tricky to identify the sites mentioned in the records because they were made before recorders started giving grid references, latitude and longitude or postcodes. The records were even made before the spelling of local place-names was standardised, and some of the places named no longer exist and cannot be easily mapped. These are challenging problems and they require a multidisciplinary approach involving both ecology and history (see e.g. Sibbald 2020: ix).

Luckily, our early modern sources were aware of these problems and sometimes attempted to pre-empt them. Often our sources provide multiple names for the species and sites they include, and anatomical descriptions and plates showing more unusual species, as well as maps and localities of sites. These features are especially valuable when the records from different sources are collected together, since it is increasingly likely that at least one author will have described the species or site. In addition to this, Linnaeus actually listed some synonyms in *Systema Naturae X* (1758), so certain names used by our early modern British and Irish sources have been identified by Linnaeus as being equivalent to the standard Linnaean name. Finally, later editors of the early modern sources – local historians and modern naturalists – have identified a good deal of the names used. Many of the local names for species have also been identified by the editors of larger dictionaries, especially the *Oxford English Dictionary, Dictionary of the Scots Language, Geiriadur Prifysgol Cymru* and *Dwelly's Illustrated Gaelic to English Dictionary*. Generally, these sources in combination are enough to identify most of the names used. Where the names cannot be securely identified (e.g. where species names are generic, or where the site name is uncertain), this is explained in the 'Recognition' section for each species, and the record is shown on the map as a diamond.

VERNACULAR LANGUAGES

The sources used for this atlas are almost exclusively written in either Early Modern English or New Latin, with a few accounts in Middle Scots and Welsh. However, in the early modern period, Latin (particularly the very academic form New Latin) was only known by the best educated members of society, and Early Modern English was not always the ordinary vernacular language in use in communities. The sources thus often add other vernacular names which the species were actually known by in the places that the authors were writing. The languages used include: Welsh, Irish, Scottish Gaelic, Norn, Late Cornish, Jèrriais and Middle Scots, as well as various dialects of Early Modern English.

This atlas lists all the relevant vernacular terms found in the early modern sources under the 'Recognition' section for each species. For ease of use, the terms are also all listed alphabetically and identified in the index. However, it is also worth briefly making some notes about each language here:

- **Welsh** was widely spoken in Wales in the early modern period, and local and national sources (especially Lhuyd 1909, 1910, 1911; Dodsley 1775; Morris 1747) provide around one hundred (mainly northern) Welsh names of animals. Some of those not referring to species featured in this atlas are (tentatively): *carlwm* for the Stoat, *soccen yr eira* for the Redwing, *tylluan gorniog* for the Long-eared Owl, *tylluan wen* for the Barn Owl, *darfen* for the Dace and *swttan*, *cod lwyd* and *bacod y melinydd* for the Pouting.

- **Middle Scots** was very widely spoken in early modern Scotland and the Ulster Plantation in Ireland. More than a dozen authors offer species names in the language (Sibbald 1803, esp. 2020; Mitchell 1906; Monro 1774; Brand 1701). However, it is difficult to calculate the number of Scots terms provided in sources written in English because there was (and is) considerable overlap between standard Scottish English and Scots itself. For this atlas, terms have been identified as Scots only when they are both contained in the *Dictionaries of the Scots Language* and are consistently spelled in a different way to any corresponding English term. Using this definition, there are over 80 Scots animal names in the database. This is actually a conservative estimate. Although it might include some terms that could be Scottish English (e.g. *teil* for the Teal) it will miss many more terms which are either shared between Scots and English, or have had their spelling anglicised to fit into English texts. Some of the species not included are (tentatively): *quhitred* for the Stoat, *kae* for the Jackdaw, *lipper-jay* for the Jay, *stronachie* or *heckleback* for the Sea Stickleback, *etterpyle* for the Three-spined Stickleback, *chuik* for the Prawn and *sand-lowper* for the Sand-hopper.

- **Irish** was the most commonly spoken language on the island of Ireland in the early modern period, but most of our sources were written in English, which was the language of much of the gentry during the Protestant Ascendancy following the establishment of the Plantations of Ireland. Two sources (O'Sullivan 2009; K'Eogh 1739), both written by educated men from Catholic Irish families, provide the vast majority of the approximately 70 Irish terms provided in our sources. Both the authors were familiar with Munster Irish. Some of the terms for the species not included in this volume are (tentatively): *grannoig* for the Hedgehog, *gealbhuin* for the Sparrow, *druid breac* for the Starling, and *loin* and *lan-duf* for the Blackbird.

- **Scottish Gaelic** was spoken especially but not exclusively in the Highlands and Islands of Scotland and local authors there (esp. Martin and Monro 2018; Buchan 1741) attest around a dozen names of species, mostly seabirds. This includes *bouger* and *albanich* for the Puffin, *scraber* for the Manx Shearwater, *trilichan* for the Oystercatcher and *lair igigh* for woodpeckers.

- **Norn** was the language descended from Old Norse spoken on early modern Orkney and Shetland. Norn animal terms are provided by a few authors (especially Gifford 1879; Monteith 1711; Brand 1701; Wallace 1700). These include *hoas* and *hoaskers* (Nynorn *hå*) for sharks, *silluk* for the Saithe, *dunter* for the Eider and *lyar* for the Manx Shearwater.

- **Late Cornish** was spoken as a vernacular language in early modern Cornwall and a few animal terms are provided by Tonkin (ed. Dunstanville 1811). These include *hernan* for the Pilchard, *hernan-gwidn* (literally 'white pilchard') for the Herring and *keligen* for the Razorshell. Borlase (1758) adds *padzher pou* (literally four-paws) for the Viviparous (Common) Lizard, and Willughby & Ray (Willughby & Ray 1686) add *morgay* for dogfish.

- Falle (1694) provides a few words of **Jèrriais**, the language descended from Norman French spoken on the Isle of Jersey. This includes *vrac* for the Ballan Wrasse, *gronnard* for the Gurnard and *lançon* for the Sandeel or Sprat.

- **Other languages** spoken by communities and individuals in early modern Britain and Ireland included the predecessor to British Sign Language spoken in Deaf communities, Manx spoken on the Isle of Man, Shelta spoken by Irish Travellers and Beurla Reagaird spoken by the Highland Scottish Travellers. There were also established communities in early modern Britain and Ireland of Huguenots who often spoke French, workers and merchants from the Low Countries who spoke Dutch and Flemish, and 'Egyptians' (the predecessors of the modern Gypsy communities) who spoke Romani. Most likely there would have been visitors or communities of people from other neighbouring countries as well as from the Ottoman Empire, the Persian Empire and free and enslaved people originally from West Africa. These people would have remembered additional names for the wildlife of Britain and Ireland, but generally these are not recorded in our sources. Some early modern sources do collect species names in other national languages, but the sources for these names seem to be international dictionaries rather than local informants so they are considered outside of the scope of this volume.
- Most species could also be identified using international **scholarly languages**. By far the most important of these was New Latin, which had a term for almost every species included in this atlas, but some authors added terms in other scholarly languages, including Ancient Greek, Classical Hebrew, Arabic, Aramaic and Middle Persian. Only the Latin terms have been included in this volume.

TREND SINCE 1772

The trend since 1772 is listed at the start of each species account, and summaries can be found in the Conclusions (p. 355). The trend is identified based on comparing distribution data from our early modern sources with twenty-first-century ecological handbooks (Crawley et al. 2020; Wood 2018; Henderson 2014; Balmer et al. 2013; Beebee 2013; Harris and Yalden 2008; Davies et al. 2004). It can be one of six values:

- **Certainly declined.** This value is given to species that were recorded in one or more regions in the early modern period where they are now believed to be absent.
- **Certainly increased.** This value is given to species only recorded as absent (not just unrecorded) in one or more early modern regions where they are now believed to be present.
- **No change.** This value is given to species that were recorded in every region in the early modern period and which are still widespread in every region today.
- **Probably no change.** This value is given where a species is recorded in the same regions in the early modern period as it is recorded in today, but not in every region. It is not as certain as 'No change' because the species might possibly have been present but unrecorded in the early modern period in the regions where it is now absent (an overlooked decline).
- **Probably increased.** This value is given when a species is unrecorded in some early modern regions where it is now known to live. Only used when statistical analysis shows the reason for the lack of records from the regions where this species now occurs is not just a lack of survey effort (see 'Identifying absence', p. 16). This value is also given where birds that seem to have had a coastal distribution in the early modern period have expanded to living in inland parts of Britain and Ireland. There is no 'probably decreased' value because the distribution of species today is so well known that no atlas species has gone from being present to just being unrecorded and likely absent across an entire region.
- **Uncertain.** This value is given where it is not clear that there has been an increase or a decline in distribution, or where both have taken place.

To help contextualise the trend since 1772, the atlas also lists the modern conservation status of each species, including its international status on the IUCN Red List and the national status of the species according to UK and Republic of Ireland's red lists (e.g. King et al. 2011). These categories reflect more short-term data (changes within the last decade), and species are ordinarily categorised based on changes in abundance rather than shifts in distribution. The categories used for modern conservation status are, from lowest to highest extinction risk: Least Concern, Near Threatened, Vulnerable, Endangered, Critically Endangered and Extinct (IUCN 2012). Some species (notably the invertebrates and marine fishes and, in the UK, the freshwater fishes) have not yet been evaluated. A second, more detailed system is in use for birds, which also incorporates some data about distribution changes, and historical population data going back to 1800. The categories used in this system are, from lowest to highest conservation concern: green list, amber list and red list (Stanbury et al. 2021; Gilbert, Stanbury & Lewis 2021).

GEOGRAPHICAL BIAS OF TOP-QUALITY RECORDS

Individual records can show us the local presence or absence of a species within a region in the early modern period. We can also look beyond these, however, at the pattern of the points on the map. This pattern can tell us about the distribution of the species across the whole of early modern Britain and Ireland. Where it is possible to reconstruct the early modern distribution of a species, this is especially valuable in giving clues about the native range of a species within the Holocene. After all, the data in this atlas pre-dates the revolutions in agriculture, transport and industry, and the popularisation of the game shooting industry, which led to declines in our wildlife before the start of modern recording schemes.

Unfortunately, drawing conclusions about local distribution based on the maps in our atlas is difficult because there is a considerable spatial bias in survey effort. Some counties were extensively surveyed in the early modern period, but others produced no records at all. We can track this bias by separating out a subset of the 'top-quality records' which are unique, reliable, site-level, presence records of the species in the atlas. This is around 30% of the total records included in the database (as shown on Figure 2). It does not include records of species not included in the atlas, repeat-records, country- or county-level records, uncertain records, or records of absence. The density of these records can be shown using a heat map (Figure 4).

There are some minor inconsistencies with this map. Most obviously, some of the key sources come from counties that have few top-quality records (Co. Clare and Co. Waterford in Munster, Suffolk in South England, Anglesey in Wales), meaning that they either only recorded nature in a small number of very biodiverse locations, or that all their records were provided at county level. For the most part, though, this map allows us to divide up Britain and Ireland into areas in which we can assume more or less effort has been put into surveying:

AREAS OF GREATER SURVEY EFFORT (MORE RECORDS)

1. South West England and southern Wales: Cornwall, Devon, Dorset, Wiltshire, Gloucestershire, Glamorgan, Pembrokeshire.
2. Munster: Co. Wexford, Co. Cork and Co. Kerry.
3. Eastern England: Norfolk, Essex, Northamptonshire, Lincolnshire, Yorkshire and Northumberland.
4. The counties surrounding the Irish Sea: Lancashire, Westmorland, Cumberland and Galloway round to Co. Antrim, Co. Down, Co. Dublin.
5. Northern Scotland: Shetland, Sutherland, Ross, Inverness, Aberdeenshire, Argyll and the Hebrides.

Figure 4 Heat map showing how many top-quality records are contained in each county. The key sources have been displayed on the map as rings.

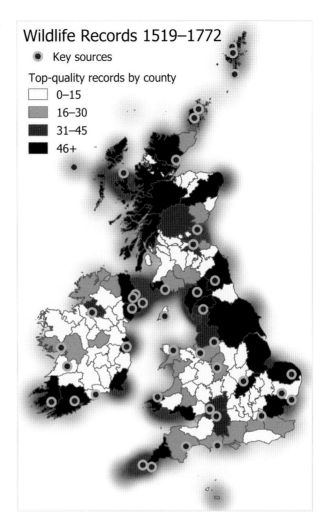

AREAS OF LESSER SURVEY EFFORT (FEWER RECORDS)

1.   Much of the English Midlands and Welsh Border counties.
2.   The inland counties of the island of Ireland.

For each map in the main species accounts of this atlas, the areas of lesser survey effort have been striped to indicate that we would not necessarily expect records from these areas, even if the species was common there.

If we generalise the top-quality records from county level to regional level, we can tabulate them as shown in Table 1. The regional biases in the recording are clearer from this table – the Highlands of Scotland (including the Northern and Western Isles) is by far the best-recorded region, and Connacht is the worst-recorded region.[2] These biases are accounted for to some extent by the relative size of each region: although there are around 16 top-quality records per 1,000 km^2 in the Highlands of Scotland

2   The lack of records from Connacht is unsurprising. The region was less developed than its neighbours and was the subject of major upheaval in the 1650s due to the confiscation of all land in Connacht, and its reassignment to transplanted Catholic landowners and soldiers (Cunningham 2011).

Table 1 Count of reliable, site-level, presence records of species in this atlas ('top-quality records') by region.

Region	Number of records	% total (rounded)	Records per 1,000 km²
Lowland Scotland	481	16%	12
Highland Scotland	629	21%	16
Ulster	256	8%	11
Connacht	59	2%	3
Leinster	155	5%	8
Munster	145	5%	6
Wales	237	8%	11
South West England	275	9%	11
South England	248	8%	6
Midlands of England	152	5%	5
North England	410	14%	11

compared to only three records per 1,000 km² in Connacht, Ulster and Leinster have an equal or higher number of records per km² than the South and Midlands of England.

IDENTIFYING ABSENCE

The recorder effort problem refers to the way that ecological surveyors can bias their data by putting more or less effort into certain areas (Balmer et al. 2013; Hill 2012; Prendergast et al. 1993). A key consequence of this is that it can be difficult to distinguish when a species is absent in an area from when insufficient effort has been applied to record it (a 'false absence'). A good example of this is the early modern records identified as *carp* (Figure 5). This included both the Crucian Carp (*Carassius carassius*) and the Common Carp (*Cyprinus carpio*).

The early modern records of carp are almost all from the south of Britain and Ireland. Assuming they are correct, these records would appear to suggest that the species were absent from Ulster and Highland and Lowland Scotland, and very rare in Connacht and North England during the early modern period. But this does not line up with their modern distributions: while the Crucian Carp has a limited distribution, the Common Carp is now widely recorded in, for example, North England and Lowland Scotland. So, have the two carp species expanded their distributions or were they present in the north of early modern Britain and Ireland but just under-recorded there?

We can answer this question statistically, because, as explained in the previous section ('Geographical bias of top-quality records', p. 14), we can estimate the level of survey effort which was applied to each region of Britain and Ireland, based on how many records there are from each region. For instance, 21% of our records overall come from Highland Scotland, and 2% come from Connacht. This implies that many times more survey effort was put into recording in Highland Scotland than recording in Connacht (Table 2).

Excluding county records and absence records, there are only 14 top-quality records of carp species from early modern Britain. This makes it hard to know if the distribution is due to chance. However, if the carp were generally distributed across Britain and Ireland, and equally common in each region, we might expect each region to have a larger or smaller share of those 14 records, based on how much survey effort was applied to each region. But if our hypothesis is correct, and the carp were actually confined to the south of Britain and Ireland in the early modern period, they will not be recorded in the north, no matter how much survey effort is applied there.

We can test the difference between the expected and actual distribution objectively using a goodness of fit statistical test. Goodness of fit tests allow us to compare expected data with the actual data to determine whether the two are significantly different.

- **Hypothesis:** The expected and observed values are significantly different.
- **Null Hypothesis:** There is no significant difference between the expected and observed values.

The distribution pattern of every species in the atlas with five or more top-quality records has been statistically tested with a goodness of fit test. To make this book accessible, the test performed and p-values have not been given in the main text of the atlas. However, for most of the species, the test used was the Exact Multinomial Goodness of Fit. A significant difference is defined for each test as a p value of less than 0.05 (i.e. a less than 5% likelihood that the difference is due to chance).

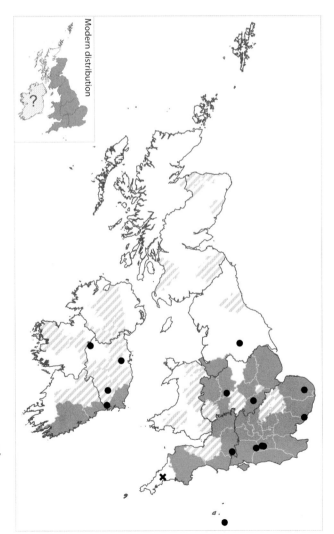

Figure 5 Map of Britain and Ireland showing where carp species were recorded between 1519 and 1772. There are records from the South, South West and Midlands of England, and part of North England. There are also records from Munster and Leinster in Ireland. Some records show captive populations only, and the counties these are in have not been shaded.

Table 2 Predicted vs actual distribution of carp records.

Region	% of all top-quality records (comparative survey effort)	Predicted distribution of the 14 carp records	Actual distribution of the 14 top-quality carp records
Lowland Scotland	16%	2	0
Highland Scotland	21%	3	0
Ulster	8%	1	0
Connacht	2%	0	0
Leinster	5%	1	3
Munster	4%	1	1
Wales	8%	1	0
South West England	9%	1	2
South England	8%	1	5
Midlands of England	5%	1	2
North England	14%	2	1

The packages used were *Exact Multinomial Test: Goodness of Fit Test for Discrete Multivariate Data version 1.2* (Menzel 2021) and *XNomial – Exact Test for Multinomial version 1.0.4* (Engels 2015). For the species with over 30 top-quality records, and a few species with a smaller number of records, the Monte Carlo method was used to reduce computational demand.

Returning to the two carp species, based on an Exact Multinomial Goodness of Fit Test, the distribution of early modern records is significantly different (p<0.05) from what we would expect based on the assumed survey effort across the different regions of Britain and Ireland. In this case, absence of evidence has become evidence of absence: the lack of carp records from Highland and Lowland Scotland and from Ulster is statistically significant and unlikely to be due to chance. The most likely reason for the lack of records in the north of Britain and Ireland is that carp were in fact absent from these regions in the early modern period (as explained in 'Carp', p. 210). Based on this test, the carp is identified as one of the species that has 'probably increased' in its trend since 1772.

In other cases, where the distribution of top-quality records for a species is a better fit with the known levels of recorder effort (i.e. there is a higher p-value than 0.05 on the goodness of fit test), we do not find evidence enough to accept the hypothesis, and so we accept the null hypothesis. In these cases, we can say that there is no clear evidence that the species was not generally distributed. This is not the same as saying that the species was generally distributed: it may or may not have been widespread.

The advantage of this method is that it allows some flexibility about the detection rate and even the survey effort applied to individual species. As long as a species was recorded at least five times, the survey effort is assumed to have been high enough to meet the species' detection rate and so the distribution of its records can be analysed. However, the test is not as reliable as a modern presence/absence analysis, and has the following issues which can sometimes undermine its reliability:

1. There may have been local interest in some species, which meant that uneven survey effort was employed to record them (e.g. the Chough in South West England).
2. Species that are only absent from poorly recorded regions where less survey effort was expected can appear to be a good fit with the survey effort, despite actually being absent from some regions (e.g. the Grass Snake, which is unrecorded in Ireland).

3. Species that were entirely absent or very poorly recorded in Britain and Ireland are invisible to the analysis (e.g. the Pool Frog).
4. The test sometimes reflects regional abundance rather than regional presence/absence (e.g. for the Trout and the other most-recorded species).

Where the test result is likely to reflect cultural interest or local abundance rather than local absence, and where the test has likely missed a regional absence, this is explained in the relevant species account.

For species where the expected and observed distribution of records are significantly different, one additional analysis has been carried out to help determine whether the species is likely to be locally absent from some regions. Modern data on topography, average summer and winter temperatures and average winter precipitation were collected for each site where these species were recorded in the early modern period (Fick & Hijmans 2017; GEBCO Compilation Group 2022). This data was used to develop a basic habitat suitability model, to establish the area of Britain and Ireland where modern conditions match those now found within the species' recorded range. This model was employed to assess how plausible it is that the species was formerly restricted in its range due to specialisation in a certain area. The model is not entirely reliable, because it is based on modern data whereas climatic conditions are known to have been different during the Little Ice Age (see 'Problems with using the Little Ice Age as a baseline', p. 357). It is used based on the assumption that even if conditions were different 250–500 years ago, the comparative conditions would still have been the same during the early modern period as they are today. At its simplest this means assuming that, whatever the exact temperatures were in the past, the north east of Scotland still had the coldest winters, the south east of England still had the warmest summers, and a species recorded in both areas is likely to have had a broad tolerance to the temperatures across early modern Britain and Ireland.

COMPARISON WITH MODERN DATA

It is worth emphasising some of the differences between the early modern data collected in the atlas database (Raye 2023) and modern biodiversity records data. The data collected for this atlas include well over 10,000 records of wildlife, of which almost a third are top-quality records of the 151 species covered in this atlas. In contrast, there are now almost 5 million records in the *Biodiversity Maps* database for the Republic of Ireland, and the UK's *National Biodiversity Network Atlas* had over 197 million records in June 2021, of which 107 million are at 1 km² resolution or better (National Biodiversity Data Centre 2022; JNCC 2021), perhaps because of the increased population in the region. While Ireland continues to be much more poorly recorded than Britain, within Britain there is now also a bias in recording towards the South of England (not one of the best-recorded early modern regions) (Hassall and Thompson 2010). Lowland Scotland is now more recorded than Highland Scotland, which was the most-recorded area in the early modern period. Because of the huge quantity of data, modern atlases can be much more temporally specific: whereas this atlas collects records from 250 years, atlas projects based on modern data can sometimes take their scope as just a small number of years. For instance, the latest British & Irish ornithological atlas (Balmer et al. 2013) looks only at records from 2007 to 2011.

In terms of species, while in the early modern period there was a bias towards recording any species which could be exploited for sport and profit (including fish, birds, mammals and marine invertebrates), there is now an overwhelming bias towards recording birds. The majority of the National Biodiversity Network's animal records refer to birds (Dickie et al. 2021: 5; Burnett et al. 1995: 19). The top ten best-recorded species in the early modern period (as shown in this atlas) were the Salmon,

Eel, Trout, Rabbit, Red Deer, Herring, Pike, Oyster, Fallow Deer and Cod. The top ten best-recorded species in the modern period would likely all be birds, especially waterfowl.

HOW TO READ THE MAPS

READER BEWARE: The maps in this atlas show where the wildlife of Britain and Ireland was recorded, not necessarily where it actually occurred. Then, as now, it is always much easier to demonstrate presence than to prove absence.

All of the species featured in this atlas have their own entry. They are listed in taxonomic order and divided into mammals, birds, fishes, amphibians and reptiles and invertebrates. Each species map in the atlas shows Britain and Ireland, from Shetland in the north to the Channel Islands in the south. The map is divided into regions (see 'Scope', p. 1), and symbols are used to indicate the different kinds of records from each area. These record types are not to be confused with the overall 'Trend since 1772' figures (see 'Trend since 1772', p. 13).

1. **Present.** Dots indicate a site or town where a species was recorded.
2. **Absent.** Crosses indicate where a recorder tells us that a species was absent from the site, town, county or region.
3. **Uncertain.** A diamond indicates a possible record that is ambiguous in terms of location or species.
4. **Present in county.** Shading indicates that a species is recorded as living wild in a historic county.
5. **No data.** Diagonal striping indicates the worst-recorded counties, meaning that records cannot be expected here even if the species was present (see 'Geographical bias of top-quality records', p. 14).
6. **Not recorded.** A blank space indicates a county that is well recorded but without any records of this species. These sometimes reflect absences and sometimes oversights (see 'Identifying absence', p. 16).

Numbers 5 and 6 here both reflect a lack of records. They are distinguished on the maps to show where survey effort is known to have been exerted – and therefore where an absence of records *might* reflect an absence of the species.

In the case of the Fox, as explained in the species account (p. 40), the animal seems to have been widespread in Highland and Lowland Scotland. There are absence records from the Northern Isles, the Outer Hebrides, the Isle of Man and Jersey; so the Fox was not present on these islands. The island of Ireland is worse recorded: most of it is striped to indicate there is no data for these counties. However, there are Fox records at local and county level in every region, so Foxes may have been widespread. Co. Kerry and Co. Cork are white, not striped, because the fauna there was relatively well recorded yet, despite this, there are no Fox records from these areas. The same is true of parts of the South, South West and the Midlands. In this case though, the lack of records does not seem to be statistically significant (see 'Identifying absence', p. 16), so the Fox may well have been widely present.

Figure 6 Map of Britain and Ireland showing where Foxes were recorded between 1519 and 1772. There are records from every region of mainland Britain and Ireland but only absence records from the Outer Hebrides, Northern Isles and the Channel Islands.

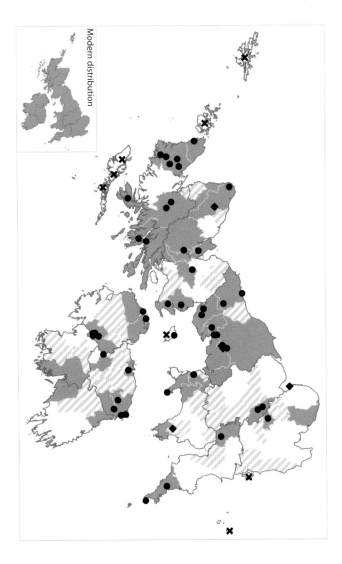

Modern distribution

RABBIT

Oryctolagus cuniculus

NATIVE STATUS Non-native (naturalised)

MODERN CONSERVATION STATUS	
World	Near Threatened
UK	Not Evaluated
ROI	Least Concern
Trend since 1772	Probably increased

In the early modern period, Rabbits were commonly kept in artificial warrens for meat and fur. The warrens seem to have been especially common on the coasts and islands of Ireland and western Britain facing the Atlantic and the Celtic Sea, as well as Orkney, but could also be found inland, especially in southern England and in the Lake District.

RECOGNITION

The Rabbit was most often called *coney* in Early Modern English, but the term *rabbet* was also used. The New Latin term was *cuniculus*, and we also find Scottish Gaelic *cuning* and Irish *connin*. All of these terms are specific to the Rabbit; no confusion with other species is possible. The term *warren* is not specific to *Oryctolagus cuniculus*; it could apply to any small area of land with controlled hunting rights (i.e. the area could be used for hunting the Fox or European Hare as well as the present species).

DISTRIBUTION

The distribution of top-quality records for this species is not statistically different from the known level of recorder effort. The gaps on the map could simply reflect decreased survey effort in some regions. It may have been widespread.

The Rabbit is one of the ten best-recorded species in the early modern sources. Rabbits appear to have been widespread all around the coasts of Britain and Ireland but were only found inland here and there. This finding helps reconcile the accounts of previous historians on the subject: Matheson (1941) found that Rabbits could be encountered on the islands of Wales from the thirteenth century, but only began to multiply in inland Wales in the nineteenth century. Ritchie likewise (1920: 252–3) points out that Rabbits were still considered rare and new in some parts of Scotland in the last decade of the eighteenth century and were not present in every county until the nineteenth century. However, Veale (1957) found that despite the initial bias towards coastal and island areas, Rabbits were common and cheap in England, including in the interior regions, from the sixteenth century. Our evidence reflects this difference, with early modern inland Rabbit populations found mainly in England, while the populations in Wales, Scotland and Ireland are largely coastal.

Rabbits were first widely introduced into Britain and Ireland after the Norman conquests. The inclusion of the species in a wildlife atlas is complicated by the fact that Rabbits were still generally farmed as livestock, hunting stock, and kept as pets and for their lawn management skills in the early modern period. Although Rabbits were left to graze for themselves, warrens were usually artificial.

Rabbit records, 1519–1772. The records are mainly distributed on islands and on every coast of Britain and Ireland, but there are also inland records across the North and South of England and elsewhere.

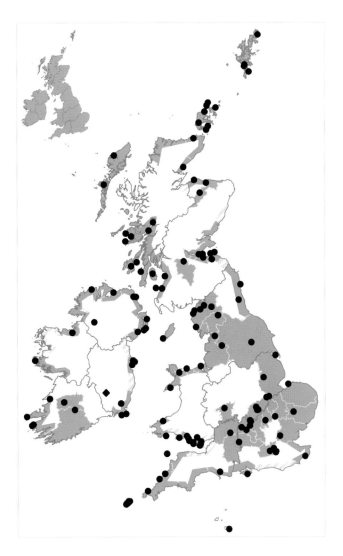

There are occasional references to people finding Rabbits unexpectedly on islands, and Stringer (1714: 161), based in Ulster, refers to hunting Rabbits in wild warrens. For the most part, though, every population of Rabbits was known and owned.

> A most delightful Airing … may be had in a Coney-Warren, sheltered all round by rising Hills, the Grass and wild Herbs of which are kept shorn as close by the Rabbits as a Scyth.
> (Harris and Smith 1744: 80)

The local distribution of this species expanded significantly after the end of the early modern period. Today, Rabbits can be found almost everywhere through both the interior and coastline of Britain and Ireland, and they seem to have been present in these areas by the start of the twentieth century (Bell 2020; Tapper 1992: 22–5). However, Rabbits have had a difficult time over the last century. In the 1950s, myxomatosis was introduced to Britain as a biological method of pest control. This has created widespread declines in Rabbit populations, but the species has since recovered to some extent. Rabbit haemorrhagic disease is also currently having a negative impact on Rabbit populations, and at present the species seems to be declining in Britain (Mathews et al. 2018: 90–8).

BROWN HARE/EUROPEAN HARE

Lepus europaeus

NATIVE STATUS Non-native (naturalised)

MODERN CONSERVATION STATUS	
World	Least Concern
UK	Not Evaluated
ROI	Not Evaluated
Trend since 1772	Uncertain

The early modern sources record Brown Hares in several locations across England and Wales, especially in the South East.

RECOGNITION

The Brown Hare was usually called just *hare* in Early Modern English and *lepus* in New Latin. Lhuyd adds Welsh *scyvarnog* to this list (ed. Campbell and Thomson 1963: 62). This leads to confusion with Mountain Hare. The records have been identified to species based on their locations (Irish and Scottish references are assumed to be Mountain Hare, English and Welsh records to be Brown Hare). In the case of Brown Hare, surprisingly this includes a few records of white hares. It is possible these white hares could represent escaped or remnant populations of Scottish Mountain Hares (who regularly turn white in winter), but they are most likely to represent Brown Hares with albinism, leucism or a medical condition which changes fur colour:

> And here I had almost forgot, though it be not too late, to mention a hare, or rather a creature partaking of both kinds, being supposed to be engendered between a hare and a white rabbit, which got loose out of one Mr Williams's house in Golours, in the parish of St. Gorran, [Cornwall]: in shape it was like a hare, but had large splats of white in its body, one of its shoulders all white, part of its head so too, with a white star in its forehead, and a long white strip over its nose, everywhere else of a hare colour.
> (Tonkin [*c.*1739], in Dunstanville 1811: 65)

DISTRIBUTION

Since the Brown Hare records have been identified in part based on location, it would be circular reasoning to attempt to reconstruct the species' distribution based on these records.

There was a great deal of interest in hares in the early modern period. Like the Rabbit, the Brown Hare provided fur for use in textiles (Lovegrove 2007: 71–2). It was rarely eaten, but it was widely hunted for sport (poorer people in early modern Britain and Ireland did not usually have the legal right to hunt hares, but often did so anyway) (Carnell 2010: 12; Griffin 2007: 110–11). In the nineteenth century the species was commonly identified as a pest and restrictions about who could hunt were relaxed in the Ground Game Act of 1880 (Lovegrove 2007: 96–7; Matheson 1941). Hundreds of hares were killed every day into the twentieth century, which seems to have led to declines in local populations.

The Brown Hare is actually an introduced species throughout the whole Britain and Ireland. It seems to have become established in England by the Iron Age. There is also an early modern record of hares on the Isle of Man, although this is from a passage of Camden's *Britannia* (ed. Gibson 1695: 1061–2) which provides other erroneous records (see 'Otter', p. 48). A second reference by Thomas Denton (2003: 496) goes some way towards confirming the record, although Denton drew heavily on Camden for his account of the Isle of Man, so this corroboration is not wholly convincing.

Brown Hare records, 1519–1772. The records come from every part of England, as well as the Isle of Man and South Wales.

The Brown Hare is usually thought to have been introduced to Scotland and Ireland after the end of the early modern period (Lever 2009: 69–71; Yalden 1999: 127–8; Tapper 1992: 26–9). It was widespread across mainland Scotland by the start of the twentieth century. The species was also introduced to Highland Scotland and the Hebrides shortly after the end of the early modern period at the end of the eighteenth century. It was then introduced to Shetland in the early nineteenth century but it died out there in the twentieth century (Yalden 1999: 230). In Ireland there were introductions from the mid-nineteenth century, of which the first successful instances seem to have been in 1876, and descendants of the introduced populations are still found in Ulster today (Lever 2009: 71; Fairley 1975: 28). The species was also introduced to Orkney in the modern period (Sharplin 2020). In Britain the Brown Hare is currently threatened by changes in agriculture (now less grazing available at certain times of year), targeted persecution, agricultural machines, motor vehicles and pesticides (Mathews et al. 2018: 105–6).

MOUNTAIN HARE/IRISH HARE

Lepus timidus

NATIVE STATUS Native

MODERN CONSERVATION STATUS	
World	Least Concern
UK	Near Threatened
ROI	Least Concern
Trend since 1772	Uncertain

The Mountain Hare was not properly distinguished from the Brown Hare in the early modern sources, but there are records that presumably refer to it from Scotland and Ireland during the early modern period.

RECOGNITION

The Mountain Hare is native to both Britain and Ireland. It either survived here through the last glacial period or arrived soon after the end thereof, and seems to have been lost from England and Wales (and presumably the south of Scotland) following the spread of woodland in the Mesolithic (Yalden 1999: 33, 127, 218; Kitchener 1998; Fairley 1975: 22). The Scottish and Irish populations have diverged and are recognised as subspecies: *Lepus timidus scoticus* (the Scottish Mountain Hare, occasionally called the Blue Hare) and *Lepus timidus hibernicus* (the Irish Hare). Unfortunately, not only were these subspecies not recognised in the early modern period, the species itself was also usually not distinguished from the Brown Hare.

Hares were referred to most often as *hare* in Early Modern English and Middle Scots. They were called *lepus* in New Latin. These terms are both generic and can be used for both the Mountain Hare and the Brown Hare. The species have been mapped here based on location: records from Scotland and Ireland are assumed to be Mountain Hare, and records from England and Wales are assumed to be Brown Hare.

K'Eogh (1739: 47) provides two Irish names: *garie* and *fiegare*. Assuming that the Brown Hare was only introduced to Ireland later, these Irish terms can be taken to normally refer to the Irish Hare.

DISTRIBUTION

Since the Mountain Hare records have been identified in part based on location, it would be circular reasoning to attempt to reconstruct the species' distribution based on these records. However, we can say that the early modern report of the absence of all hares from Shetland fits with the known introduction history of introductions there. The best presence record from Orkney is provided by Robert Sibbald, who explains:

> Lepus is noted by ordinary people as the most timid animal. It is said to sleep with its eyes open. Its running is supposed to be faster going uphill … It is found in the Orkney Islands with its hair returning to white in the winter.   (Sibbald 2020 [1684]: 11 (II:3))

The description of the white winter coat identifies the species Sibbald is describing as the Mountain Hare rather than the Brown Hare, which was introduced to Orkney in the modern period. Sibbald had multiple reliable informants on Orkney, but most other sources describing the islands say that the species was absent, so this record is surprising.

Up until the early modern period, the Irish Hare seems not to have been protected, although the use of the *milgu* (hunting hound) was especially associated with the aristocracy (Kelly 1997: 117–19).

However, in 1662 the Lord Lieutenant of Ireland dictated that the Irish Hare, the Pheasant and the Red Grouse were now the prerogative of the upper classes and could no longer be shot or trapped by ordinary people (MacLysaght 1939: 137). This seems to have been an attempt to bring Irish law closer to the English model, where Brown Hares were theoretically hunted only by the richest in society (Williamson 2013: 68–70). Yet since Irish Hare furs were exported in large numbers in the eighteenth century it may not have been especially effective (Fairley 1983).

Today the Mountain Hare continues to be present in every region of Scotland and Ireland (O'Neill and Sharplin 2020). In the nineteenth century, several attempts were made to introduce the Scottish subspecies into additional areas for shooting (Yalden 1999: 230, 244–6; Tapper 1992: 30–3). The species was introduced to the southern uplands of Scotland in the 1830s and 1840s. It was introduced to the Peak District over at least three phases in 1870–1880, and can still be seen there. It can also now be found introduced to the Isle of Man, Hebrides and Northern Isles. Finally, in the nineteenth century there were introductions to Wales, especially around Eryri. These populations died out in the twentieth century, probably as a result of changing grouse moor management strategies. In the future, the Mountain Hare is likely to decline due to habitat degradation (Mathews et al. 2018: 113–14).

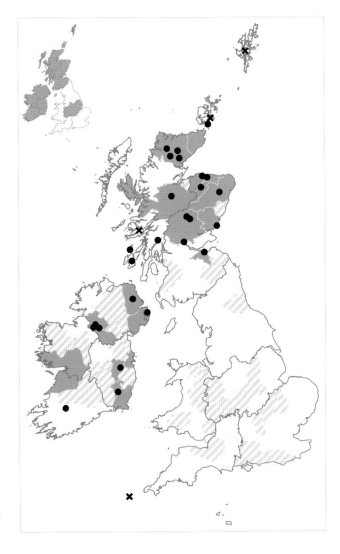

Mountain Hare records, 1519–1772.
The records come from every region
of mainland Scotland and Ireland.

RED SQUIRREL

Sciurus vulgaris

NATIVE STATUS Native to Britain, not Ireland

MODERN CONSERVATION STATUS	
World	Least Concern
UK	Endangered
ROI	Least Concern
Trend since 1772	Certainly declined

The Red Squirrel was poorly recorded in the early modern period but was mentioned by a handful of local sources from across Britain and Ireland.

RECOGNITION
The Red Squirrel is straightforward to identify in early modern texts. It is the only native animal referred to by our sources in Early Modern English as *squirrel*, in New Latin as *sciurus*, in Irish as *ira-ruo*, and in Welsh as *gwiwer*. The Grey Squirrel (*Sciurus carolinensis*), which is now the sole squirrel species found across much of England and Wales, was only introduced in the nineteenth century so cannot be indicated by any of the records on this map.

DISTRIBUTION
The distribution of top-quality records for this species is not statistically different from the known level of recorder effort. The gaps on the map may just reflect decreased survey effort in some regions. It may have been widespread.

Having said that, the Red Squirrel was probably not abundant anywhere in Britain and Ireland during the early modern period. In late medieval Europe, the Red Squirrel had produced the most popular fur of any animal. According to research by Elsbeth Veale (2003: 24, 134–9, 158–61), tens or hundreds of thousands of squirrel furs were being imported into London each year from at least the fourteenth century until the beginning of the sixteenth century. Although the most prestigious fur was the high quality *gris*, made from the paler fur of Red Squirrels living in Northern and Eastern Europe, British and Irish stocks would presumably have been targeted too, especially for less-discerning consumers.

Populations declined even further in the early modern period. The popularity of Red Squirrel fur waned, but the woodland coverage on which squirrels are dependent was reduced to a low point, which seems to have led to the Red Squirrel declining across Britain and Ireland (Williamson 2013: 20–24, 119–22; Lovegrove 2007: 96; Kitchener 1998). Local people appear to have been aware of this link. When Marchan Wood in Denbighshire was felled in around 1580, Robin Clidro, a Welsh poet, produced a humorous poem telling the story of a group of Red Squirrels who went to London with a petition in complaint (Foster Evans 2006: 77–8; Williams 1974: 87–8). By the eighteenth century, Red Squirrel populations in England seem to have stabilised with the afforestation of large estates, despite widespread continued hunting for pest-control (Williamson 2013: 119–22; Lovegrove 2007: 96). This does not seem to have been the case in Scotland and Ireland.

Red Squirrel records, 1519–1772.
There are a few records dotted
across northern and western Britain,
with one from Connacht and an
uncertain record from Dublin.

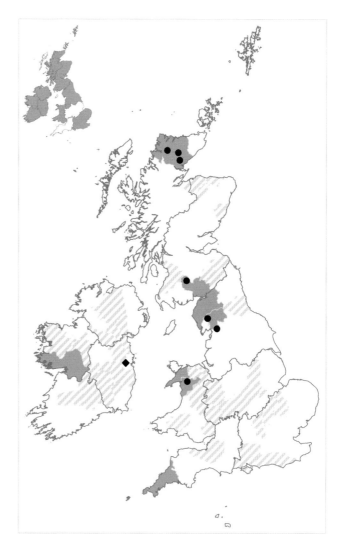

In Ireland, the Red Squirrel only appears to have been introduced in the medieval period (Barrett-Hamilton 1914: 692–5). The earliest archaeological evidence for this species is from tenth- or eleventh-century Dublin. Red Squirrels were popular animal companions in the medieval and early modern periods, and the first wild populations in Ireland might possibly have been escaped pets (Walker-Meikle 2012: 48–9; McCormick 1999). However, these initial populations seem to have declined after a few centuries. There appear to be significantly fewer records of squirrels in the early modern period than there were in the medieval period (Barrett-Hamilton 1914: 692–5). J.S. Fairley (1983), who studied exportations of furs from Ireland between 1697 and 1829, found thousands of Rabbit skins exported every year together with Fox, Otter, Hare and (probably Red) Deer, but no Red Squirrel skins. This may suggest that Ireland's Red Squirrel populations were not commercially viable by the end of the seventeenth century. The absence of Red Squirrel skins from export lists could also be explained by a ban on export, a lack of interest in squirrels or a preference for the grey-coloured Red Squirrel skins (*gris*) produced in colder parts of Europe. The most likely explanation may be indicated by what happened next: in the nineteenth century, Red Squirrels were so unknown that they had to be reintroduced to Ireland (Fairley 1975: 24; Barrington 1880), and modern Irish animals show considerable genetic diversity between populations, but little genetic diversity within populations. This is consistent with an origin with small, isolated groups being reintroduced across the island of Ireland (O'Meara et al. 2018).

In this context, the early modern Red Squirrel records from Ireland are noteworthy because they either reflect a population on its way to extinction, or a tiny remnant, later to be restocked. Apart from the uncertain record from Luttrelstown in Co. Dublin there is a county-level record from Iar Connacht and two national records. The record by O'Sullivan is especially interesting because it attests that squirrels were used for fur and possibly also kept in captivity in early modern Ireland:

> Squirrels are seen here [in Ireland]: they are smaller than martens, but larger than rats, red in colour, with hairy bodies, the tail more hairy and sickle shaped … Nevertheless, they come into the power of men, either caught down in their warrens or beaten in the chase between widely spaced trees. When kept in holes, they are amusing; when they are dead, their skin is praiseworthy. The skin is warm when added to clothes, and also it makes an unseamed purse highly esteemed by noble people, when it is adorned by silk lining inside and gold, silver or silk ribbons, as is the custom in Ireland.   (O'Sullivan 2009 [1626]: 82–3)

Red Squirrel populations responded in a similar way in Scotland. The species went extinct across the country just after the early modern period, between 1775 and 1850 (Yalden 1999: 172–3; Ritchie 1920: 290–7, 351–4). Here though, reintroduction of the species started before it went extinct, and continued through the nineteenth century. By 1900, populations look to have been stable, and by this point the Red Squirrel seems to have been once again widespread across Britain and Ireland.

But over the last century, the fortune of the Red Squirrel has reversed for a second time with the introduction of the Grey Squirrel from North America. This species outcompetes the native Red Squirrel in deciduous woodlands. It is larger, has a naturally higher population density and digests acorns much more comfortably. It also carries Squirrelpox, a disease that Grey Squirrels are resistant to, but which is deadly to the Red (Williamson 2013: 149–50; Lever 2009: 24–31; Yalden 1999: 184–90). The first Grey Squirrel populations were introduced to Britain in the second half of the nineteenth century, and to Ireland at the beginning of the twentieth century. By 1945, Grey Squirrels were well established in the wild and Red Squirrels were absent from much of South England. By 1970, Grey Squirrels had replaced Red Squirrels across most of the rest of England. By 1990, Grey Squirrels were also established across much of Wales (where they have generally replaced the Red Squirrels) and in Ulster, Leinster and the north-east of Munster, where Red Squirrels can also still be found.

At present Red Squirrels remain in parts of Wales, the Isle of Wight, Cumberland and Northumberland and across much of Scotland and Ireland (Shuttleworth 2020). Red Squirrels seem to outcompete Grey Squirrels in coniferous woodland, and possibly also in areas where the Pine Marten is still present (Sheehy and Lawton 2014), but the population continues to decline in England and Wales (Mathews et al. 2018: 121–3).

BEAVER

Castor fiber

NATIVE STATUS Native to Britain

MODERN CONSERVATION STATUS	
World	Least Concern
UK	Not Evaluated (undergoing reintroduction)
ROI	Not Present/Extinct
Trend since 1772	Uncertain

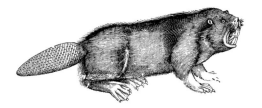

The early modern authors provided records of Beavers from Loch Ness, and (less reliably) the River Teifi.

RECOGNITION

This species was called *beaver* in Early Modern English, *bever* in Scots, and *castor* and *fiber* in New Latin. These names appear to have been specific to the species in the accounts of the early modern naturalists, although the English term must be distinguished from the word *beeves* (meaning domestic cattle).

The species was also called *avanc* in Welsh. This term is not specific to the Beaver: its more ordinary meaning referred to a water monster, and it may have only come into use to describe the Beaver in the early modern period. It is also possible that some late records might be obscured behind other animal names, as with the Lynx (p. 38) (Coles 2010).

DISTRIBUTION

Based on place-names and archaeological evidence, it seems that the Beaver was widespread in England and Wales in the first millennium CE (Coles 2006, 2010, 2019; Raye 2014). Archaeological records suggest that the Beaver continued to be found in some areas until the late medieval period. Wood gnawed by Beavers in Northumberland has been carbon dated to the late fourteenth century (Manning et al. 2014). Beavers seem to have been prized for their fur, meat and perhaps castoreum (their anal glands which had medicinal usage). They seem also to have been in decline in this time period.

By the time the early modern sources were written, Beaver populations in Britain seem to have been nearly depleted. Sixteen early modern sources refer to the Beaver, but most of these are records of past populations. There are presence records relating to two local areas, Loch Ness in Scotland, and the River Teifi in Wales.

In Scotland, beavers are referred to in the sixteenth-century *Scotorum Historia* of Hector Boece (Sutton 2010, chap. 0.18), the translation of Boece by John Bellenden (Maitland 1821: xxxiii–xxxiv), and in Raphael Holinshed's *Chronicles of England, Scotlande and Irelande* (Holinshed Project 2008a, chap. II.1.5), which appears to be another translation of Boece. This is enough to indicate that Beavers were likely present in the sixteenth century. Scotland's beavers may not have lasted long after this. Robert Sibbald, who received natural history data from correspondents across Scotland in the seventeenth century, does not seem to have been aware of any recent records from the country, reflecting: 'I don't know if it can be found now' (Sibbald 2020: 10 (II:3)).

Beavers are also regularly mentioned by the early modern authors on the River Teifi in Wales, especially around Cardigan. This follows a tradition started by Gerald of Wales, a twelfth-century scholar, who claimed that the Teifi was the only river left in Britain south of the Humber which still had Beavers (Thorpe 1978: 174). However, 11 of the 14 early modern sources discussing this population expressly described it as a former species. This is true, for example, in Joshua Childrey's natural history, *Britannia Baconia* 'In the River Teivy in times past, the Beaver (or Castor) hath been found; but now they can find none of them' (Childrey 1662: 144). Humphrey Llwyd's *Commentarioli Britannicae descriptionis fragmentum* was also influential in this regard, and is particularly reliable since his author was from Wales and his book was focused on describing Wales (Coles 2019: 34–6):

> Gerald refers to [the Teifi] as the only river in Britain abounding with beavers, but now it is not known to us what they are. Based on the bare name alone, which they say is 'avanc', it is held to be some aquatic monster.   (translated from Llwyd 1572: 63)

A contemporary English translator amusingly illustrated Llwyd's concerns by translating his word *castor* (meaning Beaver) as 'otter' (Llwyd and Twyne 1573: 76). However, Llwyd's point about the absence of the Beaver was still well enough understood by other naturalists to be cited by, for instance, John Aubrey (1847: 59). His words were echoed by Edward Lhuyd (no relation), writing a century later (Gibson 1695: 645).

Beaver records, 1519–1772.
There are two locations indicated,
including Loch Ness, and the
River Teifi in Wales (contested).

Against these 11 absence records from Wales, I am aware of only three dubious presence records in the early modern natural history sources. Edward Chamberlayne notes that 'In the River Tiver in Cardiganshire the Beaver hath been found' (1683: 253), but the reference is in the past tense ('hath been found' not 'is found'). Holinshed's *Chronicles* (this reference is from a separate section to the reference to the Beavers in Loch Ness and was written by William Harrison) refers to the Teifi as the only river still to have Beavers (Holinshed Project 2008a, chap. I.3.7); the record there is better and clearly draws on information about Beavers beyond that presented in Gerald of Wales, although he himself does not appear to have visited the River Teifi (Coles 2006: 180–1, 2019: 37–40). Harrison classifies the Beaver as a species of vermin, presumably meaning that he does not consider it a quarry species. Although he looked on it as it vermin, the Beaver was not included in the list of legal target species which parishes were required to pay bounties on in England and Wales under the 1566 Grain Act, but there is a controversial record of a bounty being paid for a 'Bever[hea]d' in Bolton Percy, North England in 1789 (Coles 2006: 187–90), which is difficult to explain based on the location and the date. The most convincing early modern record that might reflect the continued presence of the Beaver in south Britain during the early modern period is in Daniel Defoe's (1724–6) *Tour of Britain*:

> The country people told us, that they had beavers here [at Cardigan], which bred in the lakes among the mountains, and came down the stream of Tivy to feed: that they destroyed the young fry of salmon and therefore the country people destroyed them; but they could show us none of them, or any of the skins, neither could the countrymen describe them, or tell us that they had ever seen them; so that we concluded they only meant the otter, till I found after our return, that Mr Cambden mentions also, that there were beavers here seen formerly.
>
> (Defoe 1971 [1724–27]: 381)

Daniel Defoe visited this area as part of his tour. He himself seems to have mistrusted the story he heard, but this could be easily explained if Beavers were absent from much of the rest of the country, so he was not familiar with them. The local inhabitants seem to have believed there were (or had recently been) Beavers in the mountains nearby, which occasionally came downstream. Based on the overwhelming number of absence records for the area and the late date of this record, it seems most likely that this was a tall story told to travellers to explain the extinction of the famous local Beaver population. However, the record could alternatively be understood in light of a late extinction scenario for the Beaver following a period of invisible survival in the south of Britain as previously hypothesised by Bryony Coles (2006: 179–92, 2010, 2019: 81–91).

There is also one mistaken record of the Beaver living in Munster (MacBrody 1669: 956), which is quoted in the entry for the Wildcat (p. 36).

Due to accidental and purposeful introductions, the Beaver has now become officially re-established in Argyll, Tayside and in Devon. There are other temporary and permanent populations in Kent and in parts of Wales and South West England, as well as many Beavers kept in enclosures around Britain to create flood-buffer areas (Campbell-Palmer 2020).

SHIP RAT/BLACK RAT

Rattus rattus

NATIVE STATUS Non-native

MODERN CONSERVATION STATUS	
World	Least Concern
UK	Not Evaluated
ROI	Not Evaluated
Trend since 1772	Certainly declined

The Ship Rat was recorded in a few locations around early modern Britain and Ireland, especially on islands and along the coasts, but also in at least two inland counties (Oxfordshire and Staffordshire).

RECOGNITION

The Ship Rat was the only species of rat found in Britain and Ireland through most of the medieval and early modern periods. It has now been replaced across most of these islands by the Brown Rat (*Rattus norvegicus*), but the latter species was rare in Europe until the eighteenth century. Only one of our texts clearly distinguishes two kinds of rat, Rutty's *An Essay Towards a Natural History of the County of Dublin* (1772). Rutty refers to *Rattus norvegicus* as the *Norway rat, Mus aquaticus*. Since the Brown Rat arrived in these islands only around 1720, it is usually easy to identify the species intended. Uncertain records have been shown as diamonds on the map.

The Ship Rat was most often called *rat* in Early Modern English, *raton* in Irish and *Mus major* in New Latin. Occasionally, other terms were used: for instance, Lovell provides the Latin term *sorex* and O'Sullivan provides the term *Mus franci* (almost certainly as a translation for Irish *luch francach*). These other terms are less certain, but luckily always co-occur with the more certain term *rat*. The term needs to be distinguished from *rat goose* (referring to the Brent Goose) and *water rat* (meaning the Water Vole (*Arvicola amphibius*)). Lovegrove (2007: 220–1) found that the term *rat* was regularly used for bounty payments for persecution of the Water Vole in Dorset in the second half of the eighteenth century. This does not seem to be a problem for records of *rat* in the early modern natural history sources; all the reliable presence records, including those in inland counties, can be distinguished specifically because they refer to the species colonising small islands or eating stored grain alongside mice.

DISTRIBUTION

The distribution of top-quality records for this species is not statistically different from the known level of recorder effort. The gaps on the map may just reflect decreased survey effort in some regions. It may have been widespread.

Intriguingly, there are as many absence records for this species as there are presence records, and these absence records often occur close to presence records. It appears that the absence of rats may have been more noteworthy to some authors than the presence of them (which may have been embarrassing).

> I have made dilligent inquiry about Ratts in this Country [Donegal]: and find … that there is a considerable tract of Land about Donegall wherein there is not one Ratt, tho' Ballyshannon on the one hand, and Killybeggs on the other, the first ten, and the latter 12 miles distant from it, have enough [rats] to send colonies to the adjacent Countries. This is the more strange, because Donegall is a sea port town.   (Wadman in O'Flaherty 1846 [1684]: 165)

Because of this odd patchwork of presence and absence records, it is difficult to interpret the former distribution of the species. Perhaps the fairest interpretation of the data is that early modern people

Ship Rat records, 1519–1772. There are records from every region of Britain except North England and Wales, and every region of Ireland except Munster, but there are also absence records dotted across most areas.

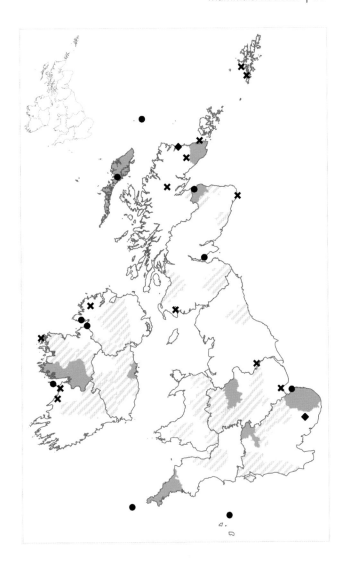

were thinking about and looking for Ship Rats in almost every region of Britain and Ireland. In the early modern period the species seems to have been common in urban areas, but less common in rural parishes (O'Connor 2017; Lovegrove 2007: 218). The species was first introduced in Britain during the Roman period, but early populations were not permanent (Yalden 1999: 125). By early modern times, however, it had been continually recorded in England (although not Scotland, Wales or the island of Ireland) for some time. In the archaeological record of medieval London and York, it is present from the tenth to the fourteenth century, and continued to be seen in these cities into the early modern period (Rielly 2010). The species is likely to be better recorded in city expense and household account books than in the early modern natural history sources used for this atlas.

From about 1720, the Ship Rat began to be replaced by the Brown Rat across much of its former range (Lever 2009: 49–50; Yalden 1999: 183–4). The decline of the Ship Rat was slowed by continuous restocking by ships in port cities, but this has been drastically reduced with improved biosecurity. The species became rare in the second half of the twentieth century and is now likely to be extinct apart from occasional transient populations brought by ships (Gurnell 2020; Mathews et al. 2018: 220–2).

WILDCAT

Felis silvestris

NATIVE STATUS Native to Britain

MODERN CONSERVATION STATUS	
World	Least Concern
UK	Critically Endangered
ROI	Extinct
Trend since 1772	Certainly declined

Wildcat populations were occasionally recorded in the early modern period in both lowland and highland Britain.

RECOGNITION

The modern binomial for this species, *Felis silvestris*, was in use centuries before Linnaeus and is the most common name encountered in the early modern period (although usually then spelled *Felis sylvestris*). Some vernacular names were also in use: *wild cat* (then two words) in Early Modern English and Middle Scots, and *cath goed* in Welsh. These obviously need to be distinguished from references to the domestic Cat (*Felis catus*), which was just called the *cat*.

The Wildcat seems to have been introduced to Ireland in the Mesolithic, but there is no evidence for a population after the Bronze Age (Montgomery et al. 2014; Kitchener and Daniels 2008; McCormick 1999). Based on this, we would not necessarily expect an Irish term for the Wildcat to be included by the naturalists. Surprisingly though, two of Ireland's national naturalists stated that the species was present and give Irish names for it, most convincingly *fiechait* provided by O'Sullivan of Beare (2009: 84–5). The same formation is later also attested in the Scottish Gaelic *fiadh-chat* (Dwelly 1988: 422). Without any local records it is hard to trust the early modern naturalists' opinion that the species was still found in Ireland (Stelfox 1965). The early modern references might easily be referring to feral domestic Cats. Sometimes historically, Irish sources have used the term *cat* to refer to the Pine Marten (Fairley 1975: 17), but both early modern authors distinguish the Pine Marten from their Wildcat. A third source, Friar Anthony Bruodin of Charles Ferdinand University in Prague, also suggested that the Wildcat was native, this time specifically in Munster, but he seems an unreliable source since he talks about the Beaver in the same sentence:

> The beauty of this land [Munster] is increased with wild quadrupeds, among which are included the red deer, fallow deer, wolves … martens, cats, beavers, hares, foxes, rabbits, ferrets, wildcats, and hundreds of other species of animals.   (trans. from MacBrody 1669: 956)

These records are surprising and fascinating but, in the absence of any more reliable evidence, it is most probable that the early modern naturalists were mistaken in suggesting that the Wildcat could be found in Ireland (Fairley 1975: 52).

DISTRIBUTION

The distribution of top-quality records for this species is statistically a poor fit with the known levels of recorder effort. It may have been locally distributed, locally abundant or of special local interest. Habitat suitability modelling based on the sites where the Wildcat was recorded in the early modern period suggests that the species may have had specific requirements. It seems to have been particularly associated with sites that today have cold winters, although this might simply be a product of its absence from milder Ireland. Statistically though, the most interesting feature of the distribution of wildcat records is the large number of local records (13!) provided in Thomas Machell's (1692–98) journal and notes on Kendal in the Lake District. Certainly, Machell had an above-average interest in Wildcats, but it also seems to be true that the Lake District acted as a stronghold for the species in the early modern period (Lovegrove 2007: 226; Langley and Yalden 1977).

Elsewhere, the Wildcat seems to have been found across much of Britain, but not Ireland in the early modern period. Apart from the Lake District, the sources attest populations in Northamptonshire, Pembrokeshire and Eryri, as well as the species' present range in the Scottish Highlands. These include some of the very last populations of Wildcats in lowland Britain. The Wildcat was regularly hunted

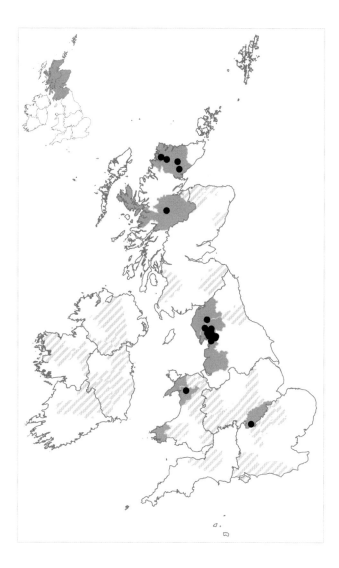

Wildcat records, 1519–1772. There are records from Highland Scotland (especially Sutherland), North England (with a cluster in the Lake District) as well as Wales and the Midlands of England.

in the medieval and early modern periods, both for sport and for its fur, and was probably already declining by the time the early modern natural histories were written:

> The wild Cat, that however [is found] of Whittlewood Forest [Northamptonshire] is generally of a larger Size, and has a Tail many Degrees bigger than the Tame. The wild Cats differ also in Colour from the common House-Cats … The She Cats at Finshed, and the like Lone-Houses do sometimes wander into the Neighbouring Woods and are gibb'd by the Wild ones there. 'Tis a difficult matter to tame the Wild Wood Cats, tho taken never so young into the House.
>
> (Morton 1712: 443)

Since the end of the early modern period, the Wildcat (now commonly called the Scottish Wildcat) has been lost from England and Wales. The decline seems to have been rapid. Langley and Yalden (1977) suggested that the species was lost first from the South, South West and Midlands of England by 1800, then from Wales, North England and most of Lowland Scotland by 1880. Lovegrove (2007: 224–5) suggested an even faster decline. In seventeenth-century England and Wales, 670 bounties were paid on Wildcats, but there were only four through the whole of the nineteenth century. The initial reason for the decline seems to have been deforestation: the Wildcat seems to have gone extinct in the open fields of the east of England before it did in the wooded counties. However, hunting for pest control may have been the final deciding factor in the extinction (Lovegrove 2007: 223–30; Yalden 1999: 175–6).

The Wildcat is no longer so commonly persecuted, but the population has fallen to the extent that it is critically endangered, lives at very low population densities and is therefore very vulnerable to hybridisation with domestic Cats (Howard-McCombe et al. 2021; Senn et al. 2019). Most of the individuals left in the wild today seem to be genetically closer to domestic Cats than they are to true Wildcats, although there is a non-hybridised population of captive animals in Scotland which could be used to re-establish the species.

LYNX

Lynx lynx

NATIVE STATUS Native to Britain

MODERN CONSERVATION STATUS	
World	Least Concern
UK	Extinct
ROI	Extinct
Trend since 1772	Certainly declined

There are some records which suggest that the Lynx could still be found in early modern Scotland.

RECOGNITION
The few records that exist refer to the species variably as *Lupus cervarius*, *luzarne*, *lynx*, or *wild cat*. The last of these is normally used for the Wildcat, so records using this term need to be treated with caution.

DISTRIBUTION
There are three records from our sources of captive Lynxes being kept in menageries at the Tower of London and Woodstock. These are omitted from the map to avoid confusion. Without these, there

Lynx records, 1519–1772. There is only one record – from Kirkcudbrightshire in south west Scotland.

are not enough local, reliable early modern records to test the goodness of fit between the recorded distribution of this species and the overall distribution of records from early modern Britain and Ireland.

The Lynx may have gone extinct in Ireland thousands of years before the historical period (Montgomery et al. 2014; Hetherington et al. 2006). In England, the species seems to have survived into the medieval period, but there is no reliable early modern evidence. However, there is some indication that the Lynx may have still been present in early modern Scotland. This is attested in two key sources. First there is a letter by Jan Boner młodśzy (Iohannis Bonarus of Balice or Hans Boner the Younger), printed in later editions of Conrad Gessner's Latin *Historiae Animalium* (Gessner 1602: 683). This letter states that the best lynx skins are those from Sweden and Scotland. By itself this text might refer to Scotland's known fur re-exportation industry (Raye 2017a), despite Topsell's (1658) paraphrase of this letter in the English translation clarifying that Lynxes lived in Scotland:

> There are Linxes in divers Countries, as in the forenamed Russia, Lituania, Polonia, Hungary, Germany, Scotland. (Topsell 1658: 382).

However, there is also a record from a traveller, Richard Pococke, from 1760 of a breeding population of large *wild cats* in the vicinity of Auchencairn in historical Kirkcudbrightshire, south west Scotland:

> They have also a wild cat three times as big as the common cat, as the pollcat is less. They are of a yellow red colour, their breasts and sides white. They take fowls and lambs, & breed two at a time. I was assured that they sometimes bring forth in a large bird's nest, to be out of the reach of dogs; and it is said they will attack a man who would attempt to take their young ones, but they often shoot … them & take the young. The county pays about £20 a-year to a person who is obliged to come and destroy the foxes when they send to him.   (Pococke 1887 [1760]: 26).

Pococke's description better fits the Lynx than any other native species and, taken together with Bonar's offhand remark and Topsell's translation, suggests a relict population of Lynxes may have still existed in early modern Scotland, despite the very low woodland coverage in the eighteenth century (Raye 2021a). If so, it is likely to have gone extinct shortly afterwards.

The Lynx is currently extinct in Britain, although there have been calls to reintroduce the species as a lost native both to help control Roe Deer and bring more ecotourism (Hetherington and Geslin 2018). Habitat-modelling work has suggested that Scotland today could support two populations of Lynxes, one in the Highlands (400 Lynxes) and one in the Southern Uplands (50 Lynxes). The 1760 record by Pococke comes from the Southern Uplands (Hetherington et al. 2008).

FOX

Vulpes vulpes

NATIVE STATUS Native to Britain, possibly not Ireland

MODERN CONSERVATION STATUS	
World	Least Concern
UK	Least Concern
ROI	Least Concern
Trend since 1772	No change

The early modern sources regularly referred to the Fox. The species seems to have been commonly found especially around the less densely populated areas of Britain and Ireland.

RECOGNITION

This species was called the *fox* in Early Modern English, and *vulpes* in New Latin. The Irish sources attest two terms *maidri rua* and *shunagh*, and the Middle Scots term was *taid*. One of Lhuyd's sources provides the North Welsh term *llwynog*. The South Welsh term *cadno* is not attested in our sources.

DISTRIBUTION

The distribution of top-quality records for this species is not statistically different from the known level of recorder effort. The gaps on the map may just reflect decreased survey effort in some regions. It may have been widespread. However, several authors attest that the Fox was absent from the Channel Islands, Outer Hebrides and the Northern Isles. The species was also usually attested as absent from the Isle of Man; the presence record appears to be a mistake, as explained under the Otter (p. 48). This approximately agrees with the modern distribution of the species, although Foxes have more recently colonised the Isle of Man (Scott 2020). There are also early modern records from the Isle of Mull, supporting the idea that Foxes were previously established there.

Until recently, the Fox was thought to be native to Ireland (Yalden 1999: 206, 218; Fairley 1975: 22). This is now being questioned (Woodman 2014). The earliest archaeological records are from the Bronze Age and some authors now suggest the species was introduced then, possibly in order that it could be hunted for its fur (Montgomery et al. 2014; McCormick 2007). Regardless of its native status, the species is now clearly naturalised across all four regions.

After deer, the Fox was one of the species most often pursued by hunters in medieval and early modern England and Wales (Poole 2015; Owen 2009; Griffin 2007: 46–7). Although initially tightly regulated through licenses, it is possible that this hunting might have had the potential to become unsustainable in some parts of England. The comparative lack of early modern records from the lowlands of England is not statistically significant, but there is some evidence that the Fox might have been locally rare or even locally absent here by the early modern period, except where populations were being protected or, especially from the eighteenth century, restocked, for hunting (Lovegrove 2007: 214–15; Thomas 1991, chap. 4.ii). Harrison, describing the population in the early sixteenth century, explained:

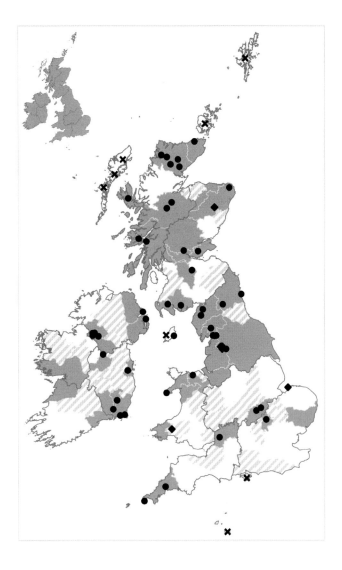

Foxes records, 1519–1772. There are records from every region of mainland Britain and Ireland but only absence records from the Outer Hebrides, Northern Isles or the Channel Islands.

> Certes if I may fréelie saie what I thinke, I suppose that these two kinds (I meane foxes and badgers) are rather preserved by gentlemen to hunt and have pastime withall at their owne pleasures, than otherwise suffered to live, as not able to be destroied bicause of their great numbers.
>
> (Holinshed Project 2008a [1577]: 1.3.7)

However, Lovegrove (2007: 210–11) notes that outside of the South and Midlands of England, Foxes were regularly targeted for pest control and despite decades of extermination (sometimes including up to 20 Foxes a year from single counties), the numbers of Foxes killed never seemed to change, which suggests that campaigns in other regions of early modern Britain were not leading to a decline in population. The Fox may have been more resilient in Wales, the North and the South West than it was in the rest of England (Lovegrove 2007: 212–13; Fairley 1983).

As well as being hunted for sport and for pest control, Foxes were also sometimes hunted for their fur in medieval and early modern Britain and Ireland. Fox furs were relatively popular, and furs were commonly exported from Ireland (Lovegrove 2007: 213–14; Fairley 1983). However, unlike other popular furs, Fox furs were not being imported into London in the period, so domestic production might have been enough to keep up with demand (Veale 1957: 58, 61).

After the end of the early modern period, the game industry began to target the Fox. Gamekeepers exterminated Foxes much more thoroughly than hunters, and the species seems to have gone extinct in eastern Scotland and east Anglia in the nineteenth century (Lovegrove 2007: 213–15; Yalden 1999: 174). At a species level, Foxes found some protection from an unlikely source. Fox hunting for sport became more popular in the nineteenth century, and hunters continued to preserve Foxes in their local areas through constant restocking. Some hunts even funded the creation of copses and Fox coverts to encourage local Fox populations.

Pressure on Fox populations eased during and after the First World War, with the reduction in the number of gamekeepers (Lovegrove 2007: 46; Yalden 1999: 174–5). In the present day, Foxes are widespread across every part of Britain and Ireland, including (increasingly) in eastern Scotland and East Anglia where the species was extinct 200 years ago (Scott 2020). Hunting continues, both for sport and for pest control, although hunting with hounds is theoretically banned in the United Kingdom. Since the 1930s and 1940s Foxes have colonised urban areas and they are now common in cities in the south of England and in Scotland (Baker and Harris 2008). This species now seems to be stable in terms of population and distribution (Mathews et al. 2018: 236–9).

WOLF
Canis lupus

NATIVE STATUS Native

MODERN CONSERVATION STATUS	
World	Least Concern
UK	Extinct
ROI	Extinct
Trend since 1772	Certainly declined

The early modern sources were very interested in the Wolf, despite it being extinct across most of Britain and Ireland. According to our sources, the Wolf was still present in some parts of Scotland and Ireland but not in England or Wales.

RECOGNITION
This species was called *wolf* in Early Modern English and Middle Scots and *lupus* in New Latin. In Irish the Wolf was called *mactiri* and once *moderalla* (literally the 'wild hound'). This last term needs to be distinguished from other kinds of *madra* including, for example, *maidri iski* (literally 'water dog', the Otter) and *maidri rua* (literally 'red dog', the Fox). While there were other species with similar names like *lupus-marinus* (the Atlantic Wolffish *Anarhichas lupus*), references to the Wolf are generally unmistakeable in the early modern natural histories.

DISTRIBUTION
The distribution of top-quality records for this species is not statistically different from the known level of recorder effort. However, in this case it is clear that the Wolf was locally distributed because there are absence records from across most of England and Lowland Scotland.

Despite the popularly held belief that the concept of extinction was only popularised in the modern period by figures like Georges Cuvier (Shubin 2019; McBrien 2016), most early modern naturalists were aware (and pleased) that the Wolf had become locally extinct in parts of early modern Britain. Over the course of the early modern period, 28 authors from the database attest that the Wolf was extinct in parts of Britain and Ireland (most often these are national absence records relating to England, where the species seems to have been hunted to extinction in the fourteenth century (Pluskowski 2010)). Only local absence records have been recorded on the map.

In Wales there are some hints that Wolves might have survived longer (Fychan 2006: 58–87; Lewis 2022). An early fourteenth-century poem from the Hendregadredd Manuscript seems to refer to a shepherd being injured by a wolf. Two satirical poems written in the fifteenth century tell a humorous but implausible story of how another poet, Tudur Penllyn, had his testicles bitten off by a Wolf. Another poet describes how Wolves howled to mourn the death of Rhys ap Llewelyn [ap Hwlcyn] of Bodychen, Anglesey in 1503, and several 'last wolf' stories postdate these records even further. These later allusions in literature and folklore are deeply dubious and perhaps constitute better evidence for the continued cultural currency of the Wolf than they do for the species' actual survival. The first early modern naturalists writing in Wales provide absence records from Pembrokeshire as well as Merionethshire in Wales (Owen 1994: 205; Gibson 1695: 655–6; Speed 1611: 117). The absence record from Merionethshire is especially valuable because this county holds part of Eryri, which contains the tallest mountains and some of the most inaccessible countryside south of the Scottish border.

Wolf records, 1519–1772. There are records from Highland Scotland, Connacht and Munster, and absence records from Lowland Scotland.

There is at least one presence record from England which goes against the narrative of a simple medieval extinction in the south of Britain. Guy Miege writes:

> I am credibly informed, that in some Places, as Warwickshire among the rest, some Wolves from time to time have been discovered. But, as it happens but seldom, so upon the least notice the Country rises amain, as it were against a common Enemy; there being such a hue and cry after the Wolf, that it is hard for him to escape the Posse Comitatus. (Miege 1691: 22)

Again, this record is not easily reconciled with others from the time period. It is possible that Miege might be describing wolf hysteria whipped up against feral dogs or animals imported for hunting (much like the records of the Wild Boar, p. 58) (see Gow 2023).

In Scotland, the Wolf appears to have been extinct in the lowlands and islands of Scotland by the time the first naturalists wrote. The Wolf's extinction in Lowland Scotland may have been relatively recent: barons were made responsible for hunting wolves on their land in an Act of James I (1427/8), and Wolf hunts in Lowland Scotland are celebrated in a Wolf-hunting poem from the mid-fifteenth century by the Gaelic poet Giolla Críost Táilliúr (Watson 1937: 178–9). However, the only Scottish county where our early modern naturalists record wolves is Sutherland, and particularly Strath Naver. These records have been criticised on the basis that Strath Naver is an unlikely place for the last wolves in Scotland to have lived (Crumley 2010, chap. 4). However, there are three records from ostensibly independent, seventeenth-century local recorders (Mitchell 1907: 454, 559; Gordon 1813: 3). The account of Robert Gordon of Gordonstoun written in around 1630–51, seems to be the most reliable:

> Ther are thrie principall forests in Sutherand (besides Scottarie, which lyeth in Strathbroray) … To witt, the forrest of Diri-chat, which is of the parish of Kildonan, wherein are conteyned the tuo hills called Bin-Ormin; the forrest of Diri-Meanigh, which is within the parish of Lairg wherein is conteyned Bin-hie, and the great hill Tain Bamd. All these forests and schases are verie profitable for feiding of bestiall, and delectable for hunting. They are full of reid deir and roes, woulffs, foxes, wyld catts.   (ed. Gordon 1813 [1630]: 3)

There is not sufficient reason to doubt this record – no other naturalist provides a contradictory absence record for early seventeenth-century Sutherland. The evidence implies that wolves were still present in Sutherland in this period. If this is the case, the presence of the Wolf there was likely short-lived. in the last quarter of the seventeenth century, all of the Scottish records of wolves on our map, and probably others, are known to have been communicated to Robert Sibbald. Sibbald seems to have concluded that these accounts were outdated. Writing in 1684, he describes the Scottish Wolf as a once-common, now extinct species (Sibbald 2020: 9 (II:3)). Pennant, writing a century later, agrees that the last Wolf in Scotland was killed in 1680 (1768: 75). Although some dubious evidence suggests that individuals and relict populations may have lingered for some time (Crumley 2010, chaps 4–5), it seems most likely that the seventeenth century saw the end of the Wolf as an ordinary Scottish species.

In Ireland, all the national authors agree that the Wolf could still be found in the early modern period. There are three county-level presence records. First, O'Flaherty's (1684) *Chorographical Description of West or H-Iar Connaught* includes the Wolf at the beginning of a list of species living in West Connacht (modern day Co. Galway) (O'Flaherty 1846: 9). Second, Bruodin (MacBrody 1669: 960) adds that although sheep need not fear Wolves in Co. Clare, they are yet not absent from that county. Finally, an anonymous account of Co. Leitrim, written in 1683 gives another possible record:

> [In Co. Leitrim] is one Deer park, viz, at Mannorhamilton w[hi]ch is walleeinging 4 miles in circuit: the wolves w[hi]ch were very numerous in this County are now very few, occationed by the care of the justices at their quarter sessions whoe always give unto those that Kill one two pence a hearth out of every hearth in the Barrony where the service is done.   (Logan 1971: 333)

It is not clear if this should be read as a presence record or a record of a population's extinction due to the new system of hearth tax bounty payments in Co. Leitrim (Hickey 2000).

There are also two site-level records from Co. Cork, both in Bishop Dive Downe's (1699–1702) *Visitation of his Diocese* (Lunham 1909), and finally one county-level absence record from early modern Ireland, relating to Co. Waterford. The population there may have been a relatively recent loss considering the records from Co. Cork, and since *The Travels of Cosmo the Third, Grand Duke of Tuscany* by Lorenzo Magalotti (1669) seems to describe Wolves as still being present in Munster and hunted with mastiffs in the seventeenth century (Magalotti 1821: 103).

The Wolf survived in Ireland for longer than it did in Britain. The species appears to have been more widespread in early modern Ireland than is represented by our natural history sources. Additional evidence can be obtained by looking at the commonwealth records (especially the interregnum period 'Commissioners of the Parliament of the Commonwealth of England for the Affairs of Ireland'). The references to Wolves in these archives were previously consulted by two nineteenth-century readers who quoted hunting licenses and bounty payments made between 1652 and 1665 from Cos

Kildare, Wicklow, Dublin, Meath, Galway, Mayo, Sligo and part of Leitrim (Prendergast 1865: xx–xxi; O'Flaherty 1846: 180–3). The relevant commonwealth records unfortunately seem to have been lost in 1922 during the Battle of Dublin (Wood 1930). Prendergast (1865: 153n) suggested some of them may have been edited as a *Book of Printed Declarations of the Commissioners for the Affairs of Ireland* by the British Museum, but I have not been able to find any surviving version of this book either. The counties mentioned by these authors cluster together. If the locations are a fair sample of the records, they suggest that the Wolf might have been especially common in Connacht and, surprisingly, Leinster in the seventeenth century. The ultimate effects of these bounties seem to have been to drive the Wolf to extinction in Ireland. Following these accounts, apart from the unpublished grand jury records apparently seen by Smith (1756: 173), the last records of Wolves in Ireland are in personal correspondence and local accounts of more questionable reliability (Fairley 1975, chap. 11). Hickey (2000), who mapped these 'last wolf' records, as well as the others mentioned here, suggests that the Wolf was hunted to extinction in Ireland within a century of our records, with the final death in 1786. Given the unreliable nature of some of this correspondence, it is difficult to be confident about the exact date of extinction, but it seems likely that the Wolf did survive into the eighteenth century in Connacht, Munster and Leinster at least. The Wolf is now of course extinct across the whole of Britain and Ireland.

BADGER
Meles meles

NATIVE STATUS Native to Britain, possibly not Ireland

MODERN CONSERVATION STATUS	
World	Least Concern
UK	Least Concern
ROI	Least Concern
Trend since 1772	No change

The Badger was recorded in every region of early modern Britain and Ireland, in both highland and lowland areas.

RECOGNITION
This species was called *meles* and *taxus* (more rarely *texon*, *tesson*, sometimes specifically *Taxus suillus*) in New Latin, *badger* and *bawson* in Early Modern English, *pry llwyd* in Welsh. The species was also called *broc* or *brock* in many of the languages of Britain and Ireland, including Irish, Scottish Gaelic, Middle Scots and Early Modern English. These names are all specific to the Badger. One name that was less specific is *grey* in Early Modern English. This was and is used for many species (even considering just those of terrestrial origin), and records using this term are only included on our map where they use one of the other terms as well.

DISTRIBUTION
The distribution of top-quality records for this species is not statistically different from the known level of recorder effort. The gaps on the map may just reflect decreased survey effort in some regions. It may have been widespread. If we include the county-level records, the Badger was recorded in every region of early modern Britain and Ireland and seems to have been widespread. This generally agrees with the modern distribution (Lewns 2020).

There are two early modern absence records included on the map. The uncertain record of absence from Devon provided by Wescote in 1630 (ed. 1845: 37) seems to just be a local record of how Badgers disappeared from a single area due to development of the land. The species may have continued to live unrecorded elsewhere in Devon. The early modern sources also attest that the Badger was absent from the Isle of Man, as it is today (Lewns 2020; Sacheverell 1859: 13; Burke 1766). There is a single unreliable presence record there which is quoted and discussed under the Otter (p. 48).

As with the Fox, the first archaeological evidence for the Badger in Ireland is from the Bronze Age. This might indicate that the Badger is not native to Ireland, although it might also simply have been overlooked (Montgomery et al. 2014; Woodman 2014; Yalden 1999: 206, 218). Either way, the records presented here suggest that the Badger was widespread in Ireland during our period.

Badgers were commonly hunted for sport as well as food, fur and medicine in medieval Britain and Ireland (Poole 2015; McCormick 1999; Kelly 1997: 282; Thomas 1991, chap. 2.i). This continued into the early modern period, as is shown in the following quotation from Robert Sibbald's *Scotia Illustrata* (1684):

Badger records, 1519–1772. There are records from every region of Britain and Ireland.

Taxus; Meles; the Brock or Badger, which is between a pig and a bear, the fox is smaller by a little in size than this fat animal.

…

Its flesh is held as a delicacy in certain regions.

Its grease also is used in place as lard as a food. Its use warm, softens, separates in areas affected by kidney stones. It soothes joint pains and softens nerve contractions.

Once a kind of clothing was made from badger pelts, which hung from the shoulders to the hips. They used to make dog collars from badger too. (Sibbald 2020: 9–10 (II:3))

In seventeenth- and eighteenth-century England and Wales bounties were regularly paid on the Badger, but the level of persecution does not seem to have been especially intensive (Lovegrove 2007: 232–4). Hunting for sport may have had more of an impact. The Badger was regularly dug out of its sett for pest control and sport, and Badgers were caught to be baited (killed by dogs). Badger populations seem to have been depleted by 1900, by which time it may have been absent from several counties (Lovegrove 2007: 230–2). Since then, populations have recovered, and the species has been increasing in abundance (Mathews et al. 2018: 247–8). Like the Fox, the Badger has started to colonise urban areas. Today it is generally distributed across Britain and Ireland except parts of the Scottish Highlands and southern uplands, the Outer Hebrides, the Northern Isles and the Isle of Man (Lewns 2020; Delahay et al. 2008). It is most common in the south and west of England, the west of Wales, and in parts of Ulster, Leinster and Munster. This is despite continuous culling across much of this area in an ineffective effort to stop the spread of bovine tuberculosis (Langton et al. 2022; Delahay et al. 2008). Despite the culls and the long history of digging and baiting Badgers in Britain and Ireland, the Badger seems to have the same distribution now as it had in the early modern period.

OTTER

Lutra lutra

NATIVE STATUS Native

MODERN CONSERVATION STATUS	
World	Near Threatened
UK	Least Concern
ROI	Least Concern
Trend since 1772	Probably no change

The Otter was recorded by early modern authors as occurring across Britain and Ireland, especially around the Scottish coasts and islands.

RECOGNITION

Most early modern authors describing this species use the English term *otter*. The authors also attest New Latin *lutra*, the Welsh *dyfrgi* and, in non-standard orthography, Irish *maidri iski* and Scottish Gaelic *madw wyske*. Some early modern sources distinguished the *otter* from the *sea otter* (which was called *Lutra marina*), but today the two populations are grouped into a single species, so records of both are considered together here. All these terms were unique to *Lutra lutra* in early modern Britain and Ireland, so records are easy to recognise.

The terms *maidri iski* and *dyfrgi* used for the Otter both mean 'water-dog'. Topsell (1658: 444) (translating Gessner) adds to these the New Latin *fluviatilis canicula*, which he translates loosely as 'dog

Otter records, 1519–1772. There are records from every region of Britain and Ireland, including the Hebrides and the Northern Isles, but there are only unreliable records from Munster.

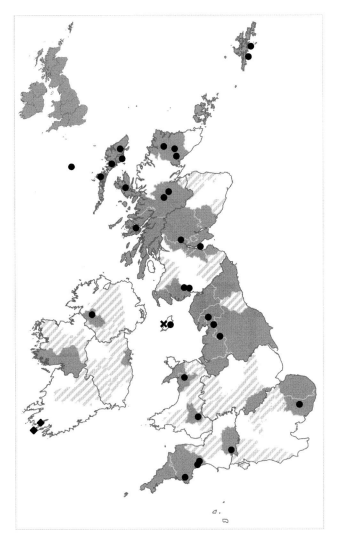

of the waters', and Botanista (1757: 109) independently attests *water-dog* as an English term. These terms, however, need to be distinguished from the Early Modern English *sea dog*, which normally referred to either the seal species or to the Spurdog (*Squalus acanthias*). Having said that, Charles Smyth (1756: 84) uses 'seals and sea-dogs' in *The Ancient and Present State of the County of Kerry* when referring to the coastal mammals of Kenmare Bay and Dursey Island in Munster. This term sounds like it was intended to refer to more than one kind of animal, so it might have included the Otter, but Smyth might also have been conscious of differences between the local Harbour and Grey Seals, or male and female seals (for comparison, early modern texts often refer to 'harts and hinds', meaning male and female Red Deer). These records are shown as diamonds on the map but are especially dubious.

DISTRIBUTION
The distribution of top-quality records for this species is not statistically different from the known level of recorder effort. The gaps on the map may just reflect decreased survey effort in some regions. It may have been widespread.

The number of records from Highland Scotland probably reflects a higher abundance of Otters there, especially on the islands and coasts (Lovegrove 2007: 248–9), but the lack of records from Ireland is somewhat surprising even if not statistically significant. There are not even any reliable county-level records from Ulster or Munster. This is strange because the Otter is native to Ireland and is present in the archaeological record from the Mesolithic period (Woodman 2014; McCormick 2007). The modern Irish Otter population is genetically diverse, with nine haplotypes (genetic groups) present (Finnegan and Néill 2010), which suggests a long history, perhaps with multiple waves of colonisation.

The Otter does not seem to be native to the Isle of Man. It is not currently found on the island (Chanin 2020), and Otters are not present in the local archaeological record. There is one dubious record from the island from the early modern sources. The author of the additions on the Isle of Man writes in the 1695 edition of Camden's *Britannia*:

> Here they have plenty of [domestic] hogs, of an ordinary bigness. There is also here great store of Otters, Badgers, Foxes, Hares and Conies. The Hares of this Island are very fat, which is a property in them not to be met with in many other Countries. There are some Deer in the Mountains, but they belong to the *Lord* of the Island and therefore none are permitted to hunt them without a licence from him under the penalty of a fine of three pounds besides imprisonment during the Lord's pleasure. (Gibson 1695: 1061–2)

This record is untrustworthy for two reasons. First, it is contradicted by two other early modern sources which say that the Otter was absent (Sacheverell 1859: 13; Burke 1766). Second, it also mistakenly attests to the presence on the early modern Isle of Man of the Fox and Badger, which also seem to have been absent there.

The Otter was hunted for its fur and for food, sport and pest control in medieval and early modern Britain and Ireland (Sibbald 2020: 10 (II:3); Owen 2009; Lovegrove 2007: 244–50; Veale 2003: 58, 61, 169; McCormick 1999; Yalden 1999: 174; Fairley 1983). In England and Wales in particular, bounties were payable for Otters under the 1566 Grain Act, although the species was not consistently targeted anywhere except in the South West. In Britain, Otter persecution intensified after the end of the early modern period, and Otter populations declined through the nineteenth century. This decline became a population crash during the 1950s–1970s, when in Britain (although not Ireland) the species was lost from most rivers in lowland England. The decline of the Otter, just like that of the Peregrine (p. 177) was probably due to organochlorine pesticides, in this case leaching into rivers (Lovegrove 2007: 247–8; Yalden 1999: 250–1). The Otter has been recovering since the 1980s, thanks especially to the ban on organochlorines and presumably helped by the legal protection of the species against hunting for sport and pest control. Today it is once again widespread in Britain (Chanin 2020; Mathews et al. 2018: 255–6).

MARTENS

Pine Marten (*Martes martes*) and possibly Beech Marten (*Martes foina*)

NATIVE STATUS Native to Britain, possibly not Ireland

MODERN CONSERVATION STATUS	
World	Least Concern
UK	Least Concern
ROI	Least Concern
Trend since 1772	Uncertain

The early modern sources recorded Pine Martens living in almost every region of Britain and Ireland.

RECOGNITION

Birks (2017: 32–5) has recently argued, based on historical sources, that as well as the Pine Marten which survives in Britain and Ireland today, the Beech (or Stone) Marten (*Martes foina*) was also present in Britain prior to 1900 (against Alston 1879). This was not recognised by the early modern authors. The only early modern local or national sources to distinguish two species of marten are those distinguishing 'foul and clean martes' (ed. Ewbank 1963: 45–9), meaning the Pine Marten from the Polecat (*Mustela putorius*); the distinction of two species of British *marten* appears to be original to Pennant's *British Zoology* (1768: 89–92). Despite publishing within the date range for this volume, Pennant is considered a 'modern' rather than early modern author, for the reasons given in the introduction (see 'Scope', p. 1), and his records are not included here. There is, however, one clear record of the Beech Marten in the database: John Wallis describes *Martes foina* as the species present in eighteenth-century Northumberland:

> Martes of Gessner … Martes aliis Foyna of Ray …
>
> The Marten is another of our mountain and wood-inhabitants, near houses. It lives upon birds, mice and other small animals … It is much esteemed for its fine fur, a deep brown, with a shade of black, bright and glossy. For shape it is as long, but slenderer, than one of our common house-cats, with shorter legs, a little peaked head, long hazel-eyes, short ears, and whiskers at the mouth, the tail as long as the whole body, very bushy, with long, thick hair; the throat whitish.
> (Wallis 1769: 412)

This description is partly borrowed from Ray's (1693: 200) worldwide handbook of mammals, amphibians and reptiles, but the fact that Wallis translated and included this description of the Beech Marten rather than the description of the Pine Marten directly above it suggests that he had identified the Beech Marten as the species in Northumberland. However, Wallis may not have come to this identification independently, since his text was published after Pennant's *British Zoology*. This means Wallis may be dependent on Pennant for his identification of the English species as the Beech Marten. Frustratingly, very few other early modern sources provide a description of the martens they are recording. It is possible that Birks (2017) is right that both martens were present in early modern Britain and that the species were just lumped together, but there is not enough early modern evidence to prove the theory.

The most common name used in the sources was *marten* or *mart* in Early Modern English and *martrick* in Middle Scots. The only possible confusion for this term comes from the *foulmart* (Polecat) but the early modern authors were clear about the difference between the two species. Our authors also attest the Welsh *bele* and the Irish *maidri criobhoig* ('tree hound', following the pattern of *madádh allaidh*, 'wild hound' = the Wolf and *maidri iski* 'water hound' = the Otter). The distinctive terms *bela goed* and *bela graig* etc. mentioned by Pennant (1768: 92–5) are not attested in the early modern texts. In later times, the term *cat* has been occasionally used to describe the Pine Marten in Ireland, but there are no examples of this in the early modern sources.

DISTRIBUTION

The distribution of top-quality records for this species is statistically a poor fit with the known levels of recorder effort. It may have been locally distributed, locally abundant or of special local interest. But habitat suitability modelling based on the sites where martens were recorded in the early modern period suggests the animal would have been comfortable in conditions across Britain and Ireland. Most likely in this case, just like with the Wildcat (p. 36), the bias seems to come from the high number of records provided by Thomas Machell's (1692–98) journal and notes on Kendal in the Lake District (ed. Ewbank 1963). Machell seems to have been more diligent in recording martens than any other early modern recorder, but it is also likely that the Lake District was a stronghold for the Pine Marten in the period. The early modern sources also highlight Highland Scotland and Eryri in Wales, which were likely to be other strongholds for the species as identified by Langley & Yalden (1977).

The Pine Marten may not have colonised Ireland naturally. The earliest remains found date from the Iron Age, and the species may have been introduced around this time, or perhaps slightly earlier

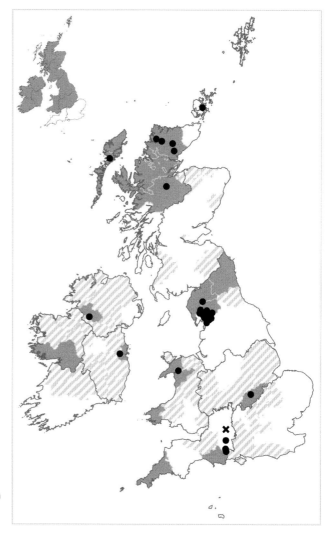

Marten records, 1519–1772. There are records from Highland Scotland, North England (with a cluster in the Lake District), the Midlands and South West of England, Wales, and Ulster, Connacht and Leinster in Ireland.

along with the Fox and Badger (Montgomery et al. 2014; Woodman 2014). However, the Pine Marten appears to have been well adapted to Ireland and was certainly long established on the island by the early modern period. It is recorded by local naturalists in Ulster, Connacht and Leinster (as shown on the map). It was also identified as one of the species of Munster by Bruodin (MacBrody 1669: 956 as quoted in Wildcat, p. 36), but this record seems less reliable as it is at regional rather than county level.

Although there are records of marten populations dotted across early modern Britain and Ireland, the species is unlikely to have been common anywhere during the period. Marten fur was very popular in medieval and early modern Britain and Ireland, and Pine Martens seem to have been persecuted wherever they could be found (Cessford 2013; Yalden 1987). Presumably domestic supply could not keep up with demand and thousands of skins of both Pine Marten and Beech Marten were imported to London from Northern and Eastern Europe in sporadic years from the fourteenth to sixteenth centuries (Veale 2003: 59, 158–61). Pine Marten skins were not exported with the Fox, Rabbit, Irish Hare and deer skins exported from Ireland between 1697 and 1819, perhaps because the species was so rare (Fairley 1983). Habitat loss is sometimes also suggested as an early reason for decline. Just like the Red Squirrel (p. 28), the Pine Marten is usually a woodland species and presumably should have been significantly affected by the deforestation of the early modern period (Sheehy and Lawton 2014; Balharry et al. 2008; Lovegrove 2007: 23–5, 203–5). On the other hand, Pine Martens in Britain and Ireland now occur in areas with woodland cover as low as 4% (Mathews et al. 2018: 257–8). The species seems to have adapted to this new habitat. In any case, an early decline is supported by the early modern sources. When Edward Topsell (1658: 387) translated the famous *Historiae Animalium* of Gessner into English, he erroneously added a note in the Pine Marten entry that the species was not found in England. However, this impression of rarity should not be overstated. While Topsell might have believed the species to be absent, naturalists still recorded local populations of martens dotted across most of the regions of early modern Britain and Ireland, as shown on the map. Langley and Yalden (1977), who studied the decline of the Pine Marten over the last two centuries, found sources suggesting that it was still widespread in England at least in 1800. Lovegrove (2007: 204–5), examining the early modern parish records from England and Wales for bounties paid under the 1566 Grain Act, concurred – although he lacked almost any records from eastern counties between Yorkshire and Bedfordshire. Most probably, martens were relatively widespread in early modern Britain and Ireland, but hunting had suppressed populations so that they existed at low density, perhaps with extinctions in a handful of counties.

The decline of the Pine Marten to its present distribution in Britain happened mainly between 1800 and 1915, with the employment of gamekeepers to shoot 'vermin'. The species went from being widespread in 1800, to extinct across much of the South and Midlands of England and southern Scotland in 1850. By 1915 it seems to have been present only in the uplands of Highland Scotland, the Lake District, Eryri in Wales, and a few other places (Lovegrove 2007: 207–8; Yalden 1999: 248; Langley and Yalden 1977). In Britain, the Pine Marten began to recover with the reduction in gamekeepers after 1914 and is now relatively widespread in Scotland north of the central belt. The species has also been reintroduced, officially and unofficially, to the Scottish border, Mid Wales and probably Shropshire, the New Forest and elsewhere (Croose 2020). It is already recorded in each of the regions where it was known in the early modern period and seems likely to expand back across Britain over the next few decades. In Ireland, the Pine Marten was not as strongly persecuted, and it survived especially in Connacht, in Cos. Galway and Clare but also in Louth, Meath, Waterford and the Slieve Bloom Mountains between Cos. Laois and Offaly (Balharry et al. 2008). It has expanded over the last 50 years or so and was reintroduced to Killarney in the 1980s. The species is now fairly widespread across Connacht, Leinster and Ulster, and expanding into the north of Munster (Lawton et al. 2020). The Beech Marten is not now present in Britain or Ireland, and Birks (2017: 49–50) argues it may have become extinct in Britain in the nineteenth century.

POLECAT
Mustela putorius

NATIVE STATUS Probably native to Britain, not native to Ireland

MODERN CONSERVATION STATUS	
World	Least Concern
UK	Least Concern
ROI	Not Present
Trend since 1772	Uncertain

The Polecat was recorded across mainland Britain in the early modern period. It was present in Eryri in Wales, in the South and South West of England, and across much of the North of England as well as in Highland Scotland where it is rarely found today, but was not recorded in Ireland.

RECOGNITION
The Polecat was called the *polecat*, *foulmart* and *fitch* in Early Modern English and *putorius* or *Mustela putorius* in New Latin. Lhuyd provides the Welsh term *ffwlbart* (Emery 1974).

DISTRIBUTION
The distribution of top-quality records for this species is statistically a poor fit with the known levels of recorder effort. It may have been locally distributed, locally abundant or of special local interest. Habitat suitability modelling based on the sites where the Polecat was recorded in the early modern period suggests that the species might not have been found in the coldest parts of Britain and Ireland. But a more important reason for the bias in the distribution pattern for this species is probably that, just like for the Wildcat (p. 36) and the martens (p. 51), there are a large number of records of the Polecat produced by Thomas Machell's (1692–98) journal and notes on Kendal in the Lake District (ed. Ewbank 1963). These likely reflect both Machell's interest in mammals and the historical importance of the Lake District for wildlife.

The Polecat's native status is uncertain. It does not occur in Ireland today, and there are no earlier historical or archaeological records to suggest the species is native there. The only early modern reference from Ireland is that Gwithers records the Polecat on a list of 'Quadrupeds common in England and rare in Ireland' (MacLysaght 1939: 416). He was presumably unfamiliar with the species since he includes it twice, once as 'polecat' and once as 'foulmart or fitchan'. The description of the species as 'rare' rather than absent is significant here, since Gwithers had a separate list for 'Quadrupeds common in England and not found in Ireland' but in the absence of any other evidence for the Polecat from Ireland it appears he was mistaken in thinking the species could be found here at all in the early modern period.

Duncan Brown (2002) has suggested that the Polecat might not be native to Britain either, but might have been introduced in the Norman period along with the Fallow Deer and the Rabbit, perhaps to help with ferreting in warrens, or for its fur. This argument is based on several pieces of evidence, perhaps most convincingly: (i) there is a break in the archaeological record between larger native Polecats early on and smaller Polecats which appear later; (ii) there are no pre-Conquest records for the Polecat from Britain; and most importantly, (iii) the Middle Welsh term for the Polecat (*ffwlbart*) does not have a native root, is borrowed from a term for Polecat fur rather than the term for the animal itself, and does not produce any place-names. These arguments are tentative, as Brown recognised, and parts of the theory need to be challenged. The distribution of Rabbits in early modern Britain is

Polecat records, 1519–1772. There are records from Wales, North England (with a cluster in the Lake District), the South and South West of England and Highland Scotland.

very well attested in the early modern sources (see p. 22). In this period, they were not yet present inland in Wales, so the presence of the Polecat in early modern Eryri does not fit with the idea that Polecats might be descended from feral strays around warrens. Although the Polecat is not recorded in any texts from before the Norman Conquest, it does appear in some prior to the fourteenth century.[3] Gerald of Wales (writing *c.*1188) remarked with characteristic smugness on the absence of Polecats (*putaciis*) from Ireland, which suggests that they were widespread in Britain at this point (O'Meara 1982: 49; Dimock 1867: 59). This is supported by a passage in the twelfth-century *Liber Eliensis* which boasted that wild animals like Otters, Weasels, Stoats and Polecats (*putesiarum*) could all be caught in traps around Ely in Cambridgeshire (Fairweather 2005: 213). The lack of a native root-word notwithstanding, it seems unlikely that the Polecat could have been so well established in the twelfth century if it was only introduced at the same time as the Fallow Deer and Rabbit. It may have been introduced earlier than this.

---

3   Early translations of medieval charters also commonly refer to the right to hunt the 'polecat' (e.g. Anon 1803: 104), but this seems to be a confused mistranslation of the Medieval Latin *catus* meaning the Wildcat, much like the mistranslation of *turdus* as 'ouzel' described in 'Ring Ouzel' (p. 193).

The Polecat was hunted for its fur in medieval and early modern Britain, but its fur was apparently never as popular as the stoat in ermine or the pine marten (Veale 2003: 61, 137). The species was more commonly hunted as pest control and was subject to intense persecution in England and Wales under the 1566 Grain Act (Lovegrove 2007: 198–200). In some counties, especially Kent, the persecution seemed to culminate between 1720 and 1760, before dropping off – possibly suggesting that early modern persecution was effective at suppressing local populations. In the nineteenth and early twentieth century, the rise in gamekeeping as a profession, on top of continued hunting for fur (Service 1891), resulted in the extermination of the Polecat (just like the Pine Marten, p. 51) across much of its range. In the case of the Polecat, the period of steepest decline seems to have been between from 1850 until 1914, when pest control eased (Yalden 1999: 174–81; Langley and Yalden 1977). In 1850 the species was still widespread across the mainland of Britain, although rare in the South, South West and Midlands of England as well as the Scottish border counties. By 1880 it was declining everywhere and had gone extinct in Kent and Essex and most of Lowland Scotland. Populations collapsed quickly after that, and by 1915 the Polecat was found only in Wales, with a few other isolated populations on the English side of the Welsh border, and possibly in North England and Highland Scotland.

The Polecat has now had a century to recover and increase its distribution from Wales since the lessening of persecution in 1915. As well as Wales, it is now present across much of England apart from Cornwall and the east coast. There are also a few populations in Scotland (Birks 2020; Mathews et al. 2018: 277–83). The species is likely to continue to expand its range across Britain in the near future since it has a broad tolerance of lowland habitats (Tapper 1999: 218–21). It remains absent from Ireland.

BROWN BEAR

Ursos arctos

NATIVE STATUS Native and Extinct

MODERN CONSERVATION STATUS	
World	Least Concern
UK	Extinct
ROI	Extinct
Trend since 1772	No change

The Brown Bear probably went extinct in Ireland at the end of the Mesolithic and in Britain in the late Neolithic or early Bronze Age, although it is possible that a small population might have survived in north Britain until the early medieval period (Montgomery et al. 2014; O'Regan 2018). The species was long extinct by the early modern period.

RECOGNITION

The species was referred to as *bear* in Early Modern English or *ursus* in Latin. These terms were generic, and used for any species of bear, but since almost every record in our sources is an absence record, we can consider that they certainly would have included the Brown Bear. The naturalists record that no bears of any species could be found wild in early modern Britain or Ireland.

DISTRIBUTION

Thirteen sources refer to the bear. Ten of these are absence records referring nationally to England and Scotland. There are also two local absence records, one referring to the Isle of Wight and one to the area from Monteith and Callander to Lochaber, which John Lesley makes the site of the pseudohistorical Caledonian Forest (Lesley 1675: 18, 1888: 29).

There are two national records of presence for the bear, one for Scotland and one for Ireland. The Irish record is in John K'Eogh's *Zoologia* (1739: 6) which includes domestic animals, so K'Eogh might well have been thinking here of the captive bears which were shipped around Europe in the early modern period for baiting and entertainment. The Scottish record is provided in an over-excited account by Nicander Nucius of Corfu, who visited in the sixteenth century:

> [Scotland] breeds in the marshes, of wild and carnivorous animals, bears [ἄρκτους] and hogs,
> besides the wolf and the fox; and of graminivorous animals, stags and hares, and others of the
> same sort. (Cramer 1841: 19)

Nucius is mistaken here. The absence of the bear in early modern Scotland is asserted by multiple Scottish accounts, including most reliably in *Scotia Illustrata* by Robert Sibbald (2020: 9 (II:3)).

Brown Bear records, 1519–1772.
There are no presence records, and
local absence records from the Isle
of Wight and from mid-Scotland.

WILD BOAR

Sus scrofa

NATIVE STATUS Native to Britain, possibly not
to Ireland

MODERN CONSERVATION STATUS	
World	Least Concern
UK	Data Deficient
ROI	Extinct
Trend since 1772	Certainly increased

The native population of Wild Boar seems to have been long extinct in early modern Britain and Ireland, but there were populations of feral Pigs and captive populations of Wild Boar.

RECOGNITION

To some extent, the early modern sources distinguished between domestic Pigs and Wild Boar just like someone might today. A domestic Pig was called a *hog*, *pig* or *swine* in Early Modern English, *sus* in Latin and *muck* in Irish. Wild Boar were called *aper* in Latin, *collagh* and *muic fhiain* in the Irish sources and in Early Modern English *wild boar* or just *boar*. These names are usually reliable guides, but some populations of pigs were not easy for early modern writers to classify.

DISTRIBUTION

There are not enough top-quality records to statistically analyse the distribution of this species.

The Wild Boar theoretically went extinct due to overhunting in Britain and Ireland in the medieval period but in practice the situation was complicated (Albarella 2010; Yalden 1999: 166–8; Kelly 1997: 281; Rackham 1986: 36–7). The Wild Boar was commonly hunted for sport and food in the medieval period, and this continued after the local extinction of the species. Wild Boar were imported by the aristocracy to restock their estates for hunting and this practice continued to a lesser extent into the early modern period. Those domestic Pig populations which were free-roaming year-round without much supervision also tended to become feral (as, for example, some of those on the Isle of Man and in Orkney). These reverted to type, becoming more like Wild Boar, and have been shown as diamonds on our map. Morton provides a good description of how the process happened in Northamptonshire:

> As to wild Hogs, there is now a Breed of them in the Right Honourable the Earl of Exeter's Purlees and the Woods adjoining. They come of a Badger-colour'd Italian Boar, and a Black Westphalian Sow that had been brought to Burleigh by the late Lord Exeter. Escaping thence into the Woods, they became wild, and continue to propagate their Kind. They are now of a Fox-colour, feed on Mast, are fierce, and disown the Government of Man.   (Morton 1712: 444)

Another population, this time described as of 'wild bores or swyne' was recorded by Leland as living on Tresco in the Isles of Scilly in the mid-sixteenth century (Smith 1907: 318). This population seems to have become extinct by the time Borlase (1756: 50) visited in the eighteenth century. It is unclear now whether this population represented feral stock which had reverted-to-type, or animals imported

Wild Boar records, 1519–1772. There is a contested presence record from the Isles of Scilly, and unreliable presence records from the Midlands, Isle of Man and Orkney, with absence records from the South and South West of England and from Wales.

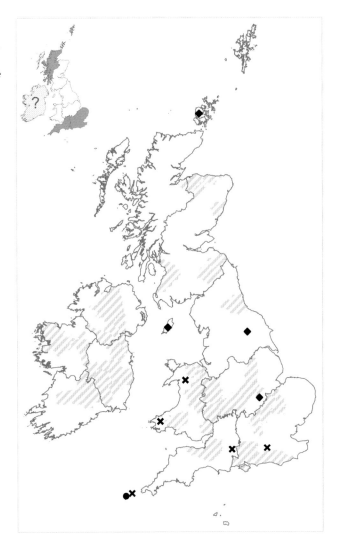

for hunting by a rich landowner. Either way, populations like this one seem to have influenced the early modern national-level naturalists to occasionally regard the species as present in Britain and Ireland (K'Eogh 1739: 10; Lovell 1660: 16). The Wild Boar seems to have been resistant to extinction.

Since the late 1990s, populations of Wild Boar have once again regularly escaped and been released from farms into the wild. The largest populations are those in the Forest of Dean and on the Kent–Sussex border, but there are also populations in Highland Scotland, South West England and possibly parts of Ireland. There seems to now be some tolerance for the species as a lost native and the Wild Boar in the Forest of Dean have been permitted to remain, with regular culling to control populations (Dutton 2020; Mathews et al. 2018: 292–7). The species will recolonise Britain and Ireland if it is allowed to do so.

RED DEER
Cervus elaphus

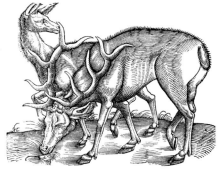

NATIVE STATUS Native to Britain, see text for Ireland

MODERN CONSERVATION STATUS	
World	Least Concern
UK	Least Concern
ROI	Least Concern
Trend since 1772	Probably increased

The Red Deer was widely recorded across early modern Britain and Ireland. There seem to have been free-roaming populations in parts of northern and western Britain and in every region of Ireland.

RECOGNITION

Ordinarily, Welsh *karw* and Early Modern English and Middle Scots *hart*, *hind* and even *stag* seem to have been usually used as specific terms for the Red Deer, although in theory only the Early Modern English term *red deere* and Irish term *fie-ruo* seems to have been unique to the species. The generic term deer was also sometimes employed but most often seems to have referred to the Fallow Deer (p. 62), so records have been included on the map for that species.

O'Sullivan and K'Eogh both attest an Irish term *carie / carrfhie*. By the early modern period this refers to the Red Deer, although it has been suggested (Scharff 1915a) that it might have originally referred to *Megaloceros giganteus* (the extinct Irish Elk).

DISTRIBUTION

The Red Deer is one of the ten best-recorded species in the early modern sources. It was also halfway between being wild and being livestock: some populations were free roaming, while others were emparked on private land for the exclusive use of the owner. For this map, shaded areas show where there is the best evidence for free-roaming populations in the early modern texts, but the distinction is not always clear. Both emparked and wild populations seem to have been exploited for sport and (to a lesser extent) food and hide in medieval and early modern Britain and Ireland (Williamson 2013: 13–15; Owen 2009; Sykes 2006; Yalden 1999: 142–56; Kelly 1997: 272–5; Fairley 1983).

The distribution of top-quality records for this species is statistically a poor fit with the known levels of recorder effort. In this case, there are more top-quality records from the Highlands of Scotland and North England than we would expect, but fewer records than expected from Ulster and Leinster. This might reflect the number of local captive populations, which were better known and thus more likely to be recorded than wild populations. Habitat suitability modelling based on the sites where the Red Deer was recorded in the early modern period suggests that the species would have been well adapted to conditions across Britain and Ireland.

There seem to have been early modern wild Red Deer strongholds in Co. Wicklow, Co. Antrim and along the west coast from Co. Donegal to Co. Kerry, as well as in Wales and across much of Highland Scotland together with the Hebrides. There may have been a population in northwest England, although the evidence for this is less clear. Curiously, based on our records, there may have been more Red Deer in some regions after they went extinct in the wild than there were before.

In Ireland, the Red Deer seems to have been able to recolonise after the last Ice Age but went extinct there during the period of colder climate known as the Younger Dryas (*c.*12,900–11,700 BP) just

Red Deer records, 1519–1772.
There are presence records
all over the map, but the only
areas that are shaded to show
free-roaming populations are
in Ireland, Wales, North England
and northern Scotland.

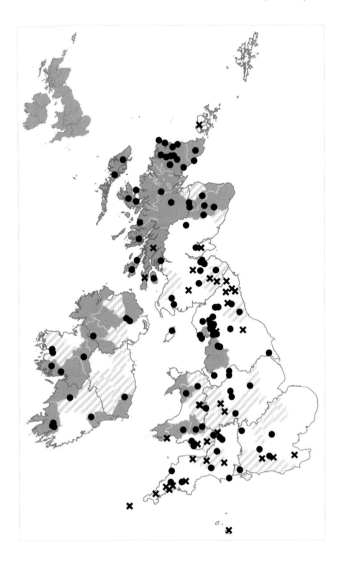

before the start of the Holocene (Montgomery et al. 2014; Carden et al. 2012). It then appears to have recolonised – or more probably have been reintroduced by humans – in the early Neolithic period. The native status of the species in Ireland therefore depends on which definition of native we use. In any case, by the early modern period the species had been present in Ireland for thousands of years and was well adapted to the environment. It went extinct again in the nineteenth century due to overhunting, except in Killarney (Fairley 1975: 22; Scharff 1918).

The Red Deer was also introduced to Orkney and the Outer Hebrides during the Neolithic or Early Bronze Age (Mulville 2010). Populations have survived in the Outer Hebrides to today, but the Orkney population seems to have died out in the early medieval period. It was attested to be absent from at least part of Orkney in the early modern sources (Pococke 1887: 136).

In Britain the Red Deer is thought to be native. As can be seen on the map, the Red Deer seems to have gone extinct as a free-roaming species across the whole of England and Lowland Scotland other than in the Lake District during the early modern period, but survived in deer parks, forests and in northern Scotland (Yalden 1999: 172).[4] Although there are still some presence records from Dartmoor local sources in the early modern period, other sources seem to suggest the species was extinct on Dartmoor and Exmoor:

> They [those with venfield rights] pay yearly four-pence … He shall also taken upon the forest [of Dartmoor] all that may do him good, except green-oak and venison, now a needless exception.
> (ed. Westcote 1845 [1630]: 85)

Populations in Devon were probably re-established in the nineteenth century when the Red Deer began to recover due to an increased interest in deer stalking and perhaps the creation of plantations. It has continued to expand its range through the twentieth century (Rose and Smith-Jones 2020; Mathews et al. 2018: 298–307; Williamson 2013: 155; Staines et al. 2008; Taylor 1948). At this point the species appears to have adapted from being predominantly a woodland dweller to being an upland specialist. At present the Red Deer is especially common across north and southwest Scotland, in Cos. Wicklow, Donegal and Kerry in Ireland and in the Lake District, East Anglia and the South West of England. It has a patchy distribution elsewhere but is quickly increasing due to escapes and releases. The species is likely to become widespread again across the uplands of Britain and Ireland in the near future. With the decline in hunting and the extinction of the Wolf, Red Deer grazing and browsing has now become a serious problem, especially in preventing the growth of new trees in ancient woodlands and the natural reafforestation of the uplands (Rao 2017; Manning et al. 2009).

FALLOW DEER

Dama dama

NATIVE STATUS Non-native (naturalised)

MODERN CONSERVATION STATUS	
World	Least Concern
UK	Not Evaluated
ROI	Least Concern
Trend since 1772	Probably increased

The Fallow Deer was widely recorded across early modern Britain and Ireland. It was naturalised in many upland areas.

RECOGNITION

The species was called *dama* in Latin and *fallow* in Early Modern English. These terms were specific to the Fallow Deer, so could not be confused. Other terms, especially *buck* and *doe* or *dae*, were used for Fallow Deer but could also apply to the Roe Deer. These records have been identified based on description.

---

4   History written about the Red Deer has been traditionally interested in distinguishing 'native' populations from those which have been restocked. This is an uncomfortable preoccupation because it suggests that populations with foreign ancestry might be less valued or less 'natural', even in areas where that species has lived for thousands of years. Extinctions and introductions are important to the history of Red Deer but need not prejudice opinions on local populations.

In Middle Scots there is a common formulaic term 'hart, hynd, dae and rae'. This is often taken to refer to the male (hart) and female (hind) Red Deer, the (female) Fallow Deer and the Roe Deer (Ritchie 1920: 286). *Dae* is occasionally used to gloss Latin *dama*, so clearly does sometimes mean the Fallow Deer. However, at other times it seems probable that it might refer to the female Roe Deer, just like *doe* in Early Modern English. Records using the term *dae* with no other clues are shown on this map as diamonds to indicate that they are uncertain.

The early modern sources use the English term *deere*, and the Middle Scots *deire* to refer generically to deer. Records using this term usually refer to deer parks, so have been included here on the map for the Fallow Deer, because the Fallow Deer was the species most commonly kept in parks by the early modern period (Williamson 2013: 14; Yalden 1999: 152–6). These records are shown as diamonds to indicate that they are unreliable.

DISTRIBUTION
The distribution of top-quality records for this species is statistically a poor fit with the known levels of recorder effort. There are fewer records than expected from Highland Scotland, Wales, Ulster and

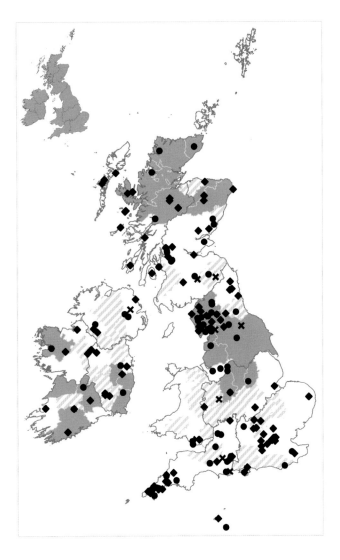

Fallow Deer records, 1519–1772. There are presence records all over the map, but the only areas that are shaded to show free-roaming populations are from the North and Midlands of England, northern Scotland, and Connacht, Munster and Leinster.

Leinster. There were more records than expected from the South West and North of England. Habitat suitability modelling based on the sites where the Fallow Deer was recorded in the early modern period suggests that the species would have been well adapted to conditions across Britain and Ireland. The distribution of the records may therefore reflect where the Fallow Deer is locally abundant. It is interesting to note that the areas with the lowest densities of emparked Fallow Deer are also the areas where Red Deer (p. 60) could still be found in the wild.

The Fallow Deer is one of the ten best-recorded species in the early modern sources. It occurred naturally in Britain in previous interglacial periods but is actually not naturally occurring in north-western Europe in the Holocene. It seems to have been introduced across the continent as part of a Europe-wide elite interest in hunting and feasting (Pluskowski 2018; Williamson 2013: 13–16; Sykes and Carden 2011; Sykes 2010; Langbein et al. 2008; McCormick 1999; Yalden 1999: 152–7; Kelly 1997: 20–22; Rackham 1986: 49–50). In England, Wales and Ireland it has been imported several times, and some populations have bred continuously here since being imported in the eleventh–thirteenth centuries, although others have been imported more recently. Initially in the medieval period, Fallow Deer were kept in royal deer parks and forests and protected for hunting, but by the early modern period records especially referred to individual herds of emparked animals which were transported and enclosed in private deer parks adjacent to stately houses. The smattering of absence records reflects where landowners less interested in Fallow Deer stopped maintaining a herd on their estates. There seem to have been emparked populations in every region of Britain and Ireland.

Although there was no Anglo-Norman invasion of Scotland, the idea of keeping deer parks was popular across Europe, and Fallow Deer seem to have been imported for this purpose in the second half of the medieval period (Lever 2009: 108; Yalden 1999: 156–7; Kitchener 1998; Ritchie 1920: 284–7). As well as emparked herds, there are also some early modern populations which were described like other species as being present (and therefore possibly free-roaming) throughout counties, rather than in particular parks. These have been indicated on the map with shading. The best example might be the description by Daniel Defoe of the landscape of Highland Scotland beyond Inverness:

> it is all one undistinguished range of mountains and woods, overspread with vast, and almost uninhabited rocks and steeps filled with deer innumerable, and of a great many kinds; among which are some of those the ancients called harts and roebucks, with vast overgrown stags and hinds of the red deer kind, and with fallow deer also.   (Defoe 1971 [1724–27]: 661)

Defoe (1971: 666–7) goes on to describe how in Caithness deer do not belong to anyone and can be killed wherever they are found. These records seem to be supported by references to Fallow Deer living in Ross-shire, Aberdeenshire and Banffshire by John Lesley (1675: 23–4) and Robert Gordon (in Blaeu 1662: 85, 104). Other possibly free-roaming populations lived in North England and the Peak District; and in Connacht, Leinster and Munster (Lever 2009: 103–13). Fallow Deer probably escaped from parks regularly and were especially known to have escaped when parks were abandoned or broken during the Wars of the Three Kingdoms (Lever 2009: 105; Ritchie 1920: 286–7). If these references are trustworthy, some of these populations, especially those in Highland Scotland, seem to have disappeared in the modern period.

The interest in keeping Fallow Deer in parks continued through the nineteenth century, but during the First and Second World Wars and the Irish War of Independence many of these parks were again broken and Fallow Deer escaped or were released into the surrounding countryside (Lever 2009: 105). They are now widespread in Wales and the South, South West and Midlands of England, with scattered populations in the North of England, Scotland and throughout Ireland (Langbein and Smith-Jones 2020; Mathews et al. 2018: 315–22; Langbein et al. 2008). The species is currently increasing and is likely to expand its distribution across the rest of Britain and Ireland where there is woodland habitat available.

ROE DEER

Capreolus capreolus

NATIVE STATUS Native to Britain, not Ireland

MODERN CONSERVATION STATUS	
World	Least Concern
UK	Least Concern
ROI	Not Present
Trend since 1772	Certainly increased

In the early modern period the Roe Deer was recorded as free-roaming in Scotland north of the central belt, including in the Outer Hebrides and North Wales, but was a rare captive species across most of England. It was absent from Ireland.

RECOGNITION

The reliable terms for this species used by our sources were *roe* and *roebuck* in Early Modern English, *rae* in Middle Scots, *iwrch* in Welsh and *capreolus* in Latin. All of these terms reliably identify the species intended.

Other less reliable terms were also commonly used. Male Roe Deer were called *bucks*, and female Roe Deer were called *does* in Early Modern English. These terms were not generally used for the Red Deer (where the males were *harts* and the females were *hinds*), but the same terms were also used for the Fallow Deer. The Latin term *caprea* was similarly used for both our species and the Goat (*Capra aegagrus hircus*). Luckily, these terms are always employed in contexts where it is obvious which species is implied.

DISTRIBUTION

The distribution of top-quality records for this species is statistically a poor fit with the known levels of recorder effort. It may have been locally distributed, locally abundant or of special local interest. In this case the early modern Roe Deer appears to have been common in Scotland north of the central belt, but absent from much of England and from Ireland. There are also absence records from Scilly, Jersey and Orkney, and national absence records for England, Ireland and (later) Wales (MacLysaght 1939: 416–17). Habitat suitability modelling based on the sites where the Roe Deer were recorded in the early modern period suggests that the species may have had specific requirements. It is only recorded on sites which today have cooler summers. However, in this case the distribution of the species is unlikely to be dictated by the climate. The early modern distribution of the Roe Deer instead suggests a species that had been hunted to extinction except in the north of Scotland.

Previous scholarship implies that the species went completely extinct in the seventeenth century in Lowland Scotland and the eighteenth century in England (Yalden 1999: 171; Ritchie 1920: 331–2). This is a dramatic change from the beginning of the medieval period, when English place-name evidence suggests the Roe Deer was fairly widespread (Yalden 1999: 135). Roe Deer were dependent on woodland and commonly hunted in medieval and early modern Britain, so deforestation and overhunting might have played a role in their decline (Yalden 1987; Ritchie 1920: 331–2).[5] The

---

5   Roe Deer, like other deer species, were the property of the monarch inside medieval English forests and strictly protected. Despite a court decision in 1338 that they should be considered beasts of warren rather than beasts of chase (Yalden, 1999: 171), they are almost never named as target species on warren licenses.

Roe Deer records, 1519–1772. There are records of free-roaming populations in Scotland north of the central belt and northern Wales, with local presence records from southern Wales, the Midlands and North of England and Lowland Scotland.

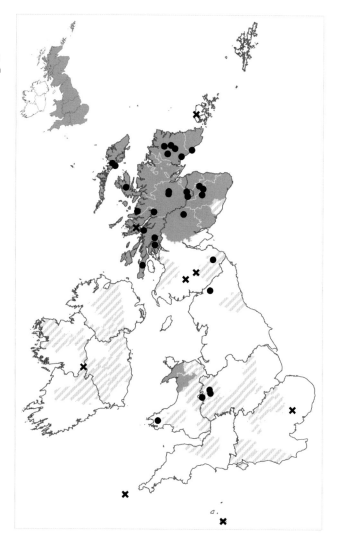

depleted early modern status of the Roe Deer corresponds approximately with the status of the Red Deer, except that the Red Deer was also recorded in captivity in hundreds of deer parks across Britain and Ireland. This does not seem to have been the case for the Roe Deer. A few captive populations are recorded, like the ones in Cumberland and Shropshire, but these are nowhere near as many or as widespread as the Red Deer references. It is possible that some additional records of Roe Deer might be hidden behind the generic references to *deer* which have been included on the map for the Fallow Deer, but for the most part it seems that Roe Deer were not as commonly kept. Yalden (1999: 152–3) observes that Roe Deer are much less suited to deer parks than Fallow Deer. Sykes (2006) likewise found that Roe Deer become less important in the archaeological record as Fallow Deer increase in the twelfth and thirteenth centuries. The early modern authors also attest that Roe Deer were absent from Ireland, meaning presumably not imported for deer parks during the period, although there were some experimental introductions in the nineteenth century (Fairley 1975: 31–2). The most likely scenario is that Roe Deer were not commonly included in early modern deer parks.

The final sources in and around Wales all come from the sixteenth century. The *Description of Cambria* written in 1584 records the Roe Deer as still present in Merionethshire and Caernavonshire:

> This [Merioneth]shire as well as Arvon, is full of cattell, foule, and fish, with great number of read
> deere and roes: but there is great scarsitie of corne.   (ed. Llwyd and Powel 1784: 9)

Many of the very few unusual records referring to Roe Deer living in deer parks also come from
around Wales in this time period. Roe Deer are described living in at least three Shropshire deer parks
in 1538–43 (Smith 1908: 80; Leland 1906: 54), and in Pembrokeshire deer parks, before 1603 (ed.
Owen 1994: 206). But when Edward Lhuyd came to edit the Welsh chapters of Camden's *Britannia*
150 years later, he listed the Roe Deer as an extinct species in Wales:

> Besides the Beaver, we have had formerly some other Beasts in Wales, which have been long
> since totally destroy'd. As first, Wolves; concerning which we read in this Author, in Meirionydh-
> shire, as also in Derbyshire and Yorkshire. Secondly, Roe-Bucks, call'd in Welsh Iyrchod; which
> have given names to several places; as Bryn yr Iwrch, Phynon yr Iwrch, Lhwyn Iwrch, &c. Thirdly,
> The Wild-Boar, whereof mention is made by Dr. Davies, at the end of his Dictionary. And lastly,
> I have offer'd some arguments to prove also that Bears were heretofore natives of this Island.
> (Gibson 1695: 644)

By 1695 there must have been few Roe Deer populations left in Britain south of the Scottish Highlands.

From the second half of the eighteenth century, the Roe Deer began to recover thanks to reafforestation
(Hewison and Staines 2008; Yalden 1999: 172; Ritchie 1920: 332). The species was reintroduced to
Lowland England at the start of the nineteenth century and had recolonised the south of Scotland
by around 1840. The species has now recovered, and is recorded in every region of Britain, although
there is still room for populations to increase in some parts, especially Wales, the Midlands and
South of England (Ward and Smith-Jones 2020; Mathews et al. 2018: 322–9; Tapper 1999: 116–19).

SEALS

Grey Seal (*Halichoerus grypus*) and Harbour Seal or Common Seal (*Phoca vitulina*) are the two species commonly seen around Britain and Ireland

NATIVE STATUS Native

MODERN CONSERVATION STATUS	
World	Least Concern
UK	Not Evaluated
ROI	Least Concern
Trend since 1772	No change

Seals were recorded all around Britain and Ireland in the early modern period.

RECOGNITION

Unfortunately, the early modern texts do not differentiate different seal species. The Early Modern English terms *seal*, *sea-calf* and *sea-dog*, the Middle Scots term *selch* (or *selchie*, hence the modern legend of the selkie, a seal that can take off its skin and become a human) and the Latin terms *phoca* and *Vitulus marinus* all seem to be used interchangeably. O'Sullivan (ed. 2009: 170) also attests the Irish term *roon* (*rón*). These terms are generally specific, except that the term sea-dog is also used for the Spurdog (*Squalus acanthias*), and perhaps once the Otter (more normally the *water-dog*).

The terms given do not distinguish different kinds of seal, either. We can identify the records as most likely to refer to either Grey Seal or Harbour Seal as these are the two species found commonly around Britain and Ireland, but it is not usually possible to distinguish between the two of them in early modern records. At times the species can be distinguished based on the description, so that, for example, Robert Sibbald's (2020: 10 (II:3)) record of 'a seal the size of a cow' is likely to refer to a Grey Seal based on the size.

DISTRIBUTION

The distribution of top-quality records for this species is statistically a poor fit with the known levels of recorder effort. It may have been locally distributed, locally abundant or of special local interest. The reason for this is the higher-than-expected number of records from the Highlands of Scotland and Connacht. Habitat suitability modelling based on the sites where seals were recorded in the early modern period suggests that these species would have been well adapted to conditions all around the coasts of Britain and Ireland, so this pattern likely reflects local abundance rather than absence. The northern and western coasts of Scotland are where seals remain most frequent today. All four records from Connacht are provided by Roderic O'Flaherty, who had an interest in seals as natural resources, as here in his description of the Barony of Moycullen:

> The countrey is generally commendable for fishing, fowling, and hunting. No river there towards the sea, and scarce any small brook, without salmons, white trouts, and eels. Many wast islands here, during the summer season, are all covered over with bird eggs, far more delicate than those laid by poultry. Here is yearly great slaughter made of seales, about Michaelmas, on wild rocks and wast islands of the sea.
> (ed. O'Flaherty 1846 [1684]: 95–6)

The number of records from Connacht probably therefore reflects both the personal interest of Roderic O'Flaherty and the local importance of the seal industry: these animals were hunted in the past for fur, meat and oil (Mac Laughlin 2010: 46–8, Yalden 1999: 170; Kelly 1997: 282–3; Gurney 1921: 87).

The demand for seal fur and oil rose after the end of the early modern period, and hunting intensified (Bolster 2012: 74–5; Yalden 1999: 170, 252–6). By 1914, the population of Grey Seals around Britain and Ireland was severely depleted. The species began to recover due to legal protection, and both the Harbour Seal and Grey Seal are now widespread around Britain and Ireland especially in the Hebrides and Northern Isles of Scotland. Until recently, due to the declines, seals were still rarely seen off the south-eastern and north-eastern coasts of England, except around in and around the Wash in East Anglia, but they are increasingly now recorded in these areas too (Thompson 2008, 2020b, 2020a; Hammond et al. 2008; Yalden 1999: 252–6). Unfortunately, at present the Harbour Seal population is declining again, perhaps due to the phocine distemper virus which has caused at least two mass mortality events.

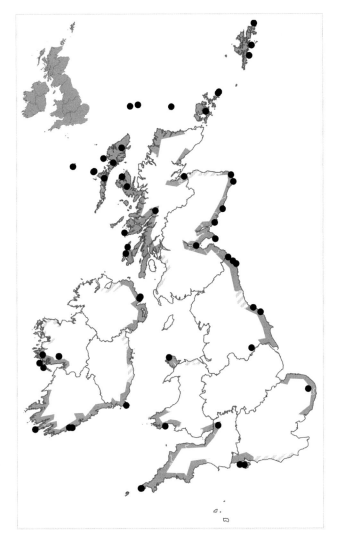

Seal records, 1519–1772. There are records of free-roaming populations around the coasts of every region of Britain and Ireland except the Midlands of England. There are also records from the Hebrides and the Northern Isles.

WALRUS
Odobenus rosmarus

NATIVE STATUS Believed to be non-native, but an occasional vagrant

MODERN CONSERVATION STATUS	
World	Vulnerable
UK	Not Present or Extinct?
ROI	Not Present
Trend since 1772	Uncertain

The early modern naturalists recorded the Walrus as a species often seen around Orkney and in the North Sea.

RECOGNITION
There was no fixed term for the species in early modern Britain and Ireland. Most early sources described it as an unknown type of *piscis* (fish), except for Sibbald (ed. 2020: 10 (II:3)) who identified it as the *walrus* and *mors*.

DISTRIBUTION
There are not enough top-quality records to statistically analyse the distribution of this species.

Hector Boece (ed. Sutton 2010, sec. 0.37) was the first early modern author to describe the Walrus. His description was soon afterward (1536) translated into Scots by Bellenden (ed. 1821: li). Boece described the Walrus as an ordinary resident of Orkney: Boece's Walrus sleeps hanging to rocks by its tusks. It can be caught if a hunter manages to bore a hole through its tail and tie a rope to it before the Walrus wakes up. Once killed, its body can be used to make oil and rope. Boece's tall tale does not inspire confidence in his record of the Walrus in Orkney, but the Atlantic Walrus was intensely hunted during the early modern period, with expeditions going from Europe to Nova Scotia for Walrus hides, ivory and oil (Bolster 2012: 72–4). In this context, the record might reflect the exploitation of a resident Orkney population by Walrus hunters. If this is the case, the population quickly became extinct. James Wallace (1693: 15), a minister resident on Orkney in the seventeenth century, attested that he had never heard of the species from anyone there.

The presence of the Walrus around Scotland, whether as a resident or as a vagrant, is supported to a limited extent by the archaeological evidence. Walrus remains have been found from Neolithic Skara Brae in Orkney and Bronze or Iron Age Jarlshof in Shetland (Yalden 1999: 229–30; Clarke 1998). Further corroborating evidence is provided by Robert Sibbald, who mentioned Boece's record, described the species in the North Sea in more detail and correctly identified it as the Walrus:

> The Walrus, or Mors, is also from the Phocidae family, another amphibious animal because it circulates the North Sea. It is a wild and powerful animal, hot to the touch, and it breathes in strongly through the nostrils … It crawls more than it moves. It has a thick, leathery skin and thick hair of a grey colour. It has a grunt just like a pig, it calls or grumbles with a strong, heavy voice. It has a very big head with two teeth thrust out, just like an elephant. It has long, wide and white tusks for defence which recline downwards across its chest.
>
> (ed. Sibbald 2020 [1684]: 10 (II:3))

Since Sibbald was based in the central belt of Scotland and was most knowledgeable about the species on the north shore of the Firth of Forth, a location has been approximated for his North Sea reference on the shore of Fife.

Walrus records, 1519–1772.
There is a contested record from
Orkney, and an uncertain record
from the North Sea off Scotland.

Today Walruses are only known around Britain and Ireland based on wandering individuals like Wally the Walrus, whom I had the pleasure of meeting on his tour of Pembrokeshire, Munster and South West England in 2021. There have been around 20 records of Walruses around Britain and Ireland (mainly around Scotland) between 1954 and 2008 (Hall 2008). They continue to be seen around Orkney, and animals were recorded there in 2013 and 2018 (Mathews and Coomber 2020).

RIGHT WHALE

Eubalaena glacialis

NATIVE STATUS Native

MODERN CONSERVATION STATUS	
World	Critically Endangered
UK	Extinct
ROI	Extinct
Trend since 1772	Uncertain

Although early modern authors often described large whales around Britain and Ireland, they provide only one possible record of the Right Whale.

RECOGNITION

The Right Whale did not have a common vernacular name in the early modern period, but it was often called the *Balaena vulgaris* (common whale) presumably because it was initially the most frequently seen of the large cetaceans (van den Hurk et al. 2022). It is now called the Right Whale because it was the right whale for whale-hunters to target, since the species is slow-moving, migrates along the coast, floats when killed and produced a great deal of oil (Kraus and Evans 2008; Yalden 1999: 169; Clark 1947). It was also the *Balaena* of Rondel or Gessner and the *Mysticetus* of Aristotle. In the most advanced species handbooks it was categorised as a baleen-whale which lacked a dorsal fin, or which had only two fins (not counting the tail). It was also often wrongly said to lack a blowhole or to have nostrils instead, since baleen whales have two blowholes in their heads. So Sibbald (1692: 64) calls the Right Whale *Balaena bipinnis caret fistula* (adapted from a title) and Ray (1713a: 6) calls it *Balaena vulgaris edentula, dorso non pinnato*.

Since the Bowhead Whale was only found far to the north of Britain and Ireland, all these terms should theoretically have been enough to distinguish the Right Whale from all other species. Linnaeus (1758: 75) understood the terms to refer to the Right Whale and Bowhead Whale (which he lumped together). However, despite having dorsal fins, other baleen species like the Blue Whale also seem to have been sometimes identified using these names, as for example the record of the 90-foot whale that stranded at Tynemouth, Northumberland in 1532, and the record of the 80-foot whale that stranded in Limekilns, Fife in 1642 (Hay 1938; Ray 1713a: 6–9 (II); Sibbald 1692: 64–6; Willughby & Ray 1686: 35–8). In practice, these terms were generic.

DISTRIBUTION

Right Whales were exploited in medieval and early modern Britain and Ireland for their baleen, oil, and likely also for food and bone (Raye 2015; Moffat et al. 2008; Gardiner 1997; Jenkins 1921: 39–40). There are two references to Right Whales from early modern England which are not immediately recognisable as misidentified larger species:

Taylor & Dale (1730: 409–10) include the Right Whale in their list of species found in Harwich and Dovercourt in Essex, but with no records of strandings and no measurements attached, which

Right Whale records, 1519–1772.
There is a single uncertain
record from Essex.

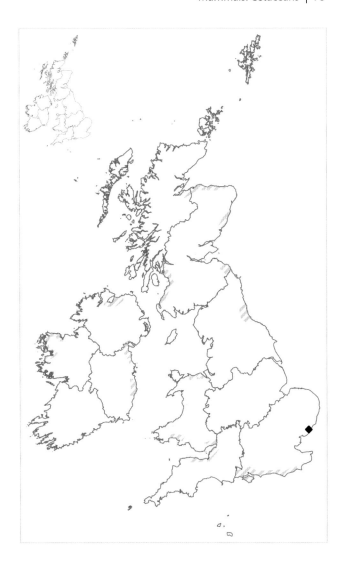

means they might have been referring to other baleen whales – especially since they cite Sibbald, and
Willughby, who describe Blue Whales using this name.

Francis Willughby and John Ray include the Right Whale on their national list of the fishes of England:

> *Balaena vulgaris* of Rondel. The Whale. Sometimes, but rarely it is found stranded on the sandy
> shore by the tide. (Willughby & Ray 1686: 25)

But again, this record does not offer a length in this section to confirm that Willughby & Ray were
not referring to the other baleen whales, and every record of this species which is provided elsewhere
in their volume (1686: 35–8) does clearly refer to a different species.

This is not to say that Right Whales were never seen. John Evelyn recorded a 58-foot animal that
seems to be this species (with 'onely two small finns') which was killed after swimming up the Thames
as far as Greenwich in London on 3 June 1658 (ed. Bray 1901: 323–4). But the Right Whale seems
to have been rare.

The fact that the name of this species is used regularly, but that almost every record using this name actually refers to a different species is striking, as is the lack of reliable records for this 'common whale' compared to, for instance, the Orca (p. 77) and the Sperm Whale (p. 74). These two facts together would fit with the theory that the movement of the whaling industry to Newfoundland and elsewhere in the sixteenth century may have been prompted by a temporary depletion of Western European populations of Right Whales, including those in British and Irish waters (Aguilar 2015; Bolster 2012: 32; Jenkins 1921: 47–8).

The Right Whale is likely to have been among the first of the cetaceans to be targeted by the industrial whaling industry (Aguilar 2015; Reeves, Smith and Josephson 2007). With the movement of the whaling industry to the West Atlantic, Right Whale stocks may have had some time to recover, but around the turn of the twentieth century, whaling resumed around Britain and Ireland, as well as in Cintra Bay in western Sahara, where the East Atlantic Right Whales may have bred (Ryan et al. 2022; Reeves 2020; Kraus and Evans 2008; Jacobsen et al. 2004; Collett 1909). This seems to have led to the final extinction of the Right Whale in the East Atlantic, and there have been only sporadic sightings for the last century. The species is also critically endangered in the West Atlantic. The remainder of the Atlantic population lives close to shore on the east coast of North America. It is threatened especially by entanglement in fishing lines, and ship strike (Kraus and Evans 2008). Right Whales from the West Atlantic are occasionally sighted around Europe and, if they can overcome the threats facing them, might one day be able to recolonise British and Irish waters.

SPERM WHALE

Physeter macrocephalus

NATIVE STATUS Native

MODERN CONSERVATION STATUS	
World	Vulnerable
UK	Not Evaluated
ROI	Not Evaluated
Trend since 1772	Uncertain

Sperm Whales were recorded most commonly on the east coast of Britain during the early modern period. The records probably represent stranded individuals rather than the normal range of the species.

RECOGNITION

This species was most reliably called the *spermaceti whale* or *cachelot* in Early Modern English and *physeter* in New Latin. These terms theoretically identify the animal to species level, but cetaceans were poorly known in the early modern period and therefore the records on this map are less reliable than records of more familiar kinds of animal.

The generic terms for large cetaceans are used by the natural history sources, including Early Modern English *whale*, Middle Scots *quhail*, Welsh *morfil*, Irish *mile-more*, Latin *cetus* or *balaena*. These records have not been included in this volume, but the map of the Sperm Whale does include records of Early Modern English *whirlepools* and *thirlpoles*. These terms often seem to refer to the Sperm

Whale (Risdon 1811: 147) but could also refer other large whale species, and therefore are also shown as diamonds, unless accompanied by more reliable terms.

The New Latin term *Balaena macrocephala* is more complicated. Robert Sibbald (1692: 30–45) distinguished three species of *Balaena macrocephala*, one with just two flippers, one with two flippers, a dorsal fin and curved teeth in the lower jaw, and one with two flippers, a dorsal fin and less curved, blunt teeth. He distinguished all of these species from his ordinary Sperm Whale. He was essentially followed in this categorisation by John Ray (1713a: 15–16 (II)), and then Linnaeus (1758: 76–7), who identified the three species as *Physeter macrocephalus*, *Physeter microps* and *Physeter tursio*. All three of these species have now been lumped together as the Sperm Whale. Records using these terms are included on the map as diamonds.

This map also does not include records of ambergris (the waxy substance secreted by Sperm Whales and used in perfume), although it was regularly recorded in the early modern period. Ambergris cannot be used to map where whales have been because it can float for years before washing up on shore.

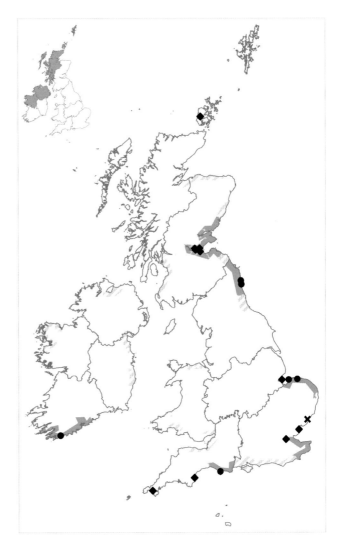

Sperm Whale records, 1519–1772. There are reliable records dotted along the North Sea coast from Kent to the Firth of Tay, with an uncertain record from Orkney. There are also reliable records from South West England and Munster.

DISTRIBUTION

The distribution of top-quality records for this species is not statistically different from the known level of recorder effort. The gaps on the map may just reflect decreased survey effort in some regions. It may have been widespread.

At time of writing, Sperm Whales are most often seen in the Atlantic Ocean and only rarely in the Celtic Sea, Irish Sea, North Sea and English Channel. The lack of reliable records from the north of Scotland and Ireland in the early modern period is therefore surprising. The most probable explanation for this is that Sperm Whales were being mapped where they were found stranded. They are a deep-water species that were probably normally not seen, and especially not identified to species level, except when they were stranded. Although the North Sea is not part of the Sperm Whale's usual range, it forms a natural 'Sperm Whale trap' and strandings have been common here over the last four hundred years. Males migrating south during the winter are especially likely to become stranded in the North Sea (Gordon and Evans 2008; Smeenk 1997).

Stranded Sperm Whales were of interest around Britain and Ireland because they were large, but also as they were commercially valuable. Spermaceti, a liquid wax found inside the head of Sperm Whales, was the most valuable part of the animal. Spermaceti was not just used for fuel; a long list of seventeenth-century medical uses is provided by Robert Sibbald:

> For colic, for anyone affected by hypochondria, for children's stomach pain, for epilepsy, for contracted nerves, for missed periods, for coagulated milk in breasts, for a blood clot, for falling from a height, for suffocating catarrh from stagnation of blood around the heart or rising to the vessels in the brain, for asthma, for a cough, hoarseness, and also for external disorders, especially for mending bone fractures and for removing smallpox scars, it has outstanding virtues. As for its use for colour and artwork I will be silent.   (translated from Sibbald 1692: 52)

Despite the value of spermaceti, Sperm Whales do not seem to have been widely exploited until around the end of the early modern period (Whitehead 2002, 2006; de Smet 1978; Jenkins 1921: 37–40, 207–26). The Sperm Whale industry had two phases. Open-boat whaling for Sperm Whales started in the eighteenth century, and reached its peak in the nineteenth century as the numbers of animals available declined. Modern whaling started in the twentieth century as the smaller population of Sperm Whales could now be commercially targeted with steam or engine-powered ships with iron hulls and harpoon guns. Sperm Whales were exploited especially intensively in the middle of the twentieth century, when stocks of the large baleen whales had been completely depleted. The sperm whale industry peaked in the 1960s, and was finally stopped (for the most part) by the international moratorium on whaling in the 1980s. The industry had probably led to worldwide Sperm Whale populations being suppressed by around 68% between 1712 and 1999 (Whitehead 2002). The species' population is now recovering.

ORCA/KILLER WHALE

Orcinus orca

NATIVE STATUS Native

MODERN CONSERVATION STATUS	
World	Data Deficient
UK	Not Evaluated
ROI	Not Evaluated
Trend since 1772	Certainly declined

The Orca was occasionally recorded around early modern Britain and Ireland, including in the South of England where it is now rare.

RECOGNITION

The Orca was the species usually intended by the terms *grampus, north-* or *sea-caper* in Early Modern English and Middle Scots, and *orca* in New Latin. However, these terms were also occasionally (confusedly?) applied to other species (Browne 1902: 91; Sibbald 1803: 117), so the records given as diamonds (which are only identified by one of the names rather than both *grampus*, and *orca*) should be considered less certain than the ordinary (dot) records.

The New Latin terms *Balaena minor utraque maxilla dentata*, *Porcus [marinus] maximus* and *Porcus marinus major* (once given in English as *great porpoise*) seem to more reliably refer to this species, as long as the latter terms are distinguished from the ordinary *Porcus marinus* (meaning the Porpoise or a dolphin).

DISTRIBUTION

The distribution of top-quality records for this species is not statistically different from the known level of recorder effort. The gaps on the map may just reflect decreased survey effort in some regions. It may have been widespread.

In the early modern period, records of Orcas are dotted around Britain and Ireland, especially on the North Sea coast. This does not match with the modern distribution of the species. Today Orca populations are based around the Northern Isles and Hebrides (the West Coast Community), with some sightings along the eastern coast of Scotland and the western coast of Ireland, and occasional sightings in South West England, Wales and the North of England (Evans and Waggitt 2020a; Boran et al. 2008). It appears that the Orca occasionally ranged much further south into the North Sea in the early modern period (Browne 1902: 34; Taylor and Dale 1730: 412–13). This is supported by the archaeological record. Orca remains have been found from excavations of early medieval Flixborough near the Humber estuary, as well as late medieval or early modern Castle of Brederode in Velsen, in the Netherlands (van den Hurk 2020; Zeiler and Kompanje 2010; Dobney et al. 2007). The change in distribution might reflect the Orca moving to adapt to changing conditions or changing climate, like the Common and Bottlenose Dolphins (p. 79).

Some of the early modern records describe the diet of the Orcas. This is important because Orca communities can be divided based on their diets, and individuals within each community seem to eat the same foods. In the modern period there are at least two communities of Orcas found around Britain and Ireland. Some Orcas are resident on the west coast of Scotland. They have been recorded eating cetaceans (including Porpoises) as well as fishes (HWDT 2018: 49–53; Beck et al.

Orca records, 1519–1772. There are reliable records from the North Sea coast between the Firth of Tay and East Anglia, as well as from South West England and Leinster. There are also unreliable records from Connacht and south-western Wales.

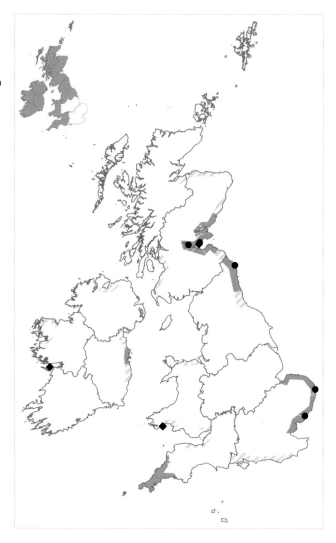

2014; Mäkeläinen et al. 2014; Foote et al. 2009, 2010; McHugh et al. 2007). Other Orcas seem to visit the Northern Isles of Scotland each summer and then migrate to Iceland and each winter. Of the Northern Isles Orcas, some eat Herring and some eat seals, possibly with little overlap between the two groups (Deecke et al. 2011). According to the early modern sources, Orcas around Britain were especially known to consume smaller cetaceans such as Porpoises (p. 81):

> Nor does [the orca] spare the porpoise which is similar to it. In fact, I dissected a porpoise this last year gone, whose eye and cheek, as it seemed, an orca had torn apart. Also, they commonly hunt porpoises and dolphins in the Firths of Forth and Tay, and they have violent fights with them, not without mutual disaster, since horrible noises precede them.
> (translated from Sibbald 1692: 18–19)

From this account, we can infer that the modern cetacean-eating West Coast Community of Orcas might be related to a now-lost community on the east coast. Accounts of marine-mammal eating are corroborated by other sources (Wallis 1769: 377; Borlase 1758: 263; Ray 1693: 15 (II)), although there are also records of Orcas hunting the Common Thresher (*Alopias vulpinus*) and Pilchard (Dunstanville 1811: 99). Some care must be taken with these sources because they echo a more influential account in the *Natural History* of Pliny the Elder (Rackham 1947, sec. IX.5), a Roman writer to whom classically educated natural history writers have always liked to refer. However, the records of the orca given in the early modern sources appears sufficiently different to provide evidence of the diet of early modern Orcas.

Today, Orcas are not doing well around Britain and Ireland. There are fewer than ten individuals left in the resident West Coast Community of Orcas, and they have not successfully bred for at least 20 years (Beck et al. 2014; Foote et al. 2010). The migratory Northern Isles Community remains larger but, like the West Coast Community, it is still likely to be threatened by declines in Herring (p. 257) and seal (p. 68) availability and pollution from toxic PCBs (polychlorinated biphenyls). PCBs were banned in the European Union in the 1980s but have persisted in the environment and tend to accumulate at the top of the food chain (Desforges et al. 2018; Jepson et al. 2016). Orcas in industrialised areas that hunt piscivorous mammals like those around Britain and Ireland seem to carry an especially high load of these chemicals. Inbreeding appears to also be an issue for the West Coast Community (Foote et al. 2021).

DOLPHINS

esp. Bottlenose Dolphin (*Tursiops truncatus*), but also often Common Dolphin (*Delphinus delphis*), White-beaked Dolphin (*Lagenorhynchus albirostris)* and Risso's Dolphin (*Grampus griseus*)

NATIVE STATUS Native

MODERN CONSERVATION STATUS (*T. truncatus*)	
World	Least Concern
UK	Least Concern
ROI	Not Evaluated
Trend since 1772	Uncertain

Dolphins were recorded by a handful of sources around early modern Britain.

RECOGNITION

Several species of dolphin are found around Britain and Ireland. The most visible is the Bottlenose Dolphin, which is resident year-round close to shore in some areas, but the Common Dolphin, White-beaked Dolphin and Risso's Dolphin are more common in some regions and would have been familiar to sailors and fishers on boats, who sometimes passed their records on to the compilers of natural histories during the early modern period.

This map primarily shows records using the English term *dolphin*, Welsh *dolffin* and Latin *delphinus*. These terms are usually used for one of the species we would still call a dolphin today but are rarely used for the Porpoise (p. 81). The Middle Scots *meerswine* and New Latin *Porcus or Sus marinus* are even less reliable terms for the dolphin species, and the related Early Modern English terms *marsuin* and Irish *mucc-marra* actually usually refer to the Porpoise but might more rarely have referred to one of the dolphins. The Middle Scots term *pellok* also usually referred to the Porpoise but was rarely used to refer to the dolphins. Lastly, the term *thornpole* is used twice in the early modern sources, both times to refer to the dolphins. All of these terms need to be treated with suspicion (Gardiner

1997). Luckily though, these records can be identified based on context: natural history texts that refer to the dolphins almost always refer to the Porpoise separately using a different term or describe the species they intend, so that records of the Porpoise can be distinguished from records of the dolphins, especially where multiple terms are used (as in the quotation from Sibbald in 'Porpoise', p. 81).

DISTRIBUTION

The distribution of top-quality records for this species is not statistically different from the known level of recorder effort. The gaps on the map may just reflect lower survey effort in some regions. It may have been widespread. This means the lack of records from Ireland, for example, is presumably just a reflection of the lack of records overall.

Dolphins were probably not as commonly exploited as Porpoises in the early modern period. Based on the number of early modern records there may even have been less interest in them. However, dolphins do still seem to have been sometimes exploited for their food, oil and bone in medieval and early modern Britain and Ireland (van den Hurk et al. 2021; Dobney et al. 2007; Gardiner 1997; Kelly 1997: 283), as in the following record by Thomas Browne of Norfolk:

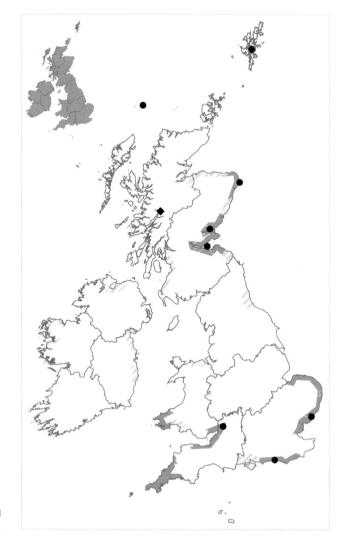

Dolphin records, 1519–1772.
There are reliable records from the South and South West of England, Wales and Lowland Scotland as well as from North Rona in the Hebrides.

The Tursio or porpose is common the Dolphin more rare though [dolphins are] sometimes taken which many confound with the porpose, butt it hath a more waved line along the skinne sharper toward the tayle the head longer and nose more extended which maketh good the figure of Rondeletius. The flesh more red & well cooked of very good taste to most palates & exceedeth that of porpose.   (ed. Browne 1902 [*c*.1662–68]: 34)

At present, there are resident Bottlenose Dolphin populations off Munster, Highland Scotland, and Wales, with smaller resident populations off Connacht and South West England (Evans and Waggitt 2020b; Wilson 2008). Other species of dolphins are also common around Ireland, in the western English Channel, and off the east coast of Scotland and North England (Evans 2008; Evans and Smeenk 2008; Murphy et al. 2008). This approximately fits with the early modern records, but dolphin distribution patterns are very likely to have dramatically shifted over the last few centuries, especially following the end of the Little Ice Age in the eighteenth century. Dolphins are known to change their distribution in response to environmental factors and prey availability. In the mid-twentieth century, the numbers of Common Dolphins and Bottlenose Dolphins sighted in the North Sea and Channel temporarily declined, but more Common Dolphins were recorded off the Netherlands and Denmark (Evans and Waggitt 2020b; Wilson 2008; Murphy et al. 2008). During this period the Cornish population of resident Bottlenose Dolphins was lost for around 20 years, but a new resident population was established there in the 1990s (Wood 1998; Tregenza 1992). Bottlenose Dolphins are also thought to have only colonised the Moray Firth in the twentieth century, where they are now resident (Wilson 1995: 23–4). Of course, populations can also change due to extinction. A thousand years ago, during the early medieval period there seems to have been a resident population of Bottlenose Dolphins in the Humber estuary in the North Sea, based on remains found at Flixborough (Dobney et al. 2007). Bottlenose Dolphins are rarely seen there today, and it seems the population might have been hunted to extinction, since mitochondrial and microsatellite analysis of the remains suggests that the animals were from a population differentiated from all known modern populations. The number of records from the east coast of Scotland during the early modern period might suggest that there was a resident population of Bottlenose Dolphins there, but might equally be explained by the presence of recorders especially interested in marine mammals.

PORPOISE
Phocoena phocoena

NATIVE STATUS Native

MODERN CONSERVATION STATUS	
World	Least Concern
UK	Not Evaluated
ROI	Not Evaluated
Trend since 1772	No change

Porpoises were recorded as living all around the coasts of Britain and Ireland during the early modern period.

RECOGNITION
The most usual New Latin term for the Porpoise was *phocaena*. In Early Modern English, authors called this species some variant of *porpasse* or *porcupiss* (some early modern authors understood this

Porpoise records, 1519–1772.
There are reliable records from
every region of Britain and Ireland.

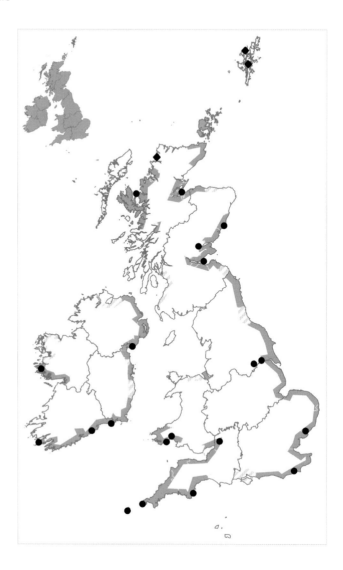

to be short for Latin 'porcus-piscis': 'pig-fish'). These terms were specific to the Porpoise, but need to be distinguished from the term *great porpoise* (and *Porcus marinus major*) which referred to the Orca.

Confusingly, the terms which translate as 'sea-pig' instead of 'pig-fish' were more ambiguous. Irish *mucc-marra* was used to refer to the Porpoise, but New Latin *Porcus marinus* (once *Sus marini*) was used for both species. Middle Scots *meerswine* also usually refers to the dolphins, but the closely related Early Modern English terms *marsuin* and *sea-swine* refer to the Porpoise. It is most probable that all the terms meaning 'sea-pig' had some degree of flexibility (Gardiner 1997).

Morris (1747) attested the Welsh terms to be *llamhidydd* and *pysgod du*. Middle Scots writers also used the term *pellok* which most usually refers to the Porpoise, but occasionally also refers to other small cetaceans.

The species intended in any given record can usually be identified where multiple terms are employed together, where the species is described, or where the source makes a distinction between the two species, as here in Robert Sibbald's *Scotia Illustrata* (1684):

> Porcus marinus chases after herrings and is believed to be the Dolphin.
>
> Phocaena, the Porpus or Porpoise, a species of dolphin. It is shorter and fatter in body.
>
> The taste of dolphin-flesh is like beef, nor is it said to provide much food.
>
> Oil (ordinarily called train-oyl) is decocted from the fatty flesh of these fish. This is used for the preparation of black soap, also in lamp-oil, and finally to anoint leather by leather- and pelt-workers. (Sibbald 2020: 23 (II:3))

Uncertain records have been shown as diamonds.

DISTRIBUTION

The distribution of top-quality records for this species is not statistically different from the known level of recorder effort. The gaps on the map may just reflect decreased survey effort in some regions. It may have been widespread.

The porpoise was regularly exploited in medieval and early modern Britain and Ireland for oil, food and bone (van den Hurk et al. 2021; Cumming 2019; Heal 2008; Sabin et al. 1999; Gardiner 1997). Carew describes how pods of what seem to be Porpoises were caught in early modern Cornwall when they chased fish into estuaries (ed. Chynoweth et al. 2004, fol. 31v). At the beginning of the early modern period, in England Porpoise meat was a high-status food which could be eaten at feasts or given as a gift, but Porpoise meat went out of fashion with Badger and Heron during the early modern period (Thomas 1991, chap. 2.i). It may never have been commonly eaten in Ireland or Scotland, based on the quotation above from Sibbald and the quotation following from John Rutty of Dublin:

> It is frequent on our coast, and is request for its Oil, Train Oil being made of it and the Seal on the coast of Youghall.
>
> The flesh is said to be fat and rank; however it was formerly frequently eaten in England, and is often eaten by Sailors. (Rutty 1772: 370)

The Porpoise is widespread around Britain and Ireland today, just as it was in the early modern period (Evans and Waggitt 2020c; Evans et al. 2008). Its population has not always been completely stable. In the 1960s/1970s, the Porpoise became rare in the southern part of the North Sea and in the English Channel, but populations recovered in the 1990s. The cause for this temporary decline is not known, but it may be connected to observed fluctuations in supply of the Porpoise's prey species (like the sandeels, p. 283, or the Herring, p. 257).

BRENT GOOSE
Branta bernicla

NATIVE STATUS Native

MODERN CONSERVATION STATUS	
World	Least Concern
UK	Amber List
ROI	Amber List
Trend since 1772	Uncertain

Several early modern sources recorded the Brent Goose around the coasts of Britain and Ireland.

RECOGNITION
This species was most reliably called *branta* in New Latin and the *brent goose*, or alternatively (Gurney 1921: 92) the *rat-, ret-* or *rout-goose* in Early Modern English and Scots. Martin Martin also attests the term *gawlin* from Uist, which could perhaps be a mistake for the word which becomes modern Scottish Gaelic *gùirnean*.

The species is occasionally referred to (mistakenly) as the *soland goose*, but this term was far more commonly used to refer to the Gannet. It is also regularly referred to as the *bernicla* in New Latin, or *barnacle* in Early Modern English. This term is unreliable and more commonly used to refer to Barnacle Goose. Records that refer to the *solan goose*, *barnacle* or *bernicla* have only been included on this map where other, more reliable terms are also used.

DISTRIBUTION
There are not enough top-quality records to statistically analyse the distribution of this species. The records suggest an early modern population wintered in the north of Scotland and Ireland and on the east coast of England.

The Brent Goose today is found overwintering around the whole coast of Britain and Ireland, but it is less commonly seen on the north and west coasts of Scotland than elsewhere. This might suggest a shift in distribution. The species seems to have been listed as present in a few texts from early modern north and west Scotland. Most clearly, 'routs' are one of the birds mentioned as present in the seventeenth-century *Genealogical History of the Earldom of Sutherland* (Gordon 1813: 3). This must have been a predecessor to the heavily persecuted population mentioned in the Moray Firth from the 1890s onward (Grundy 2007). In the North Sea there was a population decline in the mid-twentieth century, which is now thought to have been a result of the over-exploitation of eggs and unsustainable shooting of birds during nesting season (Shrubb 2013: 74, 89–90). The large flocks of northern and western Scotland were lost at this point, and the Scottish populations have not recovered to the same numbers that they had in the nineteenth century (Grundy 2007). There has also been a recent decline in Wales in particular, perhaps due to climate change (Pritchard 2021a), but the early modern evidence for a Brent Goose population in Wales is less reliable, so the population there may have been a modern colonisation.

Brent Goose records, 1519–1772.
There are reliable records from
the coasts of Ulster, Highland
Scotland and eastern England.

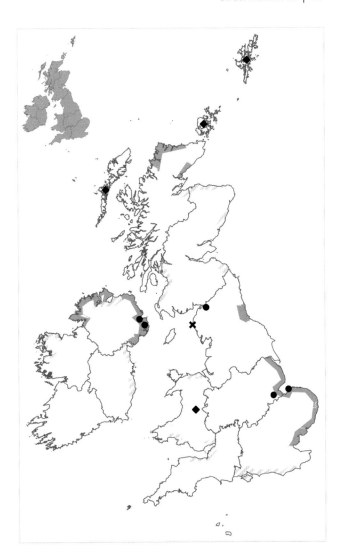

BARNACLE GOOSE
Branta leucopsis

NATIVE STATUS Native

MODERN CONSERVATION STATUS	
World	Least Concern
UK	Amber List
ROI	Amber List
Trend since 1772	Uncertain

The Barnacle Goose was recorded in a few reliable early modern sources around the coasts of Britain and Ireland. There were also many unreliable references to barnacle geese from every region of Britain and Ireland which might refer to either the Barnacle Goose or Brent Goose.

RECOGNITION

This species is reliably distinguished in Middle Scots as the *claik goose*. However, most references to this species use the New Latin *bernicla* or the English term *barnacle* (as the species is still called today). The Irish term *girrinn* (meaning barnacle) is also used. The species was called this because it was popularly believed to grow from a goose barnacle (especially the Common Goose Barnacle *Lepas anatifera*, p. 347). The terms *bernicla* and *barnacle* were doubly unfortunate because they were also applied to the Brent Goose (note that the Brent Goose is our modern *Branta bernicla*, p. 84). Records using these names are therefore not entirely reliable (and have been shown as diamonds), but are most likely to refer to the Barnacle Goose since Brent Goose had its own Early Modern English name (*brent goose*, etc.).

This species was also sometimes confusedly called the *soland goose*, but this name more usually refers to the Gannet. Both species were thought of as iconic Scottish birds, and the uncertainty seems to have come from people mixing up the names of the two. The following is perhaps the most unclear of all the references, which is shown as a diamond on our map:

> Here [in West-Connacht] is the bird engendered by the sea out of timber long lying in sea. Some call them clakes and soland-geese, some puffins, others bernacles, because they resemble them. We call them girrinn. I omit other ordinary fowl and birds, as bernacles, wild geese, swans, cocks of the wood, and woodcocks, choughs, rooks, Cornish choughs with red legs and bills, &c.
> (ed. O'Flaherty 1846 [1684]: 12–13)

DISTRIBUTION

There are not enough top-quality records to statistically analyse the distribution of this species. If the unreliable references were trustworthy, it would seem to be widespread as an overwintering bird, with the species recorded in every region and on nearly every coast.

The Barnacle Goose was of special interest to the early modern writers because it was popularly believed to grow from a Goose Barnacle (p. 347) (Pastore 2021; Marren and Mabey 2010: 59–60; Kelly 1997: 300). Like the Common Frog (p. 307), this species was thus a special case-study for the minority of early modern natural history writers who believed in spontaneous reproduction.

The Barnacle Goose has benefitted significantly from twentieth-century conservation policies. In Britain and north-west Europe, the Greenland-breeding population increased by 8× and the Svalbard-breeding population increased by 23× between the 1950s and 1990s, although some of this change is

Barnacle Goose records, 1519–1772. There are records from the coasts of every region of Britain and Ireland, but the records from Wales, the South West and Midlands of England and from Connacht and Leinster are unreliable.

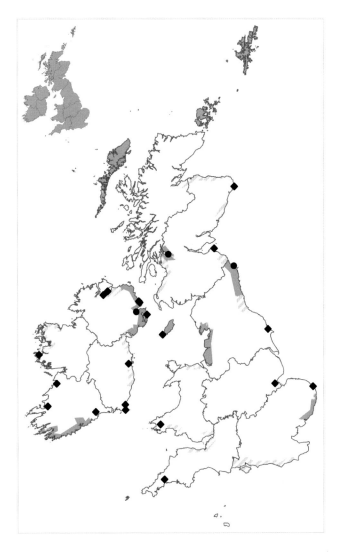

likely to be due to difference in counting methodologies (Shrubb 2013: 75). Today, the Barnacle Goose is recorded overwintering all around the coasts of Britain and Ireland, although it is less commonly recorded off the south and east coasts of Ireland, and off Wales and Cornwall, than elsewhere.

Since the 1980s, the Barnacle Goose has moved inland and also started nesting in Britain for the first time. This is partially due to a modern feral population which both winters and breeds in Britain, perhaps descended from escaped collection birds. This appears to be a modern phenomenon, although Yalden notes some inland archaeological remains from, for example, medieval Oxford and Roman York (Pritchard 2021b; Yalden and Albarella 2009; Holloway 1996: 430). Within Britain, the range of the Barnacle Goose expanded by 79% between 1981–84 and 2007–11 (Balmer et al. 2013). While there is no evidence that the Barnacle Goose has colonised new regions, it is probably much more widely found within those regions now than it was in the early modern period, when there was no evidence of inland populations.

GREYLAG GOOSE

esp. *Anser anser*

NATIVE STATUS Native

MODERN CONSERVATION STATUS	
World	Least Concern
UK	Amber List
ROI	Amber List
Trend since 1772	Certainly declined

Greylag geese were recorded in the early modern sources from every region of Britain and Ireland, inland as well as on the coasts.

RECOGNITION

This map shows the locations of what are called in Early Modern English *wild geese* and in New Latin *Anser ferus*. Despite appearance, these terms were thought of as species names used particularly for *Anser anser* – Willughby and Ray (Ray 1678: 358–61) distinguish this species from the Domestic Goose (*Anser anser domesticus*), Swan Goose (*Anser cygnoides*), Barnacle Goose (*Branta leucopsis*), Brent Goose (*Branta bernicla*), Canada Goose (*Branta canadensis*) and Spur-winged Goose (*Plectropterus gambensis*).

However, the early modern authors did not generally distinguish other less common species that resembled the Greylag Goose like the Pink-footed Goose (*Anser brachyrhynchus*) and the Bean Goose (*Anser fabalis*). This implies that these may have been lumped in with the Greylag, although the Greylag is now by far the most common of these species, and the only summer breeder, so most of the records on this map are likely to be to the Greylag. The term Greylag was not common until after our period, it may have been popularised by Pennant (Pennant 1776a: 570).

DISTRIBUTION

The distribution of top-quality records for this species is not statistically different from the known level of recorder effort. The gaps on the map may just reflect decreased survey effort in some regions. It may have been widespread. If we include the county-level records, the Greylag Goose was recorded in every region of early modern Britain and Ireland, including Connacht and Cornwall where the species is now rarer.

As with the Brent Goose, North Sea populations of Greylag Geese were reduced by overhunting in the twentieth century (Shrubb 2013: 70). South of Highland Scotland, unsustainable exploitation meant that the Greylag actually went extinct in Britain after the early modern period, and the species was reintroduced from populations in the Hebrides from the 1930s onwards (Pritchard 2021c; Holloway 1996: 68–9; Parslow 1973: 48).

The Greylag is now widespread in Britain again, although distribution remains patchier in the South West, Wales and Ireland than in England and Scotland (Balmer et al. 2013). This means that the nineteenth- and twentieth-century decline in the species is mostly invisible in our trend since 1772. In fact, due to reintroductions, the species may even have expanded its breeding range slightly since the early modern period. The only record from our early modern sources in Wales describes the bird as a migratory visitor, not a breeder:

Greylag Goose records, 1519–1772. There are records from every region of Britain and Ireland, including the Hebrides and the Northern Isles.

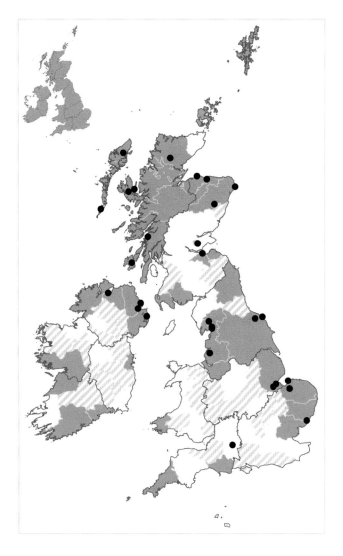

some are found always in season, as the grouse, heathcock, and wooquist, the crane, the heronshaw, the gull kept and fed, the curlew, etc. Some others are but at season, as the woodcock, the wild goose, wild duck, bittern, wild swan, etc.   (ed. Owen 1994 [1603]: 130)

This implies that the Welsh population of the Greylag may be a recent innovation. The population is expected to decline in the future due to climate change (Pritchard 2021c).

SHELDUCK
Tadorna tadorna

NATIVE STATUS Native

MODERN CONSERVATION STATUS	
World	Least Concern
UK	Amber List
ROI	Amber List
Trend since 1772	Uncertain

The Shelduck was commonly recorded in our early modern sources, especially in coastal counties to the east and south of England, Ireland and Wales.

RECOGNITION
This species was most commonly called *tadorna* and *vulpanser* in New Latin, and *shell-drake* and *burrow-duck* in Early Modern English. These terms are all specific to the Shelduck. Less specific names include the term *burranet* (which, surprisingly, is once also used to describe the Brown Shrimp) and also *burgander*, which is also used *for Mergus merganser* (Goosander).

DISTRIBUTION
The distribution of top-quality records for this species is not statistically different from the known level of recorder effort. The gaps on the map may just reflect decreased survey effort in some regions. It may have been widespread at least on the coasts of Britain and Ireland, despite the lack of records from Scotland.

The map for the Shelduck assumes that the species was only/mainly found on the coasts in early modern Britain and Ireland, except where there are site-level records to the contrary (as there are in Norfolk and Westmorland). The inland record from Wimborne St Giles in Dorset is to a captive menagerie population (ed. Pococke 1889: 138).

Today the Shelduck is recorded widely around Britain and Ireland, especially on the coasts but (increasingly in England) also sometimes inland (Balmer et al. 2013). The colonisation of inland sites in England seems to have been a twentieth-century phenomenon: almost all of the early modern records are on or near the coast, as they remain in Scotland, Wales and the island of Ireland, with the exception of a record from Norfolk. Despite the known use of Rabbit warrens by Shelducks (Holloway 1996: 72), there is not sufficient evidence to suggest that they colonised the warren sites of southern England unlike, for example, the Wheatear (p. 197). In some areas the species seems to have been rare:

> Of wild-birds, driven here by the extremity of the weather, we have all sorts … Ducks of all kinds, the true wild-duck breeding in the marsh betwixt Penzance and Marazion; widgeon, teal, woodcock, snipe, &c. The shell-drake (Tadorna Bellonii) is rare, but in the hard winter 1739, I had one brought me exactly answering the description of Ray's Willughby.   (Borlase 1758: 244–5)

In the past, the Shelduck was hunted for food and its eggs were collected (Shrubb 2013: 75, 90). This might have led to a decline in Scotland, since the bird was given legal protection, presumably because it was identified as a vulnerable species in the early modern period, and its range appears to have been limited to the north and west of the country by the eighteenth century (Patterson 2007). The Shelduck became much more widespread in Scotland in the nineteenth century, although it seems to have been declining in England in the same period (Patterson 2007; Parslow 1973).

Shelduck records, 1519–1772.
There are records from every
region of Britain south of Scotland
and from the southern and
eastern coasts of Ireland.

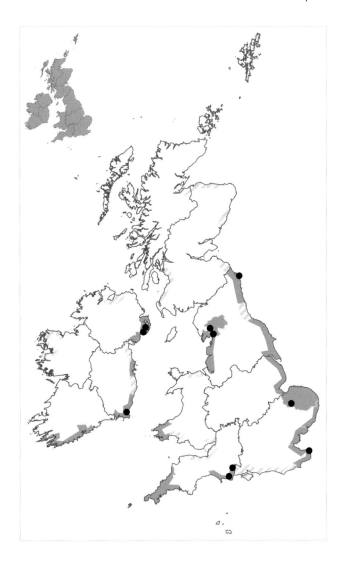

MUTE SWAN

Cygnus olor

NATIVE STATUS Native

MODERN CONSERVATION STATUS	
World	Least Concern
UK	Green List
ROI	Amber List
Trend since 1772	Uncertain

The Mute Swan was commonly mentioned in early modern natural histories from both Britain and Ireland, and authors often describe specific local populations like at Loch Spynie or Orkney.

RECOGNITION

This species was called *swan* and *signet* in Early Modern English, *cygnus* or *olor* (separately) in Latin and occasionally *Cygnus mansuetus* (literally: 'domesticated swan') as a binomial species name. Apart from the last name, these terms might hypothetically have also applied to other similar species like *Cygnus cygnus* (Whooper Swan), and *Cygnus columbianus/bewickii* (Bewick's Swan) which very closely resembles the Whooper Swan. Actually though, the Mute Swan seems to have been considered the nominal species, and the other *Cygnus* species always seem to be distinguished in our texts. The Whooper Swan by contrast was called the *elk*, *hooper* or *wild swan* in Early Modern English, and *Cygnus ferus* (literally: 'wild swan') in New Latin. These terms were used because the Mute Swan was considered to be (and arguably was) semi-domesticated.

DISTRIBUTION

The distribution of top-quality records for this species is not statistically different from the known level of recorder effort. The gaps on the map may just reflect decreased survey effort in some regions. It may have been widespread.

In medieval and early modern England, Mute Swans were highly prized. They are one of the most commonly excavated species on medieval English sites (Serjeantson 2006). Flocks of Mute Swans on the River Thames and elsewhere were famously owned, like Rabbits or Sheep, and could only be eaten by their owners – although the idea that the species was protected from harm on pain of death is a myth (Ticehurst 1926; Gurney 1921: 109–115). This level of protection seems to have ensured that Mute Swans were abundant in early modern England (Shrubb 2013: 73).

The medieval and early modern conception of the Mute Swan as domesticated has led to some question about the species' native status. In Britain they were certainly native, as Yalden has demonstrated from the archaeological record (Yalden and Albarella 2009: 81). Mute Swans are strong fliers and were clearly very well distributed in Ireland in the early modern period, so are most likely native here too.

However, surprisingly there are no early modern records of Mute Swans from Wales. The only relevant record there is of a migrating *wild swan* (ed. Owen 1994: 130). Yet this is likely not a reference to the Mute Swan but rather another species of swan, as explained above. The absence of records from Wales is not statistically significant given the small number of records overall. But there are also no archaeological remains of Mute Swans from Wales, and since Mute Swans rarely breed above 300 metres, much of Wales is unsuitable for them (Pritchard 2021d). It appears possible that the species may have been introduced to Wales more recently. It was relatively widespread in the north of the country by the end of the nineteenth century (Holloway 1996: 67).

Mute Swans remain relatively widespread and common in Britain and Ireland today, although they are not found in upland areas, so are absent from some areas like parts of the Highlands of Scotland (Balmer et al. 2013). The species is usually thought to have colonised Shetland around between 1990 and 2002 (Brown and Brown 2007). However, there is an early modern record from Shetland:

> in the Northmost end of Dunrossness, is the Lough of Flathbuster, about a Mile and a half in circumference. These Loughs are replenished with no Fishes but Trouts (whereof they have plenty) but all of them very good for Gunning, having abundance of Ducks, Teals, Swans, &c.
>
> (ed. Monteith 1711: 48)

Assuming this record is not a mistake for *Cygnus cygnus*, it implies that an earlier extinction occurred in Shetland, and that the modern population may be a recolonisation. Holloway suggests the species was absent from Shetland at the end of the nineteenth century (Holloway 1996: 67).

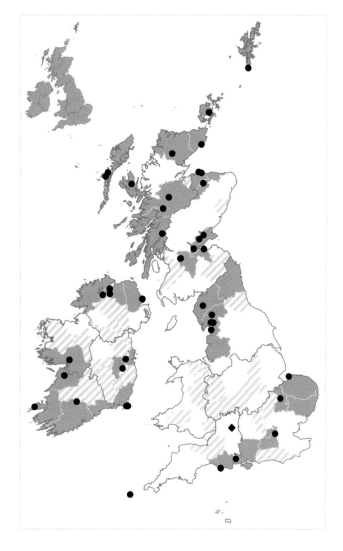

Mute Swan records, 1519–1772. There are records from every region of Britain and Ireland except Wales and the Midlands of England.

WIGEON
Anas penelope

NATIVE STATUS Native

MODERN CONSERVATION STATUS	
World	Least Concern
UK	Amber List
ROI	Amber List
Trend since 1772	Uncertain

The Wigeon was frequently recorded in our early modern sources all around Britain and Ireland.

RECOGNITION
This species was called the *widgeon* in Early Modern English. Sometimes the female bird was called the *whewer*. The species was also called the *penelope* in New Latin. The only real source of confusion can be with *Aythya farina* (Pochard). This species was occasionally called the *red-headed widgeon* or the *penelope*. Our recorders usually seem to have made the effort to distinguish the two species, and the New Latin name is never used without a vernacular name to confirm the species.

DISTRIBUTION
The distribution of top-quality records for this species is statistically a poor fit with the known levels of recorder effort. It may have been locally distributed, locally abundant or of special local interest. This species was especially commonly recorded in Ulster, which might reflect that it was more abundant or more widespread there. Habitat suitability modelling based on the sites where the Wigeon was recorded in the early modern period suggests that the species would have been well adapted to conditions all around the coasts of Britain and Ireland, as well as possibly further inland in the lowlands, except possibly in the wettest parts of Scotland.

Yalden suggests that the Wigeon may have been a common overwintering bird since the last glacial period (Yalden and Albarella 2009: 205). But there is some hint it may have been rarer in the early modern period than it is today. The Wigeon was one of the birds protected in England from 1534 due to declining numbers (Shrubb 2013: 45). However, the decline of the Wigeon cannot have been very advanced, as the early modern sources make it clear that the Wigeon continued to be widely seasonally available.

Today the Wigeon is commonly found around the whole of Britain and Ireland in winter, and is less widely recorded as a summer breeder. The bird appears to have only begun breeding in Britain and Ireland from around the middle of the nineteenth century. There are no reliable records of a breeding population in the early modern period, and several records of seasonal populations (Pritchard 2021e; Mitchell 2007; Holloway 1996: 74):

> Penelope Aldrov[andi] The Wigeon.
>
> It is a bird of passage, coming to us [Co. Dublin] with the Bernacle, and goes sooner.
>
> …
>
> It is good food, and particularly the Cock Wigeon in our Market, substituted by the Lady-fowl.
>
> It varies according to its food; for in the country where it feeds less on fish, it is commonly very sweet; else a little fishy.　　　　　　　　　　　　　　　　(Rutty 1772: 339)

Wigeon records, 1519–1772. There are records from every region of Britain and Ireland except Connacht.

The Wigeon was regularly hunted, and a close season was established in 1739, but the species appears to have continued to decline in the nineteenth century, perhaps due to habitat changes (Shrubb 2013: 45). Since then, the Widgeon has been increasing through the twentieth century. The number of wintering birds, and the distribution of nesting birds both apparently increased significantly in England and across much of the island of Ireland between the 1983–84 and 2008–09 national surveys (Balmer et al. 2013).

MALLARD
Anas platyrhynchos

NATIVE STATUS Native

MODERN CONSERVATION STATUS	
World	Least Concern
UK	Amber List
ROI	Amber List
Trend since 1772	No change

The Mallard was commonly recorded by authors from across early modern Britain and Ireland.

Mallard records, 1519–1772.
There are records from every region of Britain and Ireland including the Hebrides and the Northern Isles.

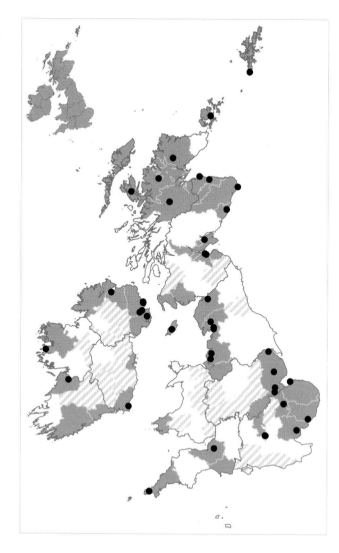

RECOGNITION

This species was called *mallard* or *Boschas major* by early modern authors. These terms were exclusively used for the Mallard. However, the species was more often described generically as *wild duck* or *drake*. Usually, this term appears to have applied to the Mallard, but there was the potential for confusion with any of the other ducks in the Anatidae family found in Britain. It does at least reliably distinguish the species from its domestic counterpart, which makes the historical records easier to interpret than the archaeological ones, where domestic birds are harder to distinguish.

> Here [the Orkney Islands] is plenty of tame and wild Fowl; they reckon they have 8 or 9 different sorts of wild Geese, and of gray Plover, Moorfowl, Wild-Duck, Swans, Teal, Whaps, or Curlew, &c. there is no place better stor'd. (Wallace 1700: 46)

DISTRIBUTION

The distribution of top-quality records for this species is not statistically different from the known level of recorder effort. The gaps on the map may just reflect decreased survey effort in some regions. It may have been widespread.

The Mallard has the same distribution now as it probably had in the early modern period. It appears to have always been widely distributed, although it does seem to have increased its abundance since 1995 (Balmer et al. 2013; Yalden and Albarella 2009: 205). In 1534, as a response to overhunting, the Mallard was given some protection in England (Shrubb 2013: 45). This appears to have been enough to protect the bird for some centuries, although a close season was (re)enacted in 1710, and some records suggest there may have been some decline and then recovery in the nineteenth century (Pritchard 2021f; Shrubb 2013: 84; Holloway 1996: 80).

TEAL

Anas crecca

NATIVE STATUS Native

MODERN CONSERVATION STATUS	
World	Least Concern
UK	Amber List
ROI	Amber List
Trend since 1772	Uncertain

The Teal was recorded in early modern sources as occurring across Britain and Ireland. It may have been widespread.

RECOGNITION

This species was called *teal* in Early Modern English, *teil* in Middle Scots, and *querquedula*, *Querquedula secunda* or *Querquedula major* in New Latin. O'Sullivan (ed. 2009: 140) also provides *plas lacha* (presumably a mistake for *pras-lacha*) in Irish. The first two terms have to be carefully distinguished from *Anas querquedula* (Garganey) which was called the *ateale* or *summer teal* in Early Modern English and occasionally *Querquedula prima* in New Latin, although it was more often called *Anas circia* in our sources. Luckily, every source using a generic term distinguishes the different birds so that the records provided on our map are reasonably secure. For example, here is the record in Smith's (1750) *Ancient and Present State of the County and City of Cork*:

Querquedula, The Teal, Aldrovandus sets down two kinds of this bird, the one larger than the other. They are the least of the duck kind. The female is distinguished from the male, as the wild-duck is from the mallard, by not having any green or red upon the head, nor black about the rump, nor those fine variegated feathers of black and white on the sides. It is by all accounted a delicate bird for the table.

(Smith 1750b: 353)

DISTRIBUTION

The distribution of top-quality records for this species is statistically a poor fit with the known levels of recorder effort. It may have been locally distributed, locally abundant or of special local interest. But habitat suitability modelling based on the sites where the Teal was recorded in the early modern period suggests that the species would have been well adapted to conditions all around the coasts of Britain and Ireland and presumably also at lowland sites inland. And at county level, the Teal was recorded in almost every region of early modern Britain and Ireland (with the exception of Connacht) and probably had a general distribution. This suggests that the pattern of records might suggest a local abundance: there are more records than expected in Ulster and fewer than expected in Highland Scotland.

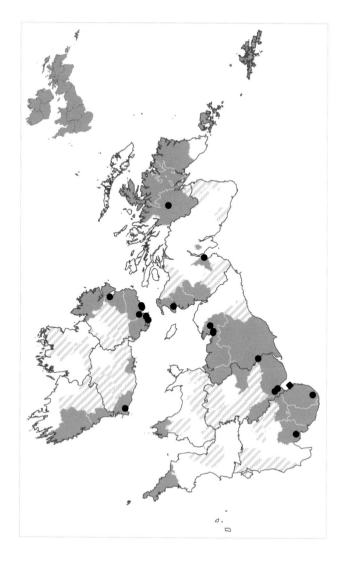

Teal records, 1519–1772. There are records from every region of Britain and Ireland except Connacht.

There seems to have been a belief in medieval Ireland that Teals either cannot or should not be hunted. However, in early modern Europe, breeding Teal were often exploited for food (Anderson 2017: 80; Shrubb 2013: 45–7, 83). They were commonly caught in nets and snares and later shot and caught in duck decoy ponds. The Teal may have been especially sought in Scotland as a close season was introduced, and its sale (for food) was banned in the middle of the sixteenth century (Baxter and Rintoul 1953: 395). A close season was also imposed in England in 1534. It is not clear that this legislation was ever widely respected, but the Teal did significantly decline due to the removal of the close season between the Game Reform Act 1831 and the Wild Birds Protection Act 1880. In Scotland, the species is thought to have declined due to land drainage projects in the eighteenth and nineteenth centuries (Lauder 2007).

The Teal is now widely distributed around lowland areas of Britain and Ireland in winter, although it is more widely distributed in the north during summer. It seems to be less common than it was in the early modern period, but there is no evidence that is has changed its distribution on a regional level.

EIDER
Somateria mollissima

NATIVE STATUS Native

MODERN CONSERVATION STATUS	
World	Near Threatened (Vulnerable in Europe)
UK	Amber List
ROI	Red List
Trend since 1772	Uncertain

The Eider was recorded by the early modern authors on a handful of islands off north Britain, including the Northern Isles, the Hebrides and most famously in the Farne Islands off Northumberland.

RECOGNITION
This bird was most commonly referred to as the *dunter* in Norn, the *colk* in early modern Scottish Gaelic, and as *St Cuthbert's duck* in Early Modern English.

DISTRIBUTION
The distribution of top-quality records for this species is not statistically different from the known level of recorder effort. The gaps on the map may just reflect decreased survey effort in some regions. It may have been widespread, although the statistical analysis does have a blind spot to species whose distribution coincides with the best recorded areas. The Eider was commonly targeted for its feathers across northern Europe in the early modern period (Shrubb 2013: 88–89), and populations regularly seem to have been suppressed (Waltho 2007).

> About Farne also lie certeine Iles greater than Farne it selfe, but void of inhabitants; and in these also is great store of puffins, graie as duckes, and without coloured fethers, sauing that they haue a white ring round about their necks. There is moreouer another bird, which the people call saint Cuthberts foules, a verie tame and gentle creature, and easie to be taken.
>
> (Holinshed Project 2008b [1577]: 44 (1.1.10))

Eider Duck records, 1519–1772.
There are records from the Hebrides,
the Northern Isles, and islands off
Lowland Scotland and North England.

The populations referred to by the early modern authors in the Northern Isles, the Hebrides, in the Firth of Forth and off Northumberland still exist today, but the species declined significantly in the first half of the nineteenth century in Northern Europe and came close to going locally extinct (Shrubb 2013: 90; Holloway 1996: 94). From the second half of the nineteenth century, it recovered in Scotland and even began breeding on the coasts of the mainland.

The pattern of records recorded from the early modern period is approximately comparable with the modern breeding range of the species, except that Eiders now breed around Ulster, and since 1968–72 they have started breeding in North England and on the Isle of Man. Over the winter, Eiders are now widespread around the coasts of Britain, and are also seen around much of the coastline of Ireland, especially in the north but also in the west and south. The Eider's extended winter range towards the south of the British and Irish mainlands reflects a modern recolonisation. In Wales, the Eider was rarely seen in the nineteenth century, but numbers began to increase from 1950 (Pritchard 2021g). The species is now declining again in Britain and Ireland.

CAPERCAILLIE

Tetrao urogallus

NATIVE STATUS Native

MODERN CONSERVATION STATUS	
World	Least Concern
UK	Red List
ROI	Extinct
Trend since 1772	Certainly declined

The Capercaillie was recorded by a handful of early modern authors in Scotland and Ireland before it went extinct.

RECOGNITION

The most common name for this species in Middle Scots was *capercailȝe* (the letter <ȝ> (yogh) is pronounced like the letter <y>). In our sources though, the name is normally spelled *capercalze* because early modern printers often did not have access to a separate character to represent the yogh. The name comes from Gaelic *capull-coille*, although this original term is not used in any early modern texts. It was specific to *Tetrao urogallus*.

More ambiguous terms were also common. The form *cock of the wood* is specific to this species although it has to be carefully distinguished from *woodcock* (=*Scolopax rusticola*) (Hall 1981). Similarly, the terms *urogallus* and *Tetrao major* are also specific, but have to be carefully distinguished from the *tetrao* and *Urogallus minor* (=the Black Grouse).

DISTRIBUTION

There are not enough top-quality records to statistically analyse the distribution of this species. However, the distribution of county-level records and the species' recorded absence from England show that it was regionally distributed. In Ireland, the bird's native status has occasionally been doubted, but its presence and continued survival into the medieval period has now been proven in the archaeological record, and there is good historical evidence for its former occurrence (Yalden and Albarella 2009; D'Arcy 1999: 99–100; Hall 1981; Shrubb 2013: 99–100; Kelly 1997: 300–1; Hall 1981; Gurney 1921: 66–7).

Our sources attest that the Capercaillie was already extinct in England by this period. In the medieval archaeological record, it is occasionally recorded in northern England but never southern England (Serjeantson 2006). It was also thought of as declining even in Scotland and Ireland, possibly due to destruction of its habitat and the colder weather in the Little Ice Age (Shrubb 2013: 99–100; Moss 2007). At the time, it was also commonly hunted. As Smith explains for Munster:

> This bird is not found in England, and now rarely in Ireland since our woods have been
> destroyed. The flesh is highly esteemed. (Smith 1750b: 334)

This statement is echoed by Lesley in Highland Scotland:

> This most rare bird the capercalze, i.e. the commonly-called woodland horse, is frequently found in Ross-shire and also Lochaber, and even other mountain areas which do not lack pine trees. It is even smaller than the raven. It greatly appeals to the tastes of those who eat it with its very lovely flavour.
> (translated from Lesley 1675: 24)

The Capercaillie was protected by law in Scotland from 1621, and in Ireland from 1711 (Shrubb 2013: 44–45; Hall 1981). However, its decline was terminal. It appears to have gone extinct in both Scotland and Ireland in the eighteenth century (Shrubb 2013: 99–101; Moss 2007; D'Arcy 1999: 105; Hall 1981). There were attempts to reintroduce the species to Scotland and Ireland in the nineteenth century, and in 1837–38 those in Scotland proved successful. In the final quarter of the nineteenth century, the Capercaillie was again recorded across most of eastern Scotland (Holloway 1996: 138–9). The bird was enthusiastically welcomed by gamekeepers, but was persecuted as a pest to forestry; it also likely suffered due to the destruction of lekking sites and the reduction in gamekeeper predator control over the twentieth century (Tapper 1992: 54–5). A close season was enforced in 1981, and a ban on shooting in 1990. Unfortunately, the species has continued to decline since the legislation was passed (Balmer et al. 2013). It is now confined to five locations in the east of Scotland, and the islands of Loch Lomond. It was never successfully reintroduced to Ireland and may soon become extinct in Britain once again.

Capercaillie records, 1519–1772. There are records from Highland and Lowland Scotland as well as Connacht and Munster.

BLACK GROUSE
Lyrurus tetrix

NATIVE STATUS Native in Britain, possibly native in Ireland

MODERN CONSERVATION STATUS	
World	Least Concern
UK	Red List
ROI	Extinct/Not Present
Trend since 1772	Certainly declined

The Black Grouse was regularly recorded by early modern authors in the uplands of Britain and, surprisingly, in parts of Ireland.

RECOGNITION
This species was commonly called the *black-cock* and the *heath-hen*, *heath-cock* or *heath-game*. In New Latin it is referred to as the *Urogallus minor* of Aldrovandi or the *tetrao*. Turner's description of the *attagen* matches the Black Grouse, but this name more usually refers to the Red Grouse. The Welsh name *ceiliog du* is also in use in the early modern period.

DISTRIBUTION
The distribution of top-quality records for this species is not statistically different from the known level of recorder effort. The gaps on the map may just reflect decreased survey effort in some regions. In this case, however, the Black Grouse does seem to have already declined by the early modern period. In Scotland, a close season was imposed for the species from the fifteenth century onwards (Shrubb 2013: 44–5).

This is not to say that the Black Grouse was completely absent from lowland England by the early modern period. Nineteenth-century naturalists provided records from Hampshire, Berkshire, Surrey, Norfolk and much of Wales and north-east England. Some of these populations may have been introduced as part of put-and-take stocking after the end of the early modern period, but others may have been native, and lingered into the twentieth century (Holloway 1996: 136–7; Tapper 1992: 50–1).

There is evidence for a widespread lowland decline of Black Grouse in England during the nineteenth century, due to habitat loss and the enclosure of heaths and commons (Shrubb 2013: 94–9; Holloway 1996: 136). Many of these populations had only recently been introduced or reintroduced in the same century, but others, like the extinctions in the north and south of Wales, were likely to be of native populations.

A further decline has occurred in the twentieth century (Balmer et al. 2013). The populations in the Midlands of England and in south Wales have been lost. In the first 1968–72 atlas of breeding birds, the Black Grouse was still recorded on Islay and in the South West of England (especially Dartmoor and Exmoor), areas where it has now been lost, although these south-western populations might have been restocked for shooting (Tapper 1992: 50–3). The population in the Peak District also briefly went extinct in the year 2000, although reintroduction was started shortly afterwards.

Black Grouse do not currently occur in Ireland, and it is debateable whether they are native. There is only a single, ambiguous Irish archaeological record from Ballynamintra Cave, which appears to have been a mistake (Yalden and Albarella 2009: 208; Ussher and Warren 1900). However, the early

Black Grouse records, 1519–1772.
There are records from every
region of Britain except South
England, as well as from Munster
and Leinster in Ireland.

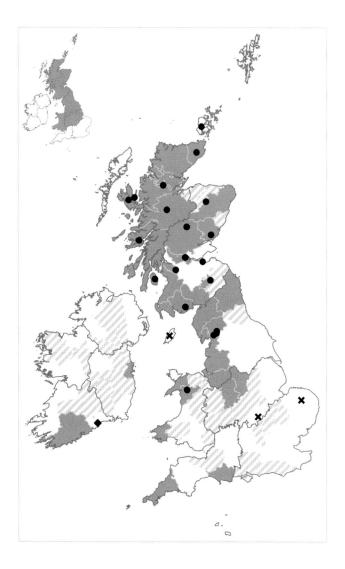

modern sources do provide a few records, which seem to suggest that the species was present in the early modern period. The most reliable of these is the account by Smith from Co. Cork:

> Tetrao seu Urogallus minor. The Heath-Cock, or Black Game, or Grouse.
>
> This species is frequent, and needs no particular description.
>
> It inhabits mountains, and is rarely seen in lower heath-grounds. The cock is almost [all] black, but the female is coloured like a Woodcock or Partridge; so that Gesner made them to be a different species of fowl.   (Smith 1750b: 334)

This record was labelled as dubious by Barrett-Hamilton on the basis that it is a translation from Willughby (Barrett-Hamilton 1899; Willughby & Ray 1676: 125). However, copying descriptions was a widespread practice during the period (see for instance the description of the Beech Marten, p. 51). The early modern sources did not give descriptions to prove they had seen the birds, but to more securely identify the species intended. This record seems actually to be trustworthy – for a few reasons. The record comes from Smith's 'Catalogue of the Birds observed in this County [Cork]', meaning that he thought of it as local. Smith takes the name of the species and its description from Willughby, but not his comment that it 'needs no particular description'. Smith was also discerning in which species he quoted. After the account of the Black Grouse, Smith skips Willughby's Hazel Grouse and Ptarmigan (which are not native to Ireland) before quoting from Willughby's description of the Red Grouse (which is native). A corroborating account of the Black Grouse can also be found in Rutty's *An Essay Towards a Natural History of the County of Dublin* (Rutty 1772: 302). There are also possible records of presence provided in Smith's account of Co. Waterford and in the national accounts of O'Sullivan's and Payne (O'Sullivan 2009: 147–8; Payne 1841: 7; Smith 1746: 336). On balance, it seems likely that the Black Grouse was present in Ireland during the early modern period, whether as an introduced sport species (like the Fallow Deer) or a now-extinct native (like the Capercaillie).

RED GROUSE/WILLOW GROUSE

Lagopus lagopus

NATIVE STATUS Native

MODERN CONSERVATION STATUS	
World	Least Concern (Vulnerable in Europe)
UK	Green List
ROI	Red List
Trend since 1772	Probably increased

The term Red Grouse described the subspecies of Willow Grouse/Willow Ptarmigan found in Britain and Ireland (*Lagopus lagopus scoticus* and *Lagopus lagopus hibernicus*). It was recorded by the early modern authors commonly in Scotland, North England, Wales and parts of Ireland, but not in South or South West England.

RECOGNITION

This species had several common Early Modern English names including *grouse*, *moorfowl*, *moor-game*. In early modern Scots it was called *muirful*. In England it is also called the *gor-cock*. It is rarely referred to in New Latin as the *Lagopus altera* of Pliny and the *Attagen* of Aldrovandi. A Welsh name is also used in our sources: *iar fynydh*. Care must be taken to avoid confusing these common names with the *moorhen* and *moorcock* (which referred and still refer to Moorhen), and the *black grouse* and *heathcock* (*Tetrao tetrix*). The term *attagen* can rarely also refer to the godwits. Records can be securely identified where they describe the species or where the other game bird species are distinguished by different names.

DISTRIBUTION

The distribution of top-quality records for this species is statistically a poor fit with the known levels of recorder effort. It may have been locally distributed, locally abundant or of special local interest. Habitat suitability modelling based on the sites where the Red Grouse was recorded in the early modern period suggests that the species may have had specific requirements. It not recorded in the areas that today have the warmest summer temperatures. There are more records than we would expect from Lowland Scotland and northern England, suggesting Red Grouse may have been more widespread or abundant in these regions than anywhere else. This fits with both the areas of optimal habitat for the Red Grouse as identified by Tapper (1999: 130–3) and the archaeological evidence that shows the species has been exclusively excavated from northern rather than southern sites in medieval England (Serjeantson 2006), which would agree with the early modern distribution. It was extensively hunted in the early modern period and various close seasons were imposed in the fifteenth and sixteenth centuries (Baxter and Rintoul 1953: 721–2).

In the modern period, the Red Grouse is mainly found on upland heather moors, where ideal conditions and predator control maintain large numbers of birds for shooting (Balmer et al. 2013). The shooting industry was facilitated by the improved travel links of the eighteenth and nineteenth centuries which enabled shooters to travel to moors from all over Britain (Shrubb 2013: 93–4). Unlike Pheasants, Red Grouse are not restocked for put-and-take shooting; instead, optimal environments are created to allow grouse numbers to increase. In order for grouse moors to be profitable, predators need to be tightly controlled, which has had and continues to have a significant effect on other wildlife species of the uplands (especially the Carnivores, p. 36, and the Raptors, p. 131) (Lovegrove 2007).

The distribution of the Red Grouse as reported by the early modern authors generally matches the current distribution of the species, except that the early modern sources were very clear that the Red Grouse did not occur on the Shetland islands during the period; it looks to have been introduced

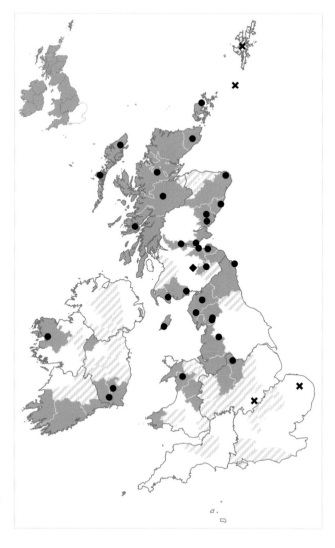

Red Grouse records, 1519–1772.
There are records from every region of Britain except the South and South West of England, and from Connacht, Munster and Leinster in Ireland.

here as well as on Dartmoor and Exmoor at the beginning of the twentieth century (Shrubb 2013: 93; Holloway 1996: 152–3; Tapper 1992: 46).

The Red Grouse increased in abundance through the first half of the twentieth century, but went through a sudden decline from the 1970s. This appears to be due to habitat changes, perhaps due in Britain to the closure of grouse moors and the reduction of predator control by gamekeepers and in Ireland to afforestation, peat extraction and misuse of burns (Pritchard 2021h; Balmer et al. 2013; Tapper 1992: 46–9). It is not clear from the evidence how widely distributed the species was in these areas in the early modern period, but at least one early modern population, now lost, was recorded in early modern Pembrokeshire for example:

> Beside these two kinds of fowl [gull and woodcock], which we account among household fare, the county yields great store of other sorts, as the [Preseli] mountains foster the grouse, heathcock, which are always in season, and the plover, both grey and russet.
> (ed. Owen 1994 [1603]: 132–3)

RED-LEGGED PARTRIDGE
Alectoris rufa

NATIVE STATUS Thought to be native in Channel Islands

MODERN CONSERVATION STATUS	
World	Least Concern
UK	Not Evaluated
ROI	Green List
Trend since 1772	Certainly increased

The Red-legged Partridge was only recorded in the Channel Islands and (surprisingly) Dublin, by our early modern sources, it became widespread after a much wider introduction in the modern period.

RECOGNITION
This species was generally called the *red-legged partridge* in Early Modern English and *Perdix russa* in New Latin. It never seems to have been called just the *partridge*, or *perdix* as these names were reserved for the Grey Partridge, which was the type species of the genus in early modern Britain and Ireland.

DISTRIBUTION
There are not enough top-quality records to statistically analyse the distribution of this species.

In the early modern period, the Red-legged Partridge was recorded by two authors in the Channel Islands (in both Jersey and Guernsey). It is likely that the species colonised these islands naturally, since they are so close to the French mainland where the Red-legged Partridge is native. Elsewhere in Britain and Ireland the species is likely to have been absent.

> We have also the Perdix russa. Willughby, or the red-legged Partridge, which has likewise red eyes, tho' the feathers are more grey than in the common. It is rare here [Dublin], but common in France, and in the islands of Guernsey and Jersey, and deemed much more delicious than the common Partridge.
> (Rutty 1772: 303)

This is not to say that there was no interest in introducing the Red-legged Partridge in the early modern period. Outside of the Channel Islands, introduction attempts seem to have begun in 1673

Red-legged Partridge records, 1519–1772. There are only records from the Channel Islands and from Dublin.

with a few individuals introduced to Windsor, and other early introductions were attempted in Wimbledon, Essex, Surrey, Suffolk and at Dungannon in Co. Tyrone (Sánchez-García et al. 2022; Lever 2009: 163; Gurney 1921: 203). The rarity record from 1772 Dublin in the quotation above is likely to reflect another release, now forgotten.

The first introduction with long-term success was at the end of the eighteenth century, and by the end of the nineteenth century the Red-legged Partridge was recorded across most of the east of England and less regularly in west England and Wales. By 1930 it was present across 30% of Britain (Sánchez-García et al. 2022; Holloway 1996: 140–1; Balmer et al. 2013; Shrubb 2013: 106; Tapper 1992: 40–1). Today this species is relatively widespread across Britain, especially to the south and to the east. Its current presence in the South West and North of England the east of Scotland and in Ulster and Leinster is new and is mainly due to releases through the second half of the twentieth century for put-and-take shooting. Many of these populations are reliant on continuous restocking and are not self-sustaining. Some, perhaps most, populations of Red-legged Partridges in these islands are actually a hybrid stock of *A. rufa* x *A. chukar*, due to interbreeding with Chukar Partridges and hybrids which were released in the 1960s and 1970s.

GREY PARTRIDGE
Perdix perdix

NATIVE STATUS Native in Britain, uncertain in Ireland

MODERN CONSERVATION STATUS	
World	Least Concern
UK	Red List
ROI	Red List
Trend since 1772	Uncertain

The Grey Partridge was widely recorded in early modern Britain and Ireland.

RECOGNITION
This species was called the *partridge* in Early Modern English and Scots, and *Perdix cinerea* in New Latin. These terms are specific to the Grey Partridge. Other species, such as the Red-legged Partridge, are always given more specific names to differentiate them (e.g. *red legged partridge*, *Perdix russa*), and were not established in Britain or Ireland until around 1770 (Lever 2009: 160–3)).

DISTRIBUTION
The distribution of top-quality records for this species is not statistically different from the known level of recorder effort. The gaps on the map may just reflect decreased survey effort in some regions. It may have been widespread.

The native status of the partridge in Ireland is uncertain. The most important evidence is phylogeographical: based on a study of variation in mitochondrial DNA, Grey Partridges can be divided into two distinct populations, the western and eastern clades (Liukkonen-Anttila et al. 2002). Almost all of the populations surveyed in Western Europe belonged to the western clade, except that the population surveyed in Ireland was of mixed origin. This suggests that the population may be descended from birds that were artificially introduced, although this could also be a result of an ancient introduction, modern restocking obscuring a previous population, or both (Potts 2012: 50). In addition to this evidence, Gerald of Wales, who visited in the twelfth century, attests that the species was absent at this point, although he was deeply unreliable when it came to the absence of Irish animals (Anderson 2017: 119; Kelly 1997: 300–1; O'Meara 1982: 47).

There are archaeological remains from Catacomb and Newhall caves, Co. Clare; Plunkett Cave in Kesh, Co. Sligo; and from the settlement at Newgrange, Co. Meath. In all cases, however, these remains seem to be modern. At all the cave sites they occur in strata alongside remains of Turkeys *Meleagris gallopavo* (which have only been brought to Ireland in the last few centuries) (Scharff et al. 1902, 1906). Ussher suggests that in the Catacomb and Newhall caves the remains were brought to the cave by rats from a nearby mansion house in the modern period, and the same may have been true at Kesh. At Newgrange, Wijngaarden-Bakker pointed out that the single bone found was smooth, unbroken and very different from the prehistoric bones of the other species; this bone likely belonged to one of the partridges that nested on site during the excavation (van Wijngaarden-Bakker 1974).

However, the partridge does seem to have been introduced to Ireland before the first records in this atlas. Payne, writing in 1589, describes Ireland (he lived in Munster) as having partridges (Payne 1841: 7). Moryson, who visited in 1617, also commented on the partridges (although he notes they

Grey Partridge records, 1519–1772.
There are records from every region of
Britain and Ireland except Connacht.
There are repeated absence records
from the Northern Isles of Scotland.

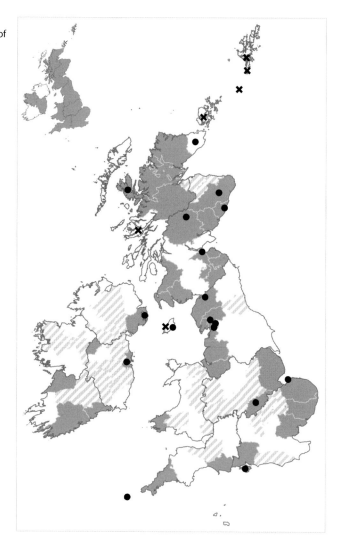

are 'somewhat rare' (Moryson 1617: 160). This means that, at the least, the Grey Partridge was established in Ireland centuries before the Red-legged Partridge.

The Grey Partridge was one of the most commonly found species in the medieval archaeological record in England and was especially favoured on elite sites (Serjeantson 2006). It became the most commonly exploited gamebird of the early modern period, although it is less well recorded than, for instance, the Red Grouse in our sources. Initially, the level of shooting in this period does not seem to have been sustainable. Grey Partridges are less common in the English archaeological record from the fourteenth century, and the bird's price rose steadily through the sixteenth century. The nests and eggs of the species were protected by an act of the Parliament of Scotland in 1457/8 (*RPS* 1458/3/32), and in 1603 the sale of partridge was banned in England (Shrubb 2013: 101; Serjeantson 2006; Baxter and Rintoul 1953: 737). In Dublin, following hunting by the Royal Irish Army and an increase in Hen Harriers, a letter from Edward Cooke, keeper of Phoenix Park, to James Butler, Lieutenant-General of Ireland, complained: 'there is scarce a partridge left about these parts' (Russell and Prendergast 1871: 192).

After the end of the early modern period, the Grey Partridge seems to have initially increased its abundance in Britain through the nineteenth century due to the increase of arable farming, the creation of hedgerows to enclose land and intense control of predators by gamekeepers (Tapper 1992: 36–9). This growing population collapsed in the twentieth century. Today the species is mainly found in lowland parts of Britain and is absent from much of the Scottish Highlands, the Lake District and Peak District, Wales and the South West of England, where habitat is suboptimal. It is also now absent from most of Ireland (Balmer et al. 2013; Tapper 1999: 146–9). The most serious change has been an abrupt decline in the population after 1952, probably following the introduction of herbicides (Balmer et al. 2013; Potts and Aebischer 1995). In the UK this decline has been 91%, and in the Republic of Ireland the Grey Partridge was almost lost entirely. There has been a distribution shift as well as an abundance change. In 1968–72 the species was still widely found across the north of Wales, in the uplands of north England, across South West England and more rarely in Ulster, Connacht and Leinster (Balmer et al. 2013). The post-1952 decline was not the first. Bag records from hunting estates show that Grey Partridges were often shot in Highland Scotland and across most of Wales between 1900 and 1938 (Tapper 1993: 36–9). Naturalists' accounts from the end of the nineteenth century confirm the Grey Partridge was common across much of Britain, and uncommon but still found across much of Ireland in this period (Holloway 1996: 142–3). The earlier declines might be due to ground-cover loss due to changes from arable to pastoral farming or the end of gamekeeper predator control in the first half of the twentieth century.

> The whole Island [of Skye] is verie fertile … with all sorts of wild foull a swans, solangeese, wildgeese, duke and drake, woodcock, heathcok, partridges, plivers, doves, hauks and hundreds of other sorts tedious to relate. (ed. Mitchell 1907 [c.1683]: 220)

Some upland and island populations of Grey Partridge might also have been established and maintained on a put-and-take basis, and therefore might have gone extinct when restocking stopped (Parish 2007).

PHEASANT
Phasianus colchicus

NATIVE STATUS Non-native

MODERN CONSERVATION STATUS	
World	Least Concern
UK	Not Evaluated
ROI	Not Evaluated
Trend since 1772	Certainly increased

The Pheasant was not native to Britain or Ireland but had escaped and become naturalised across much of Britain and Ireland by the early modern period.

RECOGNITION
Yapp has argued that the medieval Latin term *fasianus* referred to the Capercaillie rather than the Pheasant (Yapp 1981: 30–1). This was not true of the early modern period, where many of our authors made it clear that the Latin term, by then more commonly spelt *phasianus*, referred to the current species (Sibbald 2020: 16 (II:3); O'Sullivan 2009: 120–1; Smith 1750b: 332; Lesley 1675: 24; Merrett 1666: 172). Occasionally the Latin term was borrowed into the vernacular as *feasane*, but the species

was most commonly called the *pheasant* in Early Modern English. Young birds were and are still called *pouts*. The usual Irish term seems to have been *cuelagh-fa* (O'Sullivan 2009: 238–9; Scharff 1915b; K'Eogh 1739: 72), while O'Sullivan (2009: 120–1) also attests a rarer Irish term *coilleach/cearc-cruoigh*.

DISTRIBUTION

The distribution of top-quality records for this species is statistically a poor fit with the known levels of recorder effort. It may have been locally distributed, locally abundant or of special local interest. There are more records than expected from the South and South West of England, and less than expected from Highland and Lowland Scotland. Habitat suitability modelling based on the sites where the Pheasant was recorded in the early modern period suggests that the species was only found on lowland sites with mild winters. This fits approximately with the area of optimal habitat as described by Tapper (1999: 150–3).

The Pheasant was still relatively new to parts of Britain and Ireland in the early modern period. There are historical records in England from the Roman period, and the species seems to have become established here, and possibly in Wales either shortly before or shortly after the time of the Norman Conquest, but the earliest records from Scotland and Ireland are from the sixteenth century

Pheasant records, 1519–1772. There are records from every region of Britain except Highland Scotland (although there are records from the Hebrides) as well as from Munster and Leinster.

(Pritchard 2021i; Shrubb 2013: 111; Poole 2010; Lever 2009; Yalden and Albarella 2009: 208; Payne 1841: 7; Moryson 1617: 194). Throughout the early modern period (and up to the modern period), the species has continued to be recorded as absent from some areas, such as the Highlands, the Isle of Man and the Northern Isles. At regional level the Pheasant appears to have been well known and recorded across the south of Britain and (to a lesser extent) Leinster and Munster. As with the Rabbit (p. 22), the Pheasant was considered to be a poultry species at this time and was farmed for food, although some individuals appear to have escaped and have become naturalised across Britain and Ireland (Shrubb 2013: 110–11; Tapper 1992: 42–5). The wild populations do not seem to have been commonly restocked for put-and-take shooting as they are today. It is not generally possible to distinguish naturalised populations of Pheasants in the early modern records, so some of the records on the map might reflect captive populations. In sixteenth-century Scotland the bird was protected from shooting by ordinary people within a mile of royal forests and parks, which perhaps suggests it was regularly being shot even by people who did not breed it (Baxter and Rintoul 1953: 735).

Today the Pheasant is recorded all across Britain and Ireland, except that it is rare/absent from some parts of the Highlands and Hebrides (Balmer et al. 2013). This likely represents an increase from the early modern period. However, the wide distribution of the species does not necessarily indicate it is now naturalised across these islands, even today. Since the eighteenth century, the bird has become the main target for put-and-take shooting. Around 35 million birds are now released each year, of which 15 million are shot (Balmer et al. 2013; Williamson 2013).

The general absence of the species in parts of the Highlands and Hebrides is especially interesting since although there are only absence records for the Pheasant in the early modern Scottish Highlands, Martin Martin does attest a population of Pheasants in the Outer Hebrides – from context he was most probably referring to Uist:

> Here are Hawks, Eagles, Pheasants, Moor Fowls, Tarmogan, Plover, Pigeons, Crows, Swans, and all the ordinary Sea-Fowls in the West Islands. (Martin 1703: 70)

There were still birds in the Outer Hebrides in the nineteenth century (Holloway 1996: 146–7), but it is possible that this population was only viable when regularly restocked by escapes.

GREAT NORTHERN DIVER
Gavia immer

NATIVE STATUS Native

MODERN CONSERVATION STATUS	
World	Least Concern (Vulnerable in Europe)
UK	Amber List
ROI	Amber List
Trend since 1772	Uncertain

The Great Northern Diver was recorded by a few authors around the coasts of early modern Britain and Ireland.

RECOGNITION
The species has several names. In New Latin it is called both *Colymbus major* and *Colymbus maximus*, depending on who is naming it, and its most common names in English match these: *great diver*, or *great ducker*; *greatest diver*. More rarely it is called the *great loon* and *greatest loon*. In Middle Scots it is referred to as the *embergoose*. Martin Martin attests to *bonnivochill* as the Scottish Gaelic name for the species (Martin and Monro 2018: 58).

DISTRIBUTION
The distribution of top-quality records for this species is not statistically different from the known level of recorder effort. The gaps on the map may just reflect decreased survey effort in some regions. It may have been widespread.

None of the early modern records provides evidence of breeding, and some of the sources even provided far-fetched explanations as to why the Great Northern Diver is never seen sitting on a nest. The most interesting of these is probably that provided by Thomas Gifford:

> There is likewise the ember goose, which is said to hatch her eggs under her wing. This is certain, that none saw them on the land, or out of the water, and that they have a cavity, or hollow place under one of their wings, only capable of containing a large egg.
>
> (ed. Gifford 1879 [1733]: 98)

The reason that this quotation is so intriguing is because this text was written in Shetland (the northernmost part of the species' range in Britain) and in the year 1733, meaning at the end of the coldest period of the Little Ice Age (Fagan 2000: chap. 7). This would be one of the most likely situations for the bird to be found breeding in Britain and Ireland since the end of the Younger Dryas, but breeding remained so unheard of that there was a piece of folklore to explain how it must happen.

Today, the Great Northern Diver is found all around the coasts of Britain and Ireland in winter and seems to have expanded its distribution since the 1980s (Balmer et al. 2013). The species remains most common approximately in the areas where it was recorded in the early modern period: the Highlands and Islands of Scotland as well as Connacht and Munster. In the second half of the twentieth century,

Great Northern Diver records, 1519–1772. There are coastal records from the Hebrides and the Northern Isles, Lowland Scotland as well as Munster and South England.

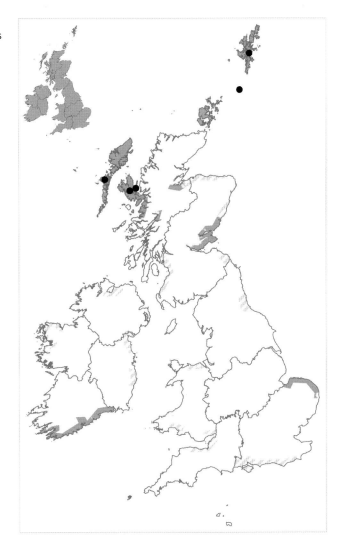

there were also some records of breeding on the coasts of the Highlands and Islands of Scotland and in Ulster and Connacht. Many of these are uncertain because Great Northern Divers do not leave to breed until after their spring moult (which can mean departing as late as April or May), passage migrants are seen in June, and some birds may return to their wintering grounds as early as the end of July (McGowan 2007; Holloway 1996: 428; Parslow 1973: 206). Only a single (unsuccessful) breeding attempt was recorded in the 2007–11 *Bird Atlas*. The birds most commonly nest in northern north America, Greenland and in Iceland in smaller numbers (Pritchard 2021j).

MANX SHEARWATER
Puffinus puffinus

NATIVE STATUS Native

MODERN CONSERVATION STATUS	
World	Least Concern
UK	Amber List
ROI	Amber List
Trend since 1772	Uncertain

The Manx Shearwater was regularly recorded by early modern authors around the coasts of Britain and Ireland.

RECOGNITION
The most common early modern name used for the Manx Shearwater was *puffin*. This is unfortunate because the Atlantic Puffin was also regularly called the *puffin* in the time period. This means that some of the records of our current species are ambiguous. However, because the descriptions are derivative, it is usually possible to identify which species is intended. For example, if a text describes a *puffin* associated with the Isle of Man and explains it is regularly eaten but has a fishy taste, the species in question is likely to be the Manx Shearwater. Where there are no clues, as for instance the records from the Farne Islands and Stromness, I have included these records as diamonds on the map.

The species was called *Puffinus anglorum* in New Latin, and *scraber* in Scottish Gaelic. Both of these names appear to be specific to the species. It was also called the *lyar* in Norn (compare Norse *líri*). Care needs to be taken with this term because it is easy to confuse with the Latin term *lyra* which was and is used to describe *Trigla lyra* (the Piper Gurnard fish).

DISTRIBUTION
The distribution of top-quality records for this species is not statistically different from the known level of recorder effort. The gaps on the map may just reflect decreased survey effort in some regions. It may have been widespread. Nevertheless, there are reliable records from the Northern Isles to Co. Cork, which suggests a wide distribution. In the medieval archaeological record, Manx Shearwaters are mainly found on northern and western sites rather than southern and eastern ones (Serjeantson 2006), which would agree approximately with the early modern record.

There is little evidence that this species has changed its distribution since the early modern period. On the map, I have indicated ambiguous records of *puffins* as diamonds, and securely identified records as spots. Because there is some confusion about where the species could be seen, compared to where it actually bred, I have not shaded any parts of the map, and have shown all county records approximately as local dot records.

Manx Shearwater records, 1519–1772. There are coastal records from every region of Britain and Ireland except Connacht.

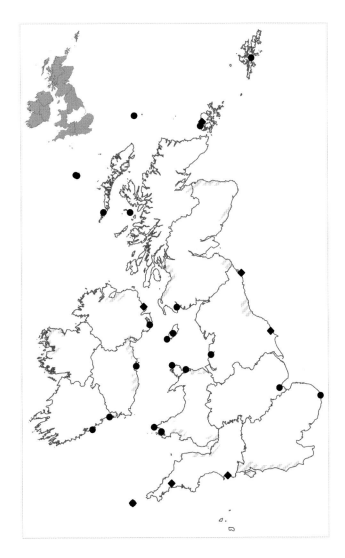

The reliable records of breeding sites mentioned in the early modern sources are:

1. The Calf of Man
2. The Isles of Scilly
3. Sula Sgeir
4. Rum
5. St Kilda
6. Ballycotton Island, Co. Cork

The best-known breeding site for Manx Shearwaters in the early modern period was the Calf of Man; this place is mentioned in nine different texts. The population on this site (and others, like the populations recorded by our authors in the Hebrides) was lost around the start of the nineteenth century when Brown Rats colonised the island, but has since begun to recover with the extermination of these rodents (Holloway 1996: 50).

At present there are no Manx Shearwater breeding sites on the east coast of Britain, nor were there any in the nineteenth century, according to Holloway (1996: 50–1). There is one ambiguous record that suggests this may not have been the case in the early modern period. Chamberlayne explains:

> In Lincolnshire and Yorkshire, near the Sea, are store of Reeves, Roughs, Gulls, and a Bird called a Stint, somewhat bigger than a Lark. Puffins and Burranets hatch in the holes of the Sea-Cliffs. Woodcocks, Sparhawks, and Fieldfares take Cornwall in their way to warmer Climates.
>
> (Chamberlayne 1683: 253)

If the *puffins* in this extract can be identified as Manx Shearwaters rather that Atlantic Puffins, this would be a record of shearwaters nesting on the mainland of east Britain. There is also a similar reference in Carew's *Survey of Cornwall* (ed. Chynoweth et al. 2004: fol. 34v). Both records are ambiguous, and it is possible the Manx Shearwater was confined to island breeding sites before the early modern period began.

WHITE STORK
Ciconia ciconia

NATIVE STATUS Possibly native to Britain

MODERN CONSERVATION STATUS	
World	Least Concern
UK	Extinct/Not Present
ROI	Not Present
Trend since 1772	Uncertain

The Stork was recorded in a few local sources from the early modern period. There were also some contemporary national records from England and Ireland.

RECOGNITION

The species is referred to as *stork* in Early Modern English and *ciconia* in New Latin. When these terms are used to describe a wild bird in early modern Britain or Ireland, they usually describe our species, although later on both terms are occasionally employed for the Grey Heron.

DISTRIBUTION

There are not enough top-quality records to statistically analyse the distribution of this species.

The best reference to the Stork from the early modern period comes from Browne in seventeenth-century Norfolk (*c.*1662–68):

> The ciconia or stork I have seen in the fennes & some haue been shot in the marshes between
> this [Norwich] and Yarmouth.   (ed. Browne 1902: 10)

The reference from Cornwall (shown with a diamond mark) is from Thomas Tonkin's notes on Carew's *Survey of Cornwall* (Dunstanville 1811: 84); it refers to a single specimen (presumably a vagrant). The other uncertain record provides a dubious account of a white vagrant bird with black wings, two yards tall, seen in Cardiganshire (Lhuyd 1712). The species is also mentioned by three national natural histories. Willughby and Ray (Ray 1678: 286) and Merrett (1666: 181) describe the species as a rare vagrant in England, and O'Sullivan (ed. 2009: 116–17) describes it as an Irish species (although there is very little other evidence to support this). As far as we can tell, the range of the bird in the early modern period was therefore not especially different from today, with it being an occasional visitor or intermittent breeder, possibly with a small year-round population in the Norfolk wetlands. This suggests that the White Stork may have become rarer since the early medieval period (Gow and Edgcumbe 2016; Serjeantson 2010; Yalden and Albarella 2009: 140; Baxter and Rintoul 1953: 333–4).[6] There is a clear record of breeding from Edinburgh in 1416 (Clarke 1919). It is also worth pointing out that the White Stork seems to be on the northern edge of its range in Britain. Our early modern

---

6   English archaeologists traditionally separate out a period before the medieval as the 'Anglo-Saxon' period but this is now recognised to be inaccurate and harmful (following Rambaran-Olm 2019).

White Stork records, 1519–1772. There are reliable records from North and South England and unreliable records from South West England and Wales.

data comes from the Little Ice Age – the coldest period of recent history – so our evidence may come from a period when the Stork was temporarily lost (Williamson 1975).

Today the White Stork is recorded occasionally in lowland areas of Britain (most commonly in England) in both winter and summer, although it does not generally breed (Balmer et al. 2013). The number of records from Britain has increased significantly over the last century; this is partially but not wholly explained by individuals from reintroduction projects and escaped collection birds in England and Europe more broadly (Green and Sandham 2021; Hogg 2007).

A reintroduction programme has recently begun in West Sussex as part of Knepp Castle Estate's rewilding project, and the first wild chicks were born there in 2020 (Green and Sandham 2021).

BITTERN

Botaurus stellaris

NATIVE STATUS Native

MODERN CONSERVATION STATUS	
World	Least Concern
UK	Amber List
ROI	Extinct
Trend since 1772	Certainly declined

Before it went extinct, the Bittern was recorded in the early modern sources in almost every region of Britain and Ireland.

RECOGNITION

This species was consistently called *Ardea stellaris* in New Latin. In Early Modern English it is called *bittern* or *bittour*, or, more rarely, *night raven* or *mire-drum*. All of these names, except the last, are specific to this species. The form *mire-drumble* is once used by Merret (1666: 181) to refer to *Ardea alba* (the Great White Egret), but this seems likely to be a mistake.

DISTRIBUTION

The distribution of top-quality records for this species is not statistically different from the known level of recorder effort. The gaps on the map may just reflect decreased survey effort in some regions. It may have been widespread. If we include county-level records, the Bittern is recorded in every region of early modern Britain and Ireland except Connacht. This is important because the Bittern went extinct as a breeder in Britain, and Ireland soon afterwards, in the nineteenth century (Anderson 2017: 302–3; Shrubb 2013: 54; D'Arcy 1999: 45). It is sometimes said to have always been absent as a breeder in Leinster, but the description of the bird in Rutty's *Essay Towards a Natural History of the County of Dublin* includes a suggestion that it is present all year-round, supporting D'Arcy's (1999: 46) view that it formerly bred in all four provinces.

> *Ardea stellaris*. The Bittern, which by some has been called the Night-raven.
>
> F[ood] Young fish or young frogs, and at the end of autumn, it goes to the woods and devours mice. (Rutty 1772: 320)

This record is interesting because migrating Bitterns are usually absent in autumn, and only arrive for the winter; Rutty's record might then imply a breeding population near Dublin in the second half of the eighteenth century.

The decline of this species in the nineteenth century was continent-wide and appears to have been a result of habitat destruction which had already begun in the early modern period, together with unsustainable hunting pressure (Shrubb 2013: 52–5; Williamson 2013; D'Arcy 1999: 46; Gooders 1983: 85–90; Gurney 1921: 182–3). In Ireland, Bitterns were used for food and medicine, and were common taxidermy trophies (D'Arcy 1999: 42–4). It has been suggested that the extinction was hastened by the cold of the Little Ice Age, but since the coldest years were arguably 1550–1700, when our sources suggest the Bittern was still widespread, this appears unlikely (Gilbert 2007). However, there is some evidence that the species was already in decline in the early modern period. Denton (ed. 2003), for example, writing in the seventeenth century, records only former populations in Cumberland, and there are no records of breeding birds in South West England in the nineteenth- and twentieth-century records

(Shrubb 2013: 53–4). Holloway suggests that it was confined to the Fens and Broads of East Anglia and only occasionally nested elsewhere by the start of the nineteenth century (Holloway 1996: 62–3).

The Bittern became re-established as a breeder in Britain in the early twentieth century, perhaps in part due to the abandonment of drainage systems during the agricultural depression period (Shrubb 2013: 54). Its numbers increased through the first half of the century, and breeders were recorded in Scotland and Wales as well as its heartland in southern England (Gooders 1983: 85–90). However, due to cold winters and poor habitat management, its population crashed in the second half of the twentieth century. It went extinct as a breeder in Wales in 1984 (Pritchard 2021k). Only 11 booming males were recorded in England in 1997. Luckily its decline was reversed once again and by 2011 that number had increased to 104 thanks to habitat restoration work (Balmer et al. 2013). Today the Bittern is widely recorded in lowland England and Wales in winter, and breeds especially in the reedbeds of the south and Midlands of England, and also in parts of Wales. There are winter records from Scotland and Ireland too, where the species remains extinct as a breeder. Despite the success of the latest conservation efforts, the Bittern likely had a broader range in the early modern period than it does today.

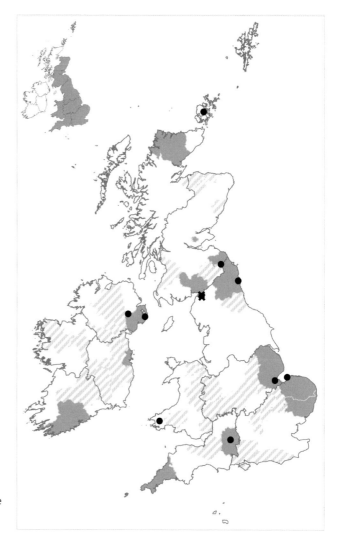

Bittern records, 1519–1772. There are reliable records from every region of Britain and Ireland except Connacht.

GREY HERON

Ardea cinerea

NATIVE STATUS Native

MODERN CONSERVATION STATUS	
World	Least Concern
UK	Green List
ROI	Green List
Trend since 1772	Uncertain

The Grey Heron was recorded in many early modern sources in almost every region of Britain and Ireland.

RECOGNITION

This species was called the *heron* or *heronshaw* in Early Modern English and *ardea* or *Ardea cinerea* in New Latin. O'Sullivan (2009: 118) attests the Irish *corr-iasc* and *corri-grieni* (possibly related to modern Scottish Gaelic terms *corra-ghrian*, *corra-grain*, *corr-ghritheach*, the last of which Dwelly (1988: 255) suggests is from Middle Irish *grith*, 'a scream').

Some of these terms are also adapted for other species. For instance, the Great White Egret was called the *white heron* or *Ardea alba*. The Bittern was called in New Latin *Ardea stellaris*. However, the Grey Heron was the generic *Ardea* species, meaning that the name was always modified before being applied to other species. The records on this map are thus likely to be reliable.

DISTRIBUTION

The distribution of top-quality records for this species is not statistically different from the known level of recorder effort. The gaps on the map may just reflect decreased survey effort in some regions. Since it was recorded in almost every region of Britain and Ireland it is likely to have been widespread.

In the early modern period, landowners would purposefully establish heronries on their land, and take Heron chicks to be fattened for the pot, or to attack them with falcons for sport (Shrubb 2013: 58–62). Wild Herons on common land were also taken by the poor. The early modern records therefore might represent some artificial heronries, established for food. In Scotland, there seems to have been such concern about the population of Grey Herons in the second half of the sixteenth century that the birds were given legal protection against hunting with bows and crossbows, and a fine of one hundred Scots pounds was imposed for anyone selling them (for food) (Baxter and Rintoul 1953: 336).

From the beginning of the nineteenth century, heronries were no longer seen as prized resources, and many were lost or destroyed as heron meat went out of fashion (Holloway 1996: 64; Thomas 1991: chap. 2). The average number of nests at heronry sites has decreased since the early modern period, which is sometimes attributed to increased persecution (Marquiss 2007). Shrubb, however, suggests this decline is most likely a result of drainage projects reducing the carrying capacity of the area where the Herons live (Shrubb 2013: 60). Since the 1980s they have generally been controlled non-lethally (Pritchard 2021l).

The Grey Heron is now found commonly across most of Britain and Ireland (Balmer et al. 2013). Its winter distribution is wider than its summer distribution, and it has not bred in recent years on Orkney or Shetland. This agrees with the eighteenth-century evidence, where the Heron was recorded in Shetland as a winter species only.

Grey Heron records, 1519–1772.
There are reliable records from
every region of Britain and
Ireland except Connacht.

There is also here [in Shetland] over winter swans, herons, wild geese of several kinds, who all
go away in the spring, and return again in autumn.   (ed. Gifford 1879 [1733]: 23)

Over the last century, persecution has not impacted Grey Heron abundance, which is increasing.
The distribution of this species has not changed at regional level since the 1970s, although it is amber
listed in Wales for local declines in breeding range (Pritchard 2021l; Balmer et al. 2013).

Herons are susceptible to harsh winters, and in the 1930s and 1940s Grey Herons in Britain were
significantly affected by the cold weather (Lamb 2011: 186; Ford 1982: 128–9; Parslow 1973). This
implies that they may also have been affected by the colder temperatures of the Little Ice Age of the
early modern period (Fagan 2000). However, Grey Herons were clearly still able to live in the north
of Scotland, including Caithness and Shetland during the Little Ice Age, so the colder weather did
not lead to wider changes in abundance during the early modern period.

GANNET
Morus bassanus

NATIVE STATUS Native

MODERN CONSERVATION STATUS	
World	Least Concern
UK	Amber List
ROI	Amber List
Trend since 1772	Uncertain

The Gannet was recorded as living around the coasts of Britain and Ireland, especially on smaller islands.

RECOGNITION
The Gannet was usually called *soland-goose* in Early Modern English, or *Anser bassanus* in New Latin after the Bass Rock in East Lothian which is still one of the biggest colonies today. These terms are almost always used for the Gannet in the period, which makes identification relatively simple. Occasionally the Brent Goose and Barnacle Goose were confusedly referred to as *soland goose*, but this is relatively rare.

DISTRIBUTION
The distribution of top-quality records for this species is not statistically different from the known level of recorder effort. The gaps on the map may just reflect decreased survey effort in some regions. It may have been widespread.

Gannets commonly nest together in large colonies called gannetries. In the early modern period, these gannetries were commonly visited in spring to take young birds as food. The most famous of these was, and perhaps continues to be, Bass Rock in the Firth of Forth, east of Edinburgh. Gannet chicks from Bass Rock were even salted and exported to England to be sold as 'geese' (Shrubb 2013: 146).

Other important gannetries were also identified in the early modern period including Ailsa Craig, St Kilda, the Skellig Rocks, Sula Sgeir and Sula Stack. All of these, as well as the Bass Rock gannetry, still exist. Most of them have been in use for over five hundred years.

In the medieval archaeological record, Gannets are mainly known from the North of England, and are rare in the Midlands, South and South West (Serjeantson 2006). Oddly, this does not agree well with the early modern distribution, since by then Gannets were commonly recorded in Cornwall, East Anglia and the Channel Islands as well as further north.

Some gannetries were lost or declined in the nineteenth century due to overexploitation for food, sport and egg-collecting. The population seems now to have recovered well, although plastic pollution has become an issue (Lindley 2021; Zonfrillo 2007; Holloway 1996: 56–7; Parslow 1973: 28). It has been suggested that the higher temperatures of the first half of the twentieth century helped make food available for the Gannet to recolonise the north of Scotland (Williamson 1975). This is surprising

Gannet records, 1519–1772. Most of the records come from the coasts of Scotland and Ireland, but there are also records dotted across England.

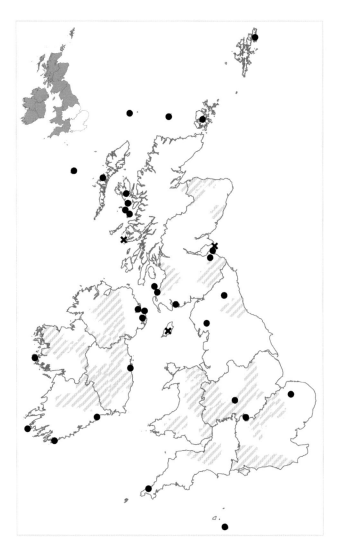

because our map seems to show that the Gannet was recorded as far north as Shetland during the Little Ice Age.

Today, the Gannet is recorded around the whole of Britain and Ireland during summer and winter, despite the migration of some individuals from the colony. However, this distribution is complicated because parents will forage at great distances from their gannetries. For this reason, our map does not include county-level records because it is not possible to separate the core nesting distribution from the much wider exploitation zone. There are also a few inland records, presumably of lost individuals.

SHAG

Phalacrocorax aristotelis

NATIVE STATUS Native

MODERN CONSERVATION STATUS	
World	Least Concern
UK	Red List
ROI	Amber List
Trend since 1772	Uncertain

The Shag was recorded by a few early modern authors as occurring around the coasts of Britain and Ireland. It was less regularly recorded than the Cormorant.

RECOGNITION

The early modern authors called this species *Corvus aquaticus minor*, to distinguish it from what they called *Corvus aquaticus* (the related Cormorant). It is also referred to in Early Modern and Modern English as the *shag*; the *Oxford English Dictionary* editors suggest this name comes from the tuft of feathers on the bird's head. Occasionally the term *coromorant* also seems to have been applied to the Shag:

> The Sea Fowls are Malls of all kinds. Coulterneb, Guillamet, Sea-Cormorant, etc. The Natives observe that the latter if perfectly Black make no good Broth, nor is its Flesh worth eating, but that a Cormorant, which has any white Feathers or Down, makes good Broth, and the Flesh of it is good Food, and the Broth is usually drunk by Nurses to encrease their Milk.
>
> (Martin 1703: 158)

The white feathers described here are more typical of the Shag. However, the Cormorant can also have white feathers. The continental *sinensis* subspecies of Cormorant commonly found inland in the South of England, is especially known for its white head although this subspecies is only occasionally found in Scotland and seems to have only colonised Britain since the 1980s (Carss and Murray 2007).

DISTRIBUTION

There are not enough top-quality records to statistically analyse the distribution of this species.

The Shag is well recorded archaeologically and seems to have been hunted for food regularly from Roman times into the twentieth century (Shrubb 2013: 145; Yalden and Albarella 2009: 210; Holloway 1996: 60). Shags were listed in the 1566 Grain Act for pest control in England and Wales, but unlike Cormorants which hunt from the shore, and even inland in rivers, Shags do not seem to have been regularly targeted (Lovegrove 2007: 100–101). They were, however, frequently hunted for sport, and declined through the nineteenth century (Holloway 1996: 60). Shooting stopped in the twentieth century, which led to some recovery and all of the populations recorded by the early modern sources exist today. The species has slightly expanded its range since the late 1990s, but at the same time the population in Britain and Ireland has declined significantly, perhaps due to extreme weather, increased predation and the decline in sandeels (Balmer et al. 2013).

Shag records, 1519–1772. There are coastal records from Highland and Lowland Scotland, the North and South West of England, and Ulster, Leinster and Munster.

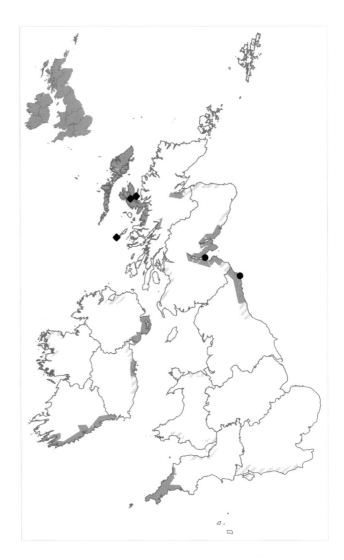

CORMORANT
Phalacrocorax carbo

NATIVE STATUS Native

MODERN CONSERVATION STATUS	
World	Least Concern
UK	Green List
ROI	Green List
Trend since 1772	No change

Many populations of Cormorant were recorded by our early modern sources, dotted around the coasts of Britain and Ireland, but also further inland.

RECOGNITION
This species was most commonly called the *cormorant* or *skart* in Early Modern English and *Corvus aquaticus* or *Corvus marinus* in New Latin. The association with the corvid genus implied by the New Latin names seems to also be made by the Irish name *fiach mairi* attested in O'Sullivan (2009: 143). The Scots term *gormaw* is also attested.

As explained in the entry on the Shag, these records are not completely reliable as some of the terms were theoretically shared between the two species and used interchangeably. However, in practice early modern naturalists usually distinguished the *shag*.

DISTRIBUTION
The distribution of top-quality records for this species is not statistically different from the known level of recorder effort. The gaps on the map may just reflect decreased survey effort in some regions. It may have been widespread.

Cormorants were listed in the 1566 Grain Act and were regularly targeted for persecution, particularly by anglers (Lovegrove 2007). They were also exploited by anglers for use in catching fish (Shrubb 2013: 32). The species was subject to especially intense persecution in the modern period, and seems to have declined in the nineteenth century, with the extirpation of many local populations (Lovegrove 2007: 102–3; Parslow 1973: 29).

Since the 1980s, Cormorants have become established not just on the coastlines, but also inland on rivers and loughs in the west of Ireland and the east and Midlands of England (Newson et al. 2007). These populations are especially but not exclusively made up of *Phalacrocorax carbo sinensis* subspecies, the continental, freshwater specialist (Balmer et al. 2013). However, inland populations, presumably of *Phalacrocorax carbo carbo* were also occasionally encountered in the past (Lovegrove 2007: 101). Holloway (1996: 58–9) found nineteenth-century inland breeding records from Connacht and Lowland Scotland and infrequent records from the South of England. It seems that Cormorants nesting inland may have been especially commonly targeted for persecution in the past (Lovegrove 2007: 102), leading to the loss of most inland colonies. For the map, all county-level records have been assumed to be to coastal populations unless the records say otherwise. Despite this, Cormorants seem to have been regularly recorded in inland England during the early modern period, although it is not clear whether these records necessarily reflect inland breeding populations.

Cormorant records, 1519–1772. There are coastal records from Highland and Lowland Scotland, Wales and every region of Ireland, but also some inland records from England.

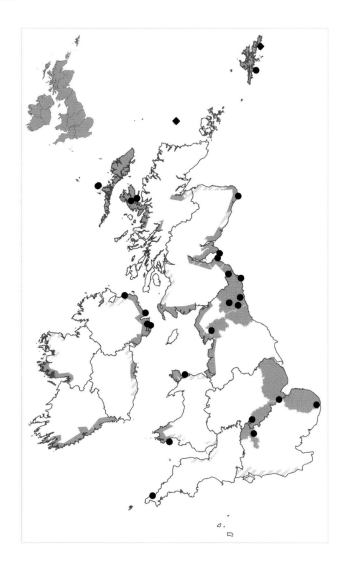

OSPREY
Pandion haliaetus

NATIVE STATUS Native

MODERN CONSERVATION STATUS	
World	Least Concern
UK	Amber List
ROI	Extinct
Trend since 1772	Certainly declined

Our early modern sources provided records of Osprey populations dotted across England and Ireland, but some of the records are unreliable.

RECOGNITION
This map shows records of what are most commonly called *ospreys*, and occasionally *sea eagles* and *bald buzzards* in Early Modern English and *haliaetus* or *ossifraga* in early modern New Latin. Lhuyd attests the Scottish Gaelic term *ean feyn* (ed. Campbell and Thomson 1963: 65–6). None of these terms were unique to our species; they were all also used for the then more common Sea Eagle, and other species in the early modern period. In some cases it is possible to distinguish reliable records based on the description of the species, or where the authors list both the Osprey and the Sea Eagle. The uncertain records are shown on our map as diamonds.

DISTRIBUTION
There are not enough top-quality records to statistically analyse the distribution of this species. The records that we do have show that the Osprey was found in the early modern period on inland wetland and coastal sites across both England and Ireland. While there are no reliable early modern records from Scotland in the database, there is later historical evidence that the Osprey was found here until the end of the period. Similarly, while there are no known Osprey records for Wales prior to the nineteenth century, the bird is assumed to have occurred there as well (Corfield 2021).

> There are in [Co. Leitrim] severall Eyeryes of Egles, as allsoe of Ospreyes, these Latter build usually on ould walls neer great rivers or Loghes and feed on fish … this is most strange that wherever they timber they have alwaies three ould ones to each Eyery: their way of fishing is they hover in the ayer when the sun shines soe as their shadow moves on ye ground under ye water which ye fish seeing are afrighted and flye to the top of ye water where ye Asprey stoopes and seises them.
> (Logan 1971 [*c.*1683]: 333–4)

The records from early modern Britain and Ireland often describe the Osprey as rare, and it was likely widely targeted for persecution (Lovegrove 2007: 107). Some populations may have been confined to islands (for example the Isle of Wight and Lambay Island in Dublin). Where a population is described on an island with no mainland population recorded, the mainland county is not shaded on our map.

The Osprey went extinct across Britain and Ireland soon after our early modern records were written. Although there are records of birds from after the end of the early modern period, these seem to have been migrating rather than breeding birds. Gurney (1921: 162) suggests the Osprey was becoming rare in England in the seventeenth century, and there are no records of bounty payments in the seventeenth–nineteenth century *Parish Records* (Lovegrove 2007: 107). However, the species may have hung on in isolated locations into the mid-nineteenth century. In Ireland, it appears to have gone extinct as a breeder due to hunting for pest control somewhere around 1750–1825 (D'Arcy 1999: 83). In Scotland, the Osprey became rare in the nineteenth century and stopped regularly breeding after 1908 (apart from occasional attempts, such as that in 1916) due to hunting for pest control and egg-collecting by naturalists (Corfield 2021; Lovegrove 2007: 108–10; Holloway 1996: 122; Dennis 1991; Gooders 1983: 108–19).

Ospreys began breeding regularly once again in Loch Garten, northern Scotland in the mid-twentieth century, following recruitment of a breeding pair from the newly recovered Scandinavian population (Dennis 1991: 9–11). Early breeding attempts resulted in collection of the eggs by egg-collectors, but in 1959 the nesting site was made a protected bird sanctuary. Since then, the Osprey has become relatively widespread in Scotland, and has even begun to recolonise North England (Balmer et al. 2013; Gooders 1983: 108–19). A population was established at Rutland Water in the Midlands of England in 1996, and birds from this program have begun to recolonise Wales and the rest of England. Migrating Ospreys have also been regularly recorded in every region of Britain and also in Ulster.

Osprey records, 1519–1772. There are reliable records from the South and South West of England and from Munster and Connacht in Ireland.

GOLDEN EAGLE
Aquila chrysaetos

NATIVE STATUS Native

MODERN CONSERVATION STATUS	
World	Least Concern
UK	Green List
ROI	Red List
Trend since 1772	Certainly declined

There were reliable early modern records of the Golden Eagle scattered across Britain and Ireland, but it does not seem to have been present in lowland England or Wales.

RECOGNITION
The Golden Eagle was not consistently distinguished from other species in the early modern period. However, this map focuses on only the few reliable records which identify the bird as the *black eagle* or *golden eagle* in Early Modern English, or which can be identified by their descriptions – this includes animals identified in New Latin as the *Aquila* of Gessner, *Chrysaetos* and *Aquila ibernica*.

Generic records of eagles are mostly included as diamonds on the map for the Sea Eagle. However, two (Castell Dinas Bran and Beaudesert Hall) have been included here as diamonds because based on the inland locations away from large lakes, they seem more likely to refer to the Golden Eagle (Yalden 2007).

DISTRIBUTION
The distribution of top-quality records for this species is not statistically different from the known level of recorder effort. The gaps on the map may just reflect decreased survey effort in some regions. It may have been widespread. Absence records suggest that the Golden Eagle was absent from lowland England in the early modern period, but present in the uplands of Wales, Scotland, Ireland and, more rarely, the uplands of England. This map approximately agrees with Tapper's (1999: 254–7) habitat suitability survey and the place-name evidence that Golden Eagles were present in Britain north and west of a line from Pembrokeshire to Yorkshire (Evans et al. 2012).

In the early modern period the species was more widespread than it is now, but it seems to have already been in decline. For instance, a letter to Lhuyd dated 1695, described a population on the edge of extinction in Eryri in Wales:

> There is scarcely anyone of the local inhabitants who has not seen the eagles; nor have three years passed from the time an eagle was observed, and even now they want one to survive. Those birds of prey were once famous for various incidents; for they are said to have snatched up not just tame birds, lambs, and sheep, but dogs, calves, and swine, and to have devoured larger animals, and, not content with just these, to have swooped on men.
>
> (trans. Emery 1974: 411)

This population was well known (the reference to the Golden Eagles of Carnedd Llewelyn in Thomas Johnson's (1641: 11) report on his botanising tour of Eryri in Wales is particularly famous). The Golden Eagles of Eryri continue to be remembered to this day, although there are few reliable records for this population from after the seventeenth century (Pritchard 2021m).

'Ernys' are included as targets under the second pest control act of James II (*RPS* 1458/3/33), and persecution may have been ongoing by the early modern period. Lovegrove (2007: 116) suggests that the persecution only began to put pressure on the species from the end of the early modern period. Records from the nineteenth century appear to agree with the early modern records and suggest that eagles were absent from Scotland south of the central belt, except a small population in Dumfries & Galloway, where they may have lingered into the middle of the century (Baxter and Rintoul 1953: 296–7). The lowland population of Golden Eagles suggested by the place-name evidence may have been already partially depleted in the early modern period (Evans et al. 2012; Holloway 1996: 120–1).

Today, the Golden Eagle is most abundant in the north of Scotland. The species stopped breeding in England in the 1780s, in Wales possibly as late as the 1850s, and in the west of Ireland at the beginning of the twentieth century (although a single Scottish pair bred in Co. Antrim from 1956 to 1960). The species has also been reintroduced to Co. Donegal in Ulster and to the southern uplands of Scotland. Until recently, birds held territory in North England. The decline seems to have been due to persecution by shepherds and especially gamekeepers (Williams et al. 2020; Anderson 2017: 310; Balmer et al. 2013; Lever 2009: 350–1; D'Arcy 1999: 68–70; Holloway 1996: 120–1). Even after

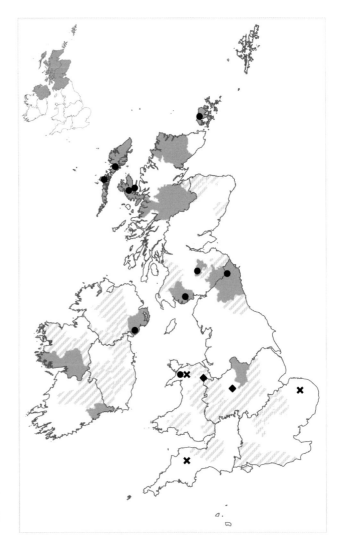

Golden Eagle records, 1519–1772. There are reliable records from Highland and Lowland Scotland, the North and Midlands of England, and Ulster, Connacht and Munster. There is also a contested record and an uncertain record from Wales.

the Golden Eagle was extinct across most of Britain and Ireland its eggs continued to demand high prices among collectors (Shrubb 2013: 120–1).

Lamb has suggested that the Golden Eagle's colonisation of the Lake District in the 1970s might have been assisted by the cooler weather (Lamb 2011: 186). It is not clear why this should have been a significant factor, considering that the Golden Eagle nests in much warmer locations than the Lake District elsewhere in the world. The Scottish population has struggled to expand due to persecution in east Scotland and lack of food in west Scotland (Balmer et al. 2013).

SPARROWHAWK

Accipiter nisus

NATIVE STATUS Native

MODERN CONSERVATION STATUS	
World	Least Concern
UK	Amber List
ROI	Green List
Trend since 1772	Uncertain

The early modern sources very commonly referred to hawk populations around Britain and Ireland, but with few species-level records of Sparrowhawks in particular.

RECOGNITION

This species was called the *sparrowhawk* or *sparhawk* in Early Modern English, *sparhalk* in Middle Scots and *Accipiter fringillarius* or *nisus* in New Latin. It is not entirely clear if these terms were specific to our species. Gurney (1921: 153) has argued that the term *sparrowhawk* might have referred at times to *Accipiter gentilis* (the Goshawk), based on Turner's description of the bird preying upon doves, pigeons and partridges. Yet Sparrowhawks do take larger prey, so this does not necessarily mean that our references are untrustworthy.

However, most of the records on this map are generic references to the *hawk*. These are uncertain references and are shown as diamonds. When used alone, the Early Modern English term *hawk*, like the Irish *sabhac/shouck* (Scharff 1915b), does often refer to the Sparrowhawk, but it could also refer to other hawks or even very different raptors like the Buzzard or the Red Kite (Lovegrove 2007: 132). It also seems likely that many of our observers might have lacked the knowledge of raptors to differentiate, for example, Sparrowhawks from Goshawks, so the diamond pins may include references to other species.

DISTRIBUTION

There are not enough top-quality records to statistically analyse the distribution of this species.

There are reliable records of Sparrowhawks from seven of the 11 regions, from Cornwall in the south, to Sutherland in the north and from Cork in the west to Angus in the east. There are also generic records of hawks all across the map. Overall, the impression of the records is that the Sparrowhawk might have been relatively widespread across Britain and Ireland, which would fit with the species' present and past distribution. The Sparrowhawk was the most commonly recorded wild falcon species in the medieval archaeological record from England (Serjeantson 2006).

Sparrowhawk records, 1519–1772.
There are records from every
region of Britain and Ireland
except the Midlands of England,
but all the records from Ulster
and Connacht are unreliable.

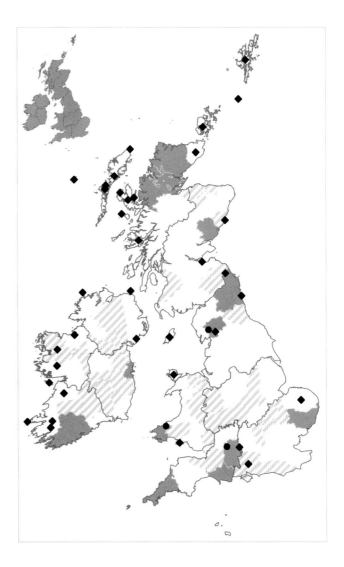

The Sparrowhawk was not mentioned in the English Grain Act of 1566, but bounties were occasionally paid for these birds (Lovegrove 2007: 132–3). From the eighteenth century onwards, they were commonly targeted by gamekeepers, and began to decline in the nineteenth century perhaps due to the development of the breach-loading gun in the middle of the century (Williamson 2013: 124; Holloway 1996: 116–17). However, they were still widespread at the turn of the nineteenth century. From this point a further decline was caused by the use of DDT from 1955. This led to the extinction of the Sparrowhawk from several eastern, arable-dominated counties in England (Parslow 1973: 53–6, 226–7). With the restriction of DDT, Sparrowhawk populations recovered, and by 1988–91 they had become widespread once again across the whole of Britain and Ireland (Balmer et al. 2013; Lovegrove 2007: 134).

The Sparrowhawk today is distributed across most of Britain except high upland areas in Highland Scotland and in Shetland, where it will spend the winter but does not breed (Balmer et al. 2013). The early modern record from Shetland of 'abundance of hawks of different kinds' (Gifford 1879: 23) need not imply a change in distribution; it may reflect overwintering Sparrowhawks or similar species which are resident there today like the Merlin (*Falco columbarius*). However, the reference from Sutherland in northern Scotland is more specific and implies a breeding population there in the seventeenth century:

> Divers kinds of halks doe breed in Southerland, such as falcons, marlions, jeyr-falcons,
> sparhalkes, &c.   (ed. Gordon 1813 [1630]: 7–8)

RED KITE
Milvus milvus

NATIVE STATUS Native

MODERN CONSERVATION STATUS	
World	Near Threatened
UK	Green List
ROI	Red List
Trend since 1772	Uncertain

The Red Kite was only rarely mentioned by the early modern sources. As a common pest species without much utilitarian value, it might have been less interesting to authors interested in cataloguing their natural resources.

RECOGNITION
The Red Kite was most reliably called *milvus* in New Latin. K'Eogh (1739: 57) attests the Irish term *pricane-nae-cark*. These terms were specific to the Red Kite, but the most common English terms were more generic (Lovegrove 1990: 13, 2007: 123). *Glede* was sometimes employed to refer to the *Circus* spp. (the harriers), and *kite* and *puttock* occasionally referred to the Buzzard. Despite this, the majority of references to this species by our sources provide multiple names and/or descriptions of the species intended, meaning that most of the records on the map are fairly secure, with the exception of Daniel Defoe's eighteenth-century record of kites in Winfrith Newburgh, Dorset (1971: 208). It also helps that the other European species of kite (*Milvus migrans*, Black Kite) was distinguished as *Milvus nigrius*, and our sources attest that it was absent from early modern Britain and Ireland.

DISTRIBUTION
There are not enough top-quality records to statistically analyse the distribution of this species.

Before the Industrial Revolution, the Red Kite was famously common in London. The level of legal protection for the species there appears to have been exaggerated, but it seems to have been well suited to cities as well as to rural areas. This included not just London but also apparently Norwich, Berwick-on-Tweed, Dundee and possibly Dublin and other cities in Britain, Ireland and in Europe more widely (Raye 2021b). In the wider countryside it was made a target for pest control in Scotland (1457/8), Ireland (1786/7), and England and Wales (1566) (Lovegrove 2007: 118–22; D'Arcy 1999: 52; Baxter and Rintoul 1953: 313).

It has sometimes been doubted that the Red Kite ever occurred in Ireland (Lovegrove 1990: 13). However, the existence of regional records from Co. Cork and Co. Waterford provides evidence of a population there prior to the species' extinction from the southern coast of Ireland (D'Arcy 1999: 50–2):

> Milvus caudâ forcipatâ. The Kite or Glead. These birds are so common that they need no
> particular description … with us [in Co. Cork] it remains all the year.   (Smith 1750b: 321)

Red Kite records, 1519–1772.
There are reliable records from
North and South England, Munster
and Leinster and from Orkney.

There are also some Irish records from the early nineteenth century indicating that Red Kites were still locally known, although D'Arcy (1999: 54) has suggested that the bird may have gone extinct as a breeder in the eighteenth century. It went extinct across most of Britain through the nineteenth century, but survived and was protected to some degree in the Welsh Valleys, although naturalist collectors continued to target Red Kite nests for eggs (Carter 2019: chap. 3; Holloway 1996: 104–5; Gooders 1983: 62–3). The initial reason for the decline appears to be that the Red Kite was widely persecuted by gamekeepers from the start of the nineteenth century (Lovegrove 1990: 43–8, 2007: 125–7; Holloway 1996: 104–5), although improved sanitation in cities may also have been important. Over the last 30 or so years, through reintroductions and the natural increase and spread of the native Welsh population, the species has begun to recover. It now nests in every region of Britain as well as in Ulster and Leinster (Carter 2019: chap. 3; Balmer et al. 2013; Lever 2009: 338–42; Carter and Whitlow 2005).

WHITE-TAILED EAGLE/SEA EAGLE

Haliaeetus albicilla

NATIVE STATUS Native

MODERN CONSERVATION STATUS	
World	Least Concern
UK	Amber List
ROI	Red List
Trend since 1772	Certainly declined

The White-tailed Eagle, also called the Sea Eagle, appears to have been regularly recorded in locations around Britain and Ireland, in both inland and as coastal populations. The early modern authors appear to attest surviving populations in every area of Britain and at least in Munster in Ireland.

RECOGNITION

This map shows records of the English *sea eagle* and the New Latin *haliaetus*. Our sources also use the terms *osprey*, *erne* and *pygargus*. These terms are difficult because they are used for the Osprey and Golden Eagle as well as the White-tailed Eagle (Love 1984: 34), but can be been identified by context if other species are distinguished or if multiple names are used.

In the past, scholars have identified records using generic terms like Early Modern English *eagle*, Welsh *eryr* and Irish *fuiller*, and the Latin term *aquila* as the White-tailed Eagle or Golden Eagle based on terrain (Williams et al. 2020; Evans et al. 2012; Yalden 2007; D'Arcy 1999: 62–3). This method is not wholly satisfactory, since the species coincide in upland areas close to water: for example, both seem to have occurred in Eryri, Wales in the seventeenth century. Because this method is not fully reliable, this map includes generic records as diamonds. It includes both records that are most likely to refer to the White-tailed Eagle and also those which are uncertain, but two generic records that are unlikely to refer to the White-tailed Eagle have been separated and are included on the Golden Eagle map.

DISTRIBUTION

The distribution of top-quality records for this species is not statistically different from the known level of recorder effort. The gaps on the map may just reflect decreased survey effort in some regions. It may have been widespread. More than half of the top-quality records come from Scotland, which might seem to fit with Love's (1984: 35) idea that the islands of Scotland were the stronghold of the species in the historical period. However, the number of Scottish records is actually not significantly different from what we might expect, considering that Highland and Lowland Scotland are some of the best-recorded areas of early modern Britain and Ireland.

For the most part, the White-tailed Eagle seems to have been a coastal species, but inland populations are also recorded around large bodies of water, including in the Lake District and around St Mary's Loch in Lowland Scotland. Our map assumes that the species had a coastal distribution in the counties where it was recorded except where there is a reliable record further inland.

It appears that previous authors have underestimated the White-tailed Eagle's distribution in England and Wales in the early modern period (Holloway 1996: 106–7; Love 1984: 110–11). In the second half of the seventeenth century, the White-tailed Eagle was still recorded in the Lake District, the Peak District, Cornwall and Norfolk. The species was also apparently known to nest along with the

Golden Eagle in Eryri in Wales in the seventeenth century (Ray 1713a: 7; Grew 1681: 56), and there are occasional vagrant records of White-tailed Eagles in Wales into the nineteenth century. This is not even to mention the historical nesting sites later remembered near Plymouth, on Lundy, the Isle of Wight and the Isle of Man, which are not mentioned in our sources (Williams et al. 2020; Yalden and Albarella 2009: 147; Love 1984: 111). Having said that, of all the early modern records from south Britain, only the records from Norfolk and Essex are confirmed by local authors (the latter seem to be vagrant records).

Outside of England the White-tailed Eagle was recorded in both the Highlands and Lowland Scotland and at least in Munster in Ireland. There are also a number of unreliable records of eagles from the coasts of Ulster and Connacht. These seem most likely to refer to this species rather than the Golden Eagle or the Osprey, and if so they would suggest a wide population in Ireland as well, agreeing with Withers' comments that the White-tailed Eagle was more common in Ireland than England (MacLysaght 1939: 416–17), and with the place-name evidence that it was formerly widespread in the north (Evans et al. 2012).

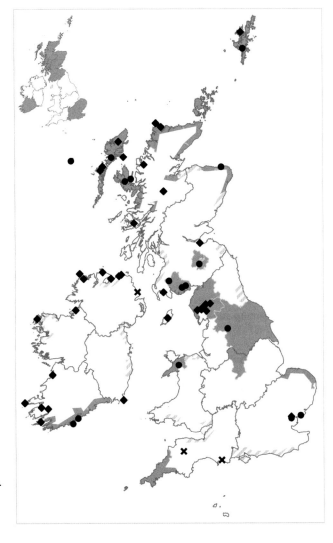

White-tailed Eagle records, 1519–1772. There are records from every region of Britain and Ireland, but the records from Ulster, Connacht and Leinster are unreliable. There are reliable inland records from the North and Midlands of England and Lowland Scotland.

When James II of Scotland passed his second pest control act in 1457/8 (*RPS* 1458/3/33), 'ernys' (eagles) were one of the targeted species (Gurney 1921: 81). Bounties were also payable in Orkney and Shetland from the seventeenth to the mid-nineteenth centuries (Love 1984: 111–12). It is not clear if this had an impact on the distribution of the Sea Eagle in early modern Scotland. Eagles were included in the English Grain Act of 1566, but Lovegrove (2007: 111–12) found no evidence that bounties were paid on the species following the Act, except in Borrowdale in the Lake District in the eighteenth century.

After the end of the early modern period, the White-tailed Eagle declined very quickly. Bounties were much more widely paid for Sea Eagles and Sea Eagle nests in Scotland in the eighteenth and nineteenth centuries (Lovegrove 2007: 113). The species seems to have been lost from England and Wales around 1800, with the last pair nesting in Kenfig, Wales, in 1816 (Williams et al. 2020; Lever 2009: 346; Holloway 1996: 106–7; Love 1984: 35–45). They were then lost from Ireland and most of Scotland by around 1900. The last pair in Britain is thought to have bred at Dunvegan Head, Skye in 1916, and the last individual was shot in Shetland in 1918. The cause of the destruction was persecution, and later the collection of eggs and specimens when the species became rare (Shrubb 2013: 189; Love 1984: 114–23; Love 2007; D'Arcy 1999: 64–5; Gooders 1983: 53–9).

The White-tailed Eagle has now breeds in Britain once more following extensive reintroduction projects starting on the Isle of Rum by the Nature Conservancy Council in 1975, and since also in Wester Ross in Highland Scotland, Fife in Lowland Scotland, Killarney National Park in Munster, and now also the Isle of Wight off the coast of South England (Lever 2009: 347–9).

GREAT BUSTARD
Otis tarda

NATIVE STATUS Uncertain

MODERN CONSERVATION STATUS	
World	Vulnerable (Least Concern in Europe)
UK	Not Evaluated
ROI	Not Present/Extinct
Trend since 1772	Certainly declined

According to the early modern recorders, before it went extinct the Great Bustard could formerly be found in the South and Midlands of England, in Lowland Scotland and in Co. Cork in Ireland.

RECOGNITION
This species was called the *bustard* in Early Modern English, the *gustard* in Middle Scots, and *otis* or *tarda* in Latin. Sometimes the Scots term was also borrowed into Latin as *gustarda*. None of these terms were applied to other species, so the references should be reliable.

DISTRIBUTION
The distribution of top-quality records for this species is statistically a poor fit with the known levels of recorder effort. It may have been locally distributed, locally abundant or of special local interest. There are far more records than we would expect from the south of England, suggesting the species was more widespread and abundant there than anywhere else. Having said that, the existence of regional records in Berwickshire and East Lothian (even if apparently nowhere else in Scotland (Grundy and McGowan 2007) and in Co. Cork, implies that the Bustard was not exclusively found in South England and the Midlands, and it was also later recorded in the Yorkshire Wolds (Gooders 1983: 48). Habitat suitability modelling based on the sites where the Bustard was recorded in the early modern period suggests that the species may have had specific requirements. The parts of lowland Scotland where it was recorded are today the warmest and driest. The colder and wetter winters of the rest of Scotland as well as Wales and South West England, especially in upland areas, might also have been less suitable for the species.

Based on the paucity of historical and archaeological records from Britain, it is usually thought that the Great Bustard might have always been very locally distributed, or even that it was only established (either through colonisation or introduction) in Britain in the medieval period (Serjeantson 2010; Yalden and Albarella 2009: 140; Gooders 1983: 46; Gurney 1921: 173–6). If this is true, the Bustard must have expanded quickly as it was surprisingly widely recorded in parts of England in the early modern period. The distribution of the species might also reflect a recent decline rather than an introduction. In the English Act against Destruction of Wildfowl (1533–4), Bustard eggs and nests were given the highest possible level of protection, equal to that of the (declining) Crane, suggesting that they were recognised as being especially vulnerable to extinction. The Bustard appears to have gone extinct as a breeder in Scotland between the accounts of John Leslie in 1578 (1675: 24) and of Robert Sibbald in 1684 (2020: 16–17 (II:3)).

Great Bustard records, 1519–1772. There are records from the South, South West and Midlands of England and from Munster, and contested records from Lowland Scotland.

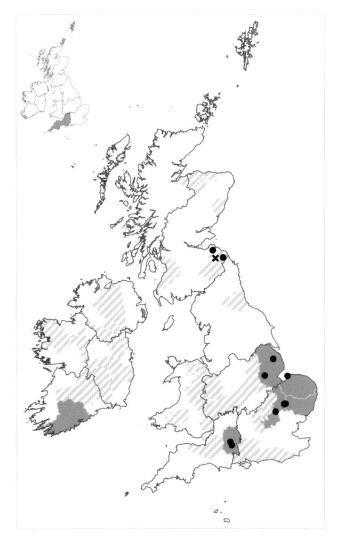

Our sources provide two records from Ireland. The most reliable is in Smith's *The Ancient and Present State of the County and City of Cork*, volume 2:

> Otis seu Tarda Avis, Will. p.129. The Bustard.
>
> It wants the back toe, by which mark alone, and by its size, it is distinguished from all other birds of the kind. They are very slow of flight, and can scarce rise off the ground, by reason of their bulk. The flesh, when in season, is very delicate.   (Smith 1750b: 334)

This reference comes from a section entitled the 'Catalogue of the Birds observed in this County' [Cork], meaning that it may be a record of a native species, just like the record of the Black Grouse (p. 103). Some further confirmation is provided by the Bustard's inclusion in O'Sullivan's *Natural History of Ireland* (2009: 122–3). These sources together suggest that, while there is little other evidence, the Bustard may have been present in Ireland, although there is not clear evidence that it bred there. It may have been rare, since it was listed as absent from Ireland in one source from 1684 (MacLysaght 1939: 416–17). I am not aware of any records detailing its later extinction, but if the species was present, it likely declined in Ireland due to overhunting, just as in Britain.

Despite two further Acts to protect the Bustard in England in 1775 and 1831, the Bustard last bred here in 1832, only 60 years after our final early modern source was written (Gooders 1983: 48–9). A few individuals survived into the 1840s, but after that the Bustard became largely a rare visitor. One of the main reasons for its extinction seems to have been overhunting for sport and food. Bustards continued to be offered in markets into the nineteenth century (Shrubb 2013: 24). In addition to this, the species likely benefitted from the open field system and so enclosure in champion counties (where there had previously been vast areas of fallow land and less hedgerow cover for predators) probably hastened the decline of the Bustard (Serjeantson 2010; Yalden and Albarella 2009: 50, 140; Lovegrove 2007: 40; Ritchie 1920: 366), although the importance of this change has been overemphasised in the past (Williamson 2013: 40–6).

Attempts were made to reintroduce the species to Salisbury Plain from the 1970s onwards (Gooders 1983: 49). These were initially unsuccessful but the latest project, begun in 2004, resulted in the first successful wild breeding in 2009 (Balmer et al. 2013). Birds from Salisbury Plain also disperse elsewhere in South and South West England over the winter months. Despite this recovery, Bustards still have a clear declining trend since 1772 due to their loss from the Midlands, south Scotland and (apparently) Munster over the last few hundred years.

CORNCRAKE

Crex crex

NATIVE STATUS Native

MODERN CONSERVATION STATUS	
World	Least Concern
UK	Red List
ROI	Red List
Trend since 1772	Certainly declined

There were early modern records of the Corncrake dotted across Britain and Ireland, and it was especially well recorded in North England.

RECOGNITION

This species was most commonly called the *rayle* or *land rail* in Early Modern English. It was also known as the *craker* or *daker-hen*, and occasionally the *corn-creek* in Ireland and Wales. Thomas attests *rhygen yr ud* (ed. Dodsley 1775: 11) as the Welsh. In New Latin it is referred to as *Rallus terrestris*, the *Ortygometra* of Aldrovandi or the *Crex*. Most of these terms are unique to the Corncrake. The term *rayle* (or *rail*, *rale* etc.) needs to be carefully distinguished from the *water rail* (still called *Rallus aquaticus*), but the Corncrake appears to have been the nominal species of rail, so the records on our map are likely to be reliable.

DISTRIBUTION

The distribution of top-quality records for this species is statistically a poor fit with the known levels of recorder effort. It may have been locally distributed, locally abundant or of special local interest. Having said that, the species was recorded in every region of Britain south of Scotland and in every region of Ireland except poorly recorded Connacht, so it may have been locally widespread.

The lack of early modern records from mainland Scotland is surprising. Habitat suitability modelling based on the sites where the Corncrake was recorded in the early modern period suggests that the species would have been well adapted to conditions across much of lowland Britain and Ireland, except perhaps north-east Scotland (where the species was actually recorded as absent in the time period), which had very cold winters. However, given that the species was one of the most widespread birds in Scotland in the nineteenth and twentieth centuries, it may have formerly been present even here (Green 2007; Holloway 1996: 158–9; Baxter and Rintoul 1953: 699–700). It is thought to have declined following the change from arable to pastoral land-use following the Highland Clearances, but these mostly took place after the end of our period. Perhaps the Corncrake was not regularly distinguished from the Water Rail in early modern Scottish sources.

John Morton was particularly interested in the Corncrake and left three records from Northamptonshire alone:

> To the same Class of Birds with a more strait Bill or Claw, belong the Poultry Kind; whereof we
> have neither the Heath-cock, or Grous, nor the Gor-cock or Moor-cock … but the Rail another

of the Birds of this Kind, that has its Name, as some suppose, from Royale, because it is a Royal or Princely Dish, we pretty often meet with, particularly in the Fields about Thengford, Northampton, Halston, and Rowel; which yet is so rare, that Turner saith, he never saw nor heard of it, but in Northumberland. (Morton 1712: 425)

The Corncrake is now red-listed in Britain because it has lost most of its habitat. It is most commonly found on the islands and peninsulas of the north and west coasts of Scotland, Ulster and Connacht, but it is poorly recorded elsewhere (Balmer et al. 2013). Several reasons have been suggested for the decline. The Corncrake was regularly hunted for food, especially from the end of the early modern period into the nineteenth century, and was one of the species licensable under the 1831 Game Reform Act in the United Kingdom (Shrubb 2013: 113). It was also affected by the change popularised in the nineteenth century in crop rotation from a fallow period to the growth of clover and turnips, although the overall effect of this change on the species may have been positive (Williamson 2013: 108–9, 111). However, the decline of the Corncrake was not rapid until the end of the nineteenth century. In 1875, the Corncrake was still common in almost every county of Britain and Ireland, except Devon, Kent, Essex and Nottinghamshire (Holloway 1996: 158–9).

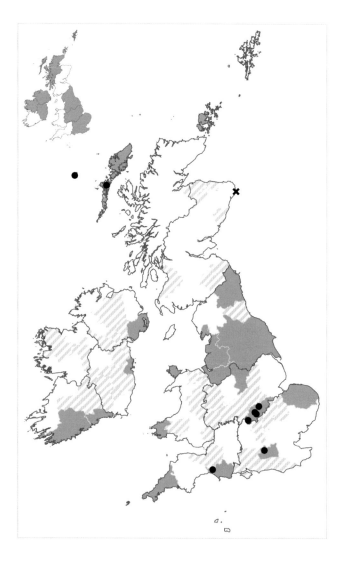

Corncrake records, 1519–1772. There are records from every region of Britain and Ireland except Lowland Scotland and Connacht.

The most important reasons for the decline in Britain (and later Ireland) following this were the introduction of mechanical mowing of meadows and the increased use of tractors (Green et al. 2021; Balmer et al. 2013; Yalden and Albarella 2009: 189; Green 2007; Parslow 1973: 71–2). In 1968–72, Corncrakes were still common in Ireland and were found dotted across Britain, especially in North England and Lowland Scotland, but by 1988–91, Corncrakes were rare. Since then, they have begun to recover in the Northern Isles and Hebrides and have been reintroduced in the Nene and Ouse Washes in Cambridgeshire.

COOT

Fulica atra

NATIVE STATUS Native

MODERN CONSERVATION STATUS	
World	Least Concern (Near Threatened in Europe)
UK	Green List
ROI	Amber List
Trend since 1772	Uncertain

The early modern authors occasionally recorded the Coot around Britain and Ireland.

RECOGNITION

This species was called *coot* in Early Modern English and *fulica* in Latin. Both these names appear to have been exclusively used for the Coot in our texts, so identification is straightforward.

DISTRIBUTION

The distribution of top-quality records for this species is statistically a poor fit with the known levels of recorder effort. It may have been locally distributed, locally abundant, or of special local interest. Habitat suitability modelling based on the top-quality records of the Coot suggests that the species may have been confined to lowland sites with cold winters, but the county-level records from Cornwall, Pembrokeshire and Co. Cork defy this model, so it is likely that another factor was responsible for determining the range and the abundance of the species. Due to the small number of records overall, the records from Harrington and Castle Ashby, stately houses in Northamptonshire, stand out (Morton 1712: 429). Presumably, these sites were colonised by the large population of Coots formerly found in the Fens of Cambridgeshire and Lincolnshire (Williamson 2013: 31–4; Gurney 1921: 12–13). Wild birds in the Fens were extensively exploited by local people as a common food resource, to the extent that the drainage of the Fens for farmland was controversial and impoverished some communities (Robson 2019: 184). Coots in particular were also frequently shot for sport and meat elsewhere in the early modern period up until the early twentieth century, sometimes in very high numbers (Shrubb 2013: 14–15, 84–5).

The record from Co. Waterford (represented with a cross) indicates that the species was not usually found there:

> Fulica, Johnst. *Mergus niger & Pullus aquaticus*, Alberti. The Coot or Bald-Coot, a particular description would be needless, being common in other places, though only seen here during the hard frost of 1739–40.   (Smith 1746: 341)

Coot records, 1519–1772. There are records from every region of Britain and Ireland except Highland Scotland, Ulster and Connacht.

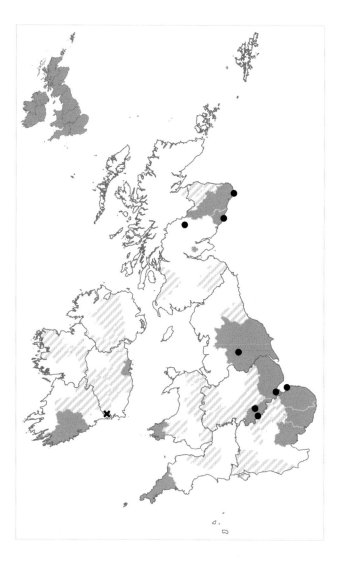

Today the Coot is widely distributed in the lowlands of Britain and Ireland, but more rarely seen in upland areas. This generally fits with the early modern pattern of records, which might suggest that the species has not changed its distribution much over the last few centuries, except that Holloway (1996: 162–3) found evidence that the Coot was common in parts of Scotland at the end of the nineteenth century, including especially Caithness, Abernethy in Perth, the Outer Hebrides and Orkney. There is evidence to suggest that the Coot's presence in these areas was the result of an expansion starting in the 1850s (Murray and Taylor 2007). For some reason, this expansion stopped in the 1940s, and from the 1950s the species retreated from the Highlands and most of the islands of Scotland. Since the 1970s it also seems to have declined and reduced its range in Ireland, here perhaps due to the loss of wetlands (Balmer et al. 2013). It has slightly reduced its range but also increased in abundance in Britain over the same period.

CRANE

Grus grus

NATIVE STATUS Native

MODERN CONSERVATION STATUS	
World	Least Concern
UK	Amber List
ROI	Extinct
Trend since 1772	Certainly declined

The Crane was recorded by early modern authors at a handful of sites from across Britain and Ireland, mainly as a migrant rather than breeder.

RECOGNITION

The Crane was generally referred to in Latin as *grus* and in Early Modern English as *crane* in the early modern accounts. Occasionally, these terms seem to have applied to the Grey Heron, or other species, but every local or county-level source in this atlas either mentions both species or describes the Crane, meaning that the records on this map are reliable.

The Irish term was *corrvona* which is distinguished from *corr-iasc* and *corri-grieni*, which were terms for the Grey Heron. K'Eogh (1739: 26) also mistakenly attests the Latin term *geranium* as a name for the Crane, presumably because several of the plants in the geranium genus were and still are called *cranesbills* in English.

DISTRIBUTION

The distribution of top-quality records for this species is not statistically different from the known level of recorder effort. The gaps on the map may just reflect decreased survey effort in some regions. It may have been widespread.

The Crane seems to have already been in decline by the start of the early modern period. This seems to have been mainly a result of overhunting for meat and falconry, not wetland drainage as has previously been argued (Stanbury and UK Crane Working Group 2011; D'Arcy 1999: 96). In the archaeological record, the species goes from being the most commonly encountered of all wild birds, to only being found at elite sites from around 1050 CE (Yalden and Albarella 2009: 36–7; Sykes 2001: 74). In the second half of the thirteenth century, Crane meat was valued at six times the cost of Grey Heron meat (Gurney 1921: 57). It continued to be bought and eaten nevertheless. The decline of the Crane may even have added to its value for conspicuous consumption (Raye 2018; Albarella and Thomas 2002). In the surviving thirteenth-century Welsh law books the Crane was protected so that it could not be hunted without a landowner's permission, and in some cases a charge was also payable to the king for each Crane killed. This essentially meant that only the rich landed gentry could hunt Cranes. There were also fines for killing Cranes in medieval Ireland, although these are usually interpreted to have been for religious reasons (D'Arcy 1999: 92). In England, the Act against Destruction of Wildfowl (1533–4), meant that Crane eggs were protected with a 20d fine (Gurney 1921: 167). Nevertheless, the breeding population finally went extinct at the end of the sixteenth century, perhaps partly due to improved gun technology (Gurney 1921: 167–8). The last record of the native population breeding is from 1603 CE:

> In the [Pembrokeshire] bogs breed the crane, the bittern, the wild duck and teal and divers others of that kind. On high trees the heronshaws, the shovelard and the woodquist [woodpigeon].
>
> (ed. Owen 1994 [1603]: 133)

In our records from 1604–1772 the Crane was still recorded as a migrant in a few areas including Co. Cork, Cornwall, the Fens in Cambridgeshire, Lincolnshire and Norfolk, the Isles of Scilly and, despite occasional suggestions that the Crane might never have been common in Scotland (Gurney 1921: 171–3), in the Northern Isles and Hebrides. All these records seem to refer to wild, seasonal populations, but Cranes were also sometimes kept as pets and for meat in the medieval period and at the beginning of the early modern period (Shrubb 2013: 68).

The Crane was lost, even as a migrant, in the second half of the eighteenth century, and became only a rare vagrant (Shrubb 2013: 66; Pennant 1776b: 14). It recolonised and began breeding again in the Norfolk Broads around 1981, in Yorkshire in 2002, and has since been reintroduced to the Somerset Levels, where it began breeding in 2013 (Balmer et al. 2013).

Crane records, 1519–1772. There are records from Highland Scotland, Ulster, Munster, Wales, and the South, South West and Midlands of England.

OYSTERCATCHER
Haematopus ostralegus

NATIVE STATUS Native

MODERN CONSERVATION STATUS	
World	Least Concern
UK	Amber List
ROI	Red List
Trend since 1772	Probably increased

The Oystercatcher was recorded by many of the early modern naturalists around the coasts of Britain and Ireland.

RECOGNITION
This species was most commonly called the *sea pie* in Early Modern English and the *sea piot* in Middle Scots. In New Latin it was occasionally called *Pica marina* (probably a calque from the vernacular term) but more commonly the *Haematopus* of Bellon. In Scottish Gaelic it was called the *trilichan*, and O'Sullivan (2009: 144) attests *realach* as the form in Irish.

Of these terms, the first two are always distinguished from the terrestrial *pie*, called the *Pica* (our Magpie), so there does not appear to have been any confusion between these two birds. However, Sibbald (ed. 2020: 18–19 (II:3)) writes a long note distinguishing our *Haematopus* from the *Himantopus* of Pliny (likely *Himantopus himantopus*, our Black-winged Stilt), but since the latter species is only a rare vagrant in Britain and Ireland this is unlikely to affect many of our records. More seriously, Merrett (1666: 182) seems to call *Tringa totanus* (Redshank) the *haemantopus*, but this appears to have been a mistake – this species is much more commonly referred to in New Latin as the *totanus*. The records on the map are thus likely to be reliable. There does not seem to be any historical support for interpreting the term *sea-pie* as referring to the *Larus* species.

There are a few other vernacular terms. Gifford (ed. 1879: 23) attests a possible Norn term *chalders* (possibly related to Nynorn *tjoga* = Guillemot/Razorbill?). Two voyagers to St Kilda also offer additional terms. Martin Martin (ed. Martin and Monro 2018: 280) refers to the species as the **t**irma, and Alexander Buchan (1741: 14) refers to the *firma*. This difference in first letter seems likely to be a copying error. Given the very similar contexts for the two texts, I would suggest Buchan copied the term from Martin. In both cases, the context makes it clear that this should be considered a Scots or English term rather than Scottish Gaelic, which is also how the *Dictionaries of the Scots Language* takes it. As we have seen, a more ordinary term in Middle Scots was *sea piot*, but *tirma* might have been a Hebridean or St Kildan term.

DISTRIBUTION
The distribution of top-quality records for this species is not statistically different from the known level of recorder effort. The gaps on the map may just reflect decreased survey effort in some regions. It may have been widespread. Based on the county-level records, the Oystercatcher appears to have

Oystercatcher records, 1519–1772.
There are coastal records from
every region of Britain and
Ireland except Connacht.

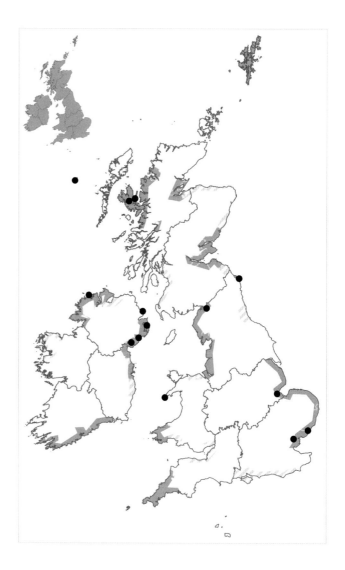

been widespread but coastal. It is recorded on the coast of every region except Connacht, which is perhaps the least recorded region in our early modern sources. It is not recorded on any site inland.

The records that do not have a county shaded beneath them come from counties where the only records come from islands. For instance, Dobbs records a population on Hulin Rocks north of Portmuck in Co. Antrim:

> On the great rock great store of fowl breed, as Gulls of several sorts and sizes, sea pys, sea parrots, directly so shaped, but Black and white with red bills, and other fowl like puffins but larger; here likewise seals and sea calves, used very much, and about Michaelmas and Allsaints day, bring out their young among these rocks. The bigest rock seems to be a little Isle; but the stones so soddened or wedged together, you cannot (tho' you would think otherwise) get one loose to throw at a fowl.
> (ed. Hill 1873 [1683]: 380)

Not shading Co. Antrim may be overly cautious since there are early modern records from both mainland Co. Down and Co. Donegal, but Dobbs' impulse to throw stones at the birds helps explain why easier-to-reach mainland populations might not have done so well in the early modern period.

Oystercatchers were also regularly shot for sport in the nineteenth and early twentieth centuries, and were affected by the increasing use of beaches for leisure in the nineteenth century. This may have suppressed the population in some areas. Oystercatchers were extinct in some counties of South and South West England by the end of the nineteenth century, and numbers recovered through the twentieth century. The bird is now widespread on all the coasts of Britain, and common inland in the north of Britain (Balmer et al. 2013; Shrubb 2013: 117; Holloway 1996: 164–5; Sharrock 1974: 207).

The Oystercatcher is now only a winter migrant along much of the coast of Munster, especially in Cos. Cork and Waterford (Anderson 2017: 132; Balmer et al. 2013). There is some evidence that Oystercatchers were formerly more common in this area. Holloway (1996: 164–5) found evidence for breeding in Co. Cork at the end of the nineteenth century. The species was recorded there in the eighteenth century, although it is not clear whether these are breeding or wintering records.

Today the Oystercatcher is recorded in every region of Britain. However, it is not as well recorded in Ireland, Wales, or in the South and South West of England as it is in Scotland and England northwards from Norfolk (Balmer et al. 2013). Until recently it was also rare on the east coast of England. In the Midlands and North of England to Lowland Scotland it has now colonised places inland. This appears to have been a modern development (perhaps comparable to the inland colonisation of the gulls), since all of the early modern records specific to site-level are situated on the coast. The colonisation of inland sites appears to have started in Scotland in the nineteenth century and begun in England later (O'Brien 2007a; Holloway 1996: 164–5). There are no reliable inland records from the early modern period, although there are some county-level records which might have included inland populations.

LAPWING

Vanellus vanellus

NATIVE STATUS Native

MODERN CONSERVATION STATUS	
World	Least Concern
UK	Red List
ROI	Red List
Trend since 1772	Uncertain

The Lapwing was recorded by early modern writers across Britain and Ireland, especially in the east of England.

RECOGNITION

This species was referred to as the *capella* and *vanellus* in New Latin, and the *lapwing* or *bastard plover* in Early Modern English. These terms allow specific identification for records from early modern Britain and Ireland.

It was also called the *puet* (and less often *tewit* or *kywit*) by our sources, but *puet or pewit* is also the most common name for the Black-headed Gull. It is usually possible to distinguish records to this species from context: the Lapwing is listed with other waders rather than with gulls. Likewise, there are occasional references to the Lapwing as the *green plover* (e.g. Morton 1712: 428), but this term was much more commonly used for the Golden Plover.

Lapwing records, 1519–1772.
There are records from every region
of Britain and Ireland except Highland
Scotland, Ulster and Connacht.

DISTRIBUTION

The distribution of top-quality records for this species is statistically a poor fit with the known levels of recorder effort. It may have been locally distributed, locally abundant or of special local interest. Habitat suitability modelling based on the sites where the Lapwing was recorded in the early modern period suggests that the species may have had specific requirements. It is only recorded on lowland sites which today have dry winters and relatively warm summers. The site-level records are confined mainly to the South and Midlands of England, suggesting the Lapwing may have been more abundant or more widely distributed there than elsewhere.

However, the Lapwing is recorded in every region of Britain and Ireland except the Highlands (although there is a possible record from Shetland), Ulster and Connacht, so it must have been relatively widespread, and is the most common wader on archaeological sites through the Roman and medieval periods (Shrubb 2013: 117). It seems to have strongly benefited from the increased agriculture of the early modern period and adapted to changes well. Before the Fens were drained, it was commonly found exploiting the wetlands but after drainage and embankment the Lapwing continued to be seen in the Fenland counties, nesting in the fields and exploiting the high water-table which pushed invertebrates to the upper soil. In traditional systems, Lapwings nested in common pastureland and ate the stubble left in the agricultural fields over winter, but when farmers later introduced the winter turnip into the field rotation and enclosed the commons, Lapwings benefitted from the increased waste grain and began to nest in fields (Williamson 2013: 10, 31–4, 41, 109). Lapwings do not seem to have been persecuted as pests and were in fact sometimes taken as pets and used to control pests in gardens in this period, although they were often hunted for food and their eggs were taken (Shrubb 2013: 117–20, 178; Lovegrove 2007: 27).

The Lapwing has been declining steadily through the modern period. It is now nearly absent as a breeder in South West England, Wales and Munster (Balmer et al. 2013). There are records from all of these areas from the early modern period, and even into the end of the nineteenth century, so this represents a modern loss (Holloway 1996: 178–9). These are not just overwintering records. The record from Co. Cork shows that the species commonly bred there:

> Capella [or] Vanellus, Willughby p.228. The Lapwing or bastard Plover. This bird is in all countries very well known. They build their nests on the ground, in fields or heath, open and exposed to view, only laying a few straws about the eggs to hide the nest. As soon as the young are hatched, they forsake the nest, and follow the old ones like chickens. These birds are very useful in gardens, as they destroy worms and insects. (Smith 1750b: 345)

The decline of the Lapwing seems to have begun in the second half of the nineteenth century, as a result of shooting, netting and egg-collecting for food (Shrubb 2013: 48, 117–20; Holloway 1996: 178). The Lapwing has also been negatively affected by the afforestation of the uplands (although it has benefitted from deforestation for agriculture), drainage of the lower moors and the abandonment of commercial reedbeds (Williamson 2013: 100, 146, 154). Most recently, changes in agricultural practice, including the change to autumn-sown crops, the loss of pasturelands and increased use of chemicals have had a negative impact on the species (Hawkins and Hughes 2021; Balmer et al. 2013).

It appears that Lapwings were occasionally seen in the Northern Isles in the nineteenth century, possibly due to the end of the Little Ice Age (Shrubb 2013: 206; O'Brien 2007b; Booth et al. 1984: 91; Parslow 1973: 76). This may actually have been a recolonisation. Gifford (ed. 1879: 73), writing in the first half of the eighteenth century, provides an uncertain record of the Lapwing as either present or possibly recently lost from Shetland. Holloway (1996: 178–9) records it as present in both Orkney and Shetland by the end of the nineteenth century.

GOLDEN PLOVER

Pluvialis apricaria

NATIVE STATUS Native

MODERN CONSERVATION STATUS	
World	Least Concern
UK	Green List
ROI	Red List
Trend since 1772	Uncertain

There were records from authors across Britain and the south coast of Ireland that were most likely to refer to the Golden Plover. The records mostly do not have an exact date, so the map here should be assumed to refer to the Golden Plover's (much wider) winter population.

RECOGNITION

This species was generally called the *green plover* in Early Modern English and *Pluvialis viridis* in New Latin. These terms normally refer to the Golden Plover but were also sometimes used for the Lapwing, so the records on this map are not altogether trustworthy.

Other names for this species are even less certain, including especially generic references to the *plover* or *Pluvialis*. These might also be records of the Lapwing or the Grey Plover (Shrubb 2013: 117; Baxter and Rintoul 1953: 590; Gurney 1921: 121–2), but are perhaps most likely to refer to the Golden Plover since it was more common than the Grey Plover and early modern sources commonly list both *plover* and *lapwing* in their records (Reyce 1902: 44; Sibbald 1892: 61). They have been included on this map as diamonds to indicate that they are less trustworthy references.

DISTRIBUTION

The distribution of top-quality records for this species is not statistically different from the known level of recorder effort. The gaps on the map may just reflect decreased survey effort in some regions. It may have been widespread.

The Golden Plover was one of the most common species found on Roman and medieval sites, especially on elite sites (Yalden and Albarella 2009: 215; Serjeantson 2006). It has significantly benefitted in the past by association with humans and the conversion of, for example, woodland into arable field and pasture, and the maintenance of young heather on grouse moors (Williamson 2013: 10, 29). It was commonly hunted in the early modern period, although perhaps not as much as the Lapwing; it is not known if this had any effect on the bird's population (Shrubb 2013: 118–19; Gurney 1921: 121–2), although in the fifteenth century the Parliament of Scotland passed an act (*RPS* 1458/3/32) to protect their nests and eggs.

> Of the cloven footed Water Fowl with slender Bills of a middle length … To these we may fitly add the true Green Plover, *Pluvialis viridis*: A Bird almost unknown to our Northamptonshire Fowlers, who usually call the Lapwing, Green Plover. Indeed it does not often occur with us. I have seen it in Oxendon Meadows amongst a Company of Lapwings in the Winter Months. We rarely see above five or six of them at a time. (Morton 1712: 428–9)

The Golden Plover remains common across Britain and Ireland as an overwintering bird; however, as a summer breeder it has been declining for just over a century, especially since the 1970s. The breeding population in South West England, for instance, formerly known from Dartmoor as well as

Golden Plover records, 1519–1772. There are records from every region of Britain and Ireland except Connacht, but the records from the South West of England and from Wales are unreliable.

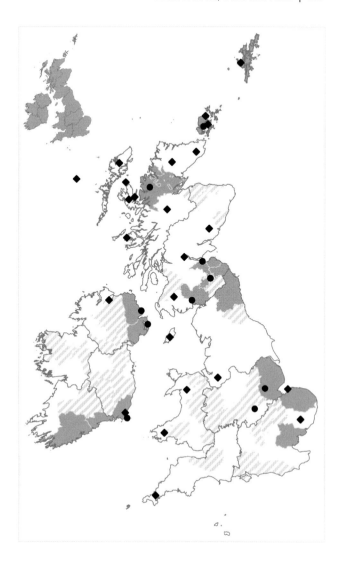

perhaps Exmoor and the Mendips, seems to now be extinct (Balmer et al. 2013; Williamson 2013: 154; Holloway 1996: 176–7). The decline may have initially been because of shooting, but more recently is due to changes in the management of the uplands (especially afforestation) and an increase in nest predation by species like Foxes and Carrion Crows (Balmer et al. 2013; Williamson 1975, 2013: 154; Thompson 2007a; Holloway 1996: 176). Climate change is also an important factor in the decline of the Golden Plover because the species is reliant on craneflies (Tipulidae spp.), and craneflies are significantly less abundant following hot summers (Pearce-Higgins et al. 2010).

GREY PLOVER

Pluvialis squatarola

NATIVE STATUS Native

MODERN CONSERVATION STATUS	
World	Least Concern
UK	Amber List
ROI	Red List
Trend since 1772	Uncertain

The Grey Plover was recorded by a few authors around the coastline of Britain and Ireland. These records mostly seem to reflect winter rather than summer birds.

RECOGNITION

This species was called the *grey plover* in Early Modern English or *Pluvialis cinerea* in the early modern texts. It can usually be distinguished from the Golden Plover because the latter was usually called the *green plover* in Early Modern English. A few records refer only generically to the *plover*. These are more difficult to identify to species level but are included on the map for the Golden Plover rather than here, because the Golden Plover is likely to have been more common.

DISTRIBUTION

The distribution of top-quality records for this species is not statistically different from the known level of recorder effort. The gaps on the map may just reflect decreased survey effort in some regions. It may have been widespread, at least on the coasts.

For this map, Grey Plover populations have been assumed to be coastal except where sites further inland are mentioned. A few authors do record Grey Plovers and Golden Plovers together on heaths near the coast. They were recorded, for example, on Thetford Heath in Norfolk and in the Lammermuir Hills by the Firth of Forth (Mitchell 1908: 170–1; Browne 1902: 20). This agrees approximately with the modern situation where Grey Plovers occasionally overwinter inland, especially in the east of Britain. In the archaeological record, Grey Plover remains are also commonly found inland in Britain, although this might reflect birds traded for food (Yalden and Albarella 2009: 215).

> Pluuialis or plouer green & graye in great plentie about Thetford & many other heaths. they breed not with us butt in some parts of Scotland and plentifully in [Iceland].
>
> (ed. Browne 1902 [*c.*1662–68]: 20)

Despite Browne's reference here, no clear local records of Grey Plover breeding survive from Scotland from the early modern period. He may have been referring only to the Golden Plovers as breeding.

Emery (1974) suggests that the reference by one of Edward Lhuyd's correspondents to the *brondhu'r twyne* in Eryri might also refer to the Grey Plover, but it seems to me that this reference is more likely to be to a Golden Plover; I have included the record on the map of that species.

Grey Plovers were occasionally hunted for food in the early modern period (Gurney 1921: 147), although perhaps not as commonly as the Golden Plover or Lapwing. They were frequently shot on the coasts of England in the late nineteenth and early twentieth centuries (Shrubb 2013: 117). This might have suppressed population levels, and the bird seems to have become rarer in Britain and Ireland after the early modern period. In Scotland, it was rarer until the 1970s, when the species

Grey Plover records, 1519–1772. There are coastal records from every region of Britain and Ireland except Connacht and South West England, and more inland records from the South and Midlands of England.

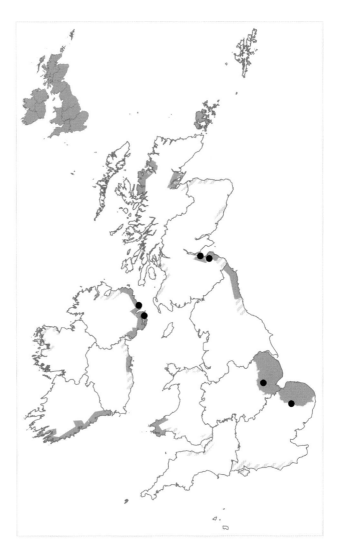

became a relatively common wintering bird (Thompson 2007b). It is now widely recorded on all coasts (Balmer et al. 2013), although it is still not a breeder. The species' distribution is stable, but the number of individuals wintering in Britain and Ireland is declining. This is partly due to a decreasing world population, but more because the Grey Plover seems to be shifting its wintering grounds due to climate change (Pritchard 2021n).

EURASIAN CURLEW
Numenius arquata

NATIVE STATUS Native

MODERN CONSERVATION STATUS	
World	Near Threatened (Vulnerable in Europe)
UK	Red List
ROI	Red List
Trend since 1772	Uncertain

The Curlew was recorded by our early modern naturalists around the coasts, and occasionally inland in Britain and Ireland.

RECOGNITION

This species was called the *whap* in Middle Scots, *crotach mhuire* in Irish, and *numenius* or *arquata* in New Latin (the two Latin names were only put together later). These terms appear to have been specific to our species. However, the most common term for this bird was *curlew* in Early Modern English. This last term, as well as probably *cuerlúin* in Irish, is not totally reliable; there may have been some confusion between this species and Whimbrel (*Numenius phaeopus*). The Stone-curlew (*Burhinus oedicnemus*) may also have caused some confusion – this last species formerly had a wider distribution including much of the south and east of England (Holloway 1996: 168–9) and possibly parts of Ireland (Smith 1750b: 344). All three of these species, as well as others, were sometimes described as types of *curlew* and although the Eurasian Curlew seems to have been the nominal species of Curlew, the other species were not always formally distinguished – as, for example:

> In Devonshire there are three sorts of Curlicus; the first as big as a Muscovie Duck, the second as big as an ordinary Duck, the third somewhat less.   (Chamberlayne 1683: 253)

Based on this record, I have shown all records that only use the English term *curlew* as diamonds on our map to show that they are less reliable, except where the other birds are distinguished by the author.

DISTRIBUTION

The distribution of top-quality records for this species is not statistically different from the known level of recorder effort. The gaps on the map may just reflect decreased survey effort in some regions. It may have been widespread.

The idea that the species might have had a coastal distribution is supported by the fact that nine of the ten top-quality records come from coastal areas. The uncertain records, if trustworthy, also support this interpretation. This will surprise many readers, given the modern summer range of the species. The county-level records are problematic in this regard, because it is often not clear whether they refer to coastal or inland populations. Based on the site-level records, I have assumed that the county-level records refer to a coastal population unless an inland site is mentioned. This is a conservative estimate and, for example, the county-level record from Devon might actually refer to the inland breeding population on Dartmoor or Exmoor which was recorded in the nineteenth century (Holloway 1996: 192–3).

The lack of records from inland areas of Britain and Ireland in the early modern period might at first sight suggest that in the past these islands were only used by overwintering coastal populations, but not by breeding inland populations. This would find some support in the archaeological evidence, which seems to imply that the Curlew was less common before the Roman period (Yalden and Albarella 2009: 217). Historical evidence from the nineteenth century suggests that the Curlew was formerly rare in the South, South West and Midlands of England (Green 2021; Holloway 1996: 192–3). There is evidence from the Roman, medieval and early modern periods suggesting that the Curlew was commonly hunted for food during the winter (Shrubb 2013: 13, 18, 25, 115; Williamson 2013: 31–2; Gurney 1921: 57); however, it was usually the most expensive wader on the market (Shrubb 2013: 117), so we might posit that hunting pressure could have been suppressing some breeding populations. In areas with small populations, egg predation by Carrion Crows and presumably other predators can also eliminate Curlews (Williamson 2013: 160; Bainbridge 2007) and prior to the rise of the gamekeeping industry, predators may have been more common in the uplands, despite the parish bounties provided under the Tudor Grain Acts. The Curlew may have been only an uncommon breeder in early modern Britain and Ireland.

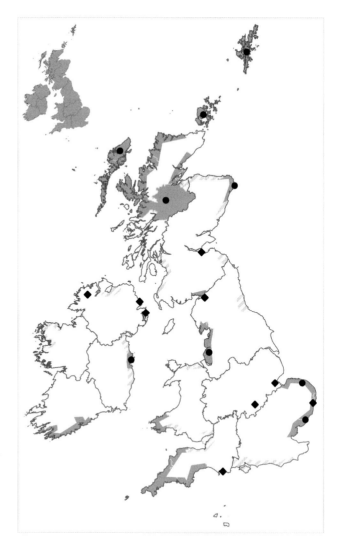

Eurasian Curlew records, 1519–1772. There are coastal records from every region of Britain and Ireland except Connacht. The records from the Midlands are unreliable, but there seem to be reliable inland records from Highland Scotland.

Yet this interpretation is not fully supported in the early modern texts. In the *Description of Pembrokeshire*, for instance, we are told that the Curlew 'continues always in this county' (i.e. all year-round), although admittedly the text also adds that 'yet [it is] never found to breed' (ed. Owen 1994: 133). Pococke (ed. 1891: 59) also notes in passing that populations of Curlews in Co. Donegal lay eggs 'in the rocks of inland mountains'. Even further from the coasts, in *The Natural History of Northampton-shire* we are informed:

> the Curlew, which tho' a Sea Fowl, usually seeking its Food upon the Sands and Ouze in the Salt Marshes, yet sometimes straggles up into our Midland Fields.   (Morton 1712: 428)

These records together suggest that the Curlew was likely still breeding in some areas of Britain. Perhaps an alternative explanation, which would make sense of the apparent coastal distribution and the above records, is to point out that the moors where Curlews traditionally bred were not as well recorded as the estuaries and coastal fields in which Curlews overwinter. A similar bias resulted in the early modern distribution of the Wheatear seeming to be focused on the Downs of south-east England (p. 199). Curlew nests are also notoriously hard to find (in Welsh and Irish folklore, the nests are sometimes said to be protected by a saint (Green 2021; Anderson 2017: 144)). We could also add that the population of Curlews is higher in winter due to the presence of migratory birds, and the birds are also more concentrated when they are on the coasts. These factors together might perhaps have biased the early modern sources towards showing winter populations, and against showing summer populations.

Today, the Curlew breeds inland on rough pasture and moors in every region of Britain and Ireland. It is has declined significantly since the 1970s. Since the 1968–72 *Bird Atlas* it has gone from being widespread in Ireland, the south of Wales and South West England to being much rarer. It may become extinct in these regions without intervention. This needs to be put into the context of an increase in abundance and distribution before that (Green 2021; Parslow 1973: 82–3). To some extent, in Britain the decline might represent a return to the pre-1860s Curlew breeding distribution (Green 2021; Bainbridge 2007; Holloway 1996: 192–3), which included only the moors and uplands (not pasture or arable land) of the Midlands, South and South West including Devon, Cornwall and Shropshire; the species was only found more generally in Wales, North England and Scotland. In the winter, most breeding Curlew populations move to the coast, and Britain and Ireland also host birds that have migrated from Scandinavia (Balmer et al. 2013).

GODWITS

Bar-tailed Godwit (*Limosa lapponica*),
Black-tailed Godwit (*Limosa limosa*)

NATIVE STATUS Native

MODERN CONSERVATION STATUS	
World	Near Threatened (Least Concern in Europe)
UK	Amber List
ROI	Red List
Trend since 1772	Probably increased

There were reliable records of godwits from the coasts of Ulster, Leinster and Munster and the southern and eastern coasts of England. There also seem to have been inland populations in the Fens.

RECOGNITION

The early modern authors did not reliably distinguish the *Limosa* species in the early modern period. On paper, Willughby and Ray distinguish the *Fedoa* of Gessner (the Bar-tailed Godwit) from the *Fedoa secunda* (the Black-tailed Godwit). The Latin term *fedoa* seems to come from an Early Modern English *ffedowe*, which is attested in the Lestrange household accounts (Gurney 1921: 127). The terms *barge* and *yarwhelp* were usually associated with the first species and the terms *stone-plover* and *totanus* were usually associated with the second species. However, in the field the terms do not seem to have been consistently differentiated by authors, including Willughby and Ray (Smith 1746: 227–8; Ray 1678: 292–3). Some of these references also use the New Latin *attagen*, which was more normally applied to the Red Grouse.

DISTRIBUTION

The distribution of top-quality records for these species is statistically a poor fit with the known levels of recorder effort. They may have been locally distributed, locally abundant or of special local interest. Habitat suitability modelling based on the sites where the godwits was recorded in the early modern period suggests that these species may have had specific requirements. They are only recorded on lowland sites which today have dry winters. There are more records than expected from the Midlands and Ulster and less than expected from Scotland. The southern and eastern counties of England seem to have been a stronghold for the godwits, both in the early modern period and according to the archaeological evidence prior to that (Yalden and Albarella 2009: 216). The only inland populations recorded in the early modern period are in the Fen counties – Lincolnshire, Cambridgeshire and possibly Norfolk – most other counties where they were present note the godwit as a coastal bird.

Godwits and their eggs were commonly collected as food in the early modern period, which might have suppressed some populations where humans could reach them (Shrubb 2013: 47–8, 114–15, 126–7; Williamson 2013: 31–2; Gurney 1921: 171). Black-tailed Godwits were so in demand they were even taken and fattened before being sold (Shrubb 2013: 126–7). Large flocks of Bar-tailed Godwits seem to have been unknown before the twentieth century (Pritchard 2021o; Insley 2007).

The lack of records from Scotland finds some justification in the nineteenth-century distribution, when the Black-tailed Godwit did not breed and the Bar-tailed Godwit was attested but in far lower numbers than are known today, or were known from Ireland at the time (Baxter and Rintoul 1953: 530–6). Both species may have declined earlier in Scotland than in the rest of these islands.

Godwits records, 1519–1772.
There are coastal records from every
region of England and from Ulster,
Leinster and Munster. There are
also inland records from the South
and Midlands of England.

After the early modern period, both species of godwit are thought to have declined due to the drainage of wetland habitats (Williamson 2013: 10). The Black-tailed Godwit, which was the only one of the two to have bred in Britain, appears to have become extinct as a breeder there between 1829 and 1835 due to the loss of this habitat (Shrubb 2013: 126; Holloway 1996: 188–9). Wintering godwits continued to be regularly hunted by wildfowlers (Shrubb 2013: 117).

At present, both species are commonly found during the winter on all coastlines around Britain and Ireland. The Black-tailed Godwit has returned as a breeder, and is sometimes found breeding, especially in the Northern Isles and in South England. The Bar-tailed Godwit has also significantly increased in abundance and now overwinters in large flocks (Pritchard 2021o; Insley 2007). The distribution of the Bar-tailed Godwit increased by 19% between the 1981–84 *Bird Atlas* and the 2007–11 *Bird Atlas*, especially due to new populations in the west of Ireland and Scotland, and in Wales and South West England (Balmer et al. 2013).

WOODCOCK

Scolopax rusticola

NATIVE STATUS Native

MODERN CONSERVATION STATUS	
World	Least Concern
UK	Red List
ROI	Red List
Trend since 1772	No change

The Woodcock was recorded as a migrant in every region of early modern Britain and Ireland but only more rarely as a breeding bird.

RECOGNITION

This species was most often and most specifically called the *woodcock* in Early Modern English and *scolopax*, or occasionally *Gallus sylvestris* in New Latin. The Irish term was *creabhar*. Thomas (ed. Dodsley 1775: 10) attests the Welsh term to be *cuffylog*. It was also called *gallinago* and *rusticola* in Latin. These terms can also refer to the snipe species, although these are usually distinguished in the early modern texts as *Gallinago minor* and *Rusticola minor*. One text refers simply to a seasonal kind of *cock* (ed. Hore 1859: 467), which seems most likely to refer to the Woodcock, but might instead mean other species. Where records are uncertain, they are shown as diamonds. All the records shown as dots appear to be reliable.

DISTRIBUTION

The distribution of top-quality records for this species is not statistically different from the known level of recorder effort. The gaps on the map may just reflect decreased survey effort in some regions. It may have been widespread.

The Woodcock was and continues to be a target for hunters. The method of hunting for sport in the early modern period may have been sustainable (Williamson 2013: 67–8), although the trapping of the bird for food may have been more impactful. The Woodcock is one of the most commonly excavated species on Roman and medieval sites, especially elite sites, and was very commonly sold in later medieval markets (Shrubb 2013: 18–21; Yalden and Albarella 2009: 108, 216; Serjeantson 2006). However, most records noted on this map and elsewhere are from overwintering populations only. There are only a few possible breeding records, which might suggest that the species was not a common breeder in the early modern period:

> Woodcocks are reckoned birds of passage, but they do not always leave the county to which they occasionally resort: Some gentlemen, hunting in the neighbourhood of Penzance, in the summer-time 1755, flushed a woodcock, surprised at seeing such a winter bird at that season of the year, they hastened to the bush and there found a nest with two eggs in it: a gentleman, more curious than the rest, carried the eggs home; and one being accidentally broke, the body of a young woodcock appeared.
> (Borlase 1758: 245)

The Woodcock began to breed more commonly across Britain and Ireland in the nineteenth century, especially from the 1820s, perhaps due to increased protection of wooded breeding areas for forestry and shooting Pheasants (Pritchard 2021p; Roberts 2007; Holloway 1996: 186–7; Tapper 1992: 56–9). Today it is found widely across Britain and (less commonly) Ireland over winter. As a breeder it is once again declining, although it remains more common than it was in the early

Woodcock records, 1519–1772.
There are records from every
region of Britain and Ireland.

modern period. The decline is probably due to changes in woodland management. It is now rare as a breeder in South West England, Wales, Ulster, Connacht and Munster and absent from Orkney, Shetland and the Outer Hebrides, although it remains widespread in all of these areas over the winter (Balmer et al. 2013).

SNIPE
Gallinago gallinago

NATIVE STATUS Native

MODERN CONSERVATION STATUS	
World	Least Concern
UK	Amber List
ROI	Red List
Trend since 1772	Uncertain

There were a few records of the Snipe from early modern Britain and Ireland, but it was usually thought of as overwintering, not as a breeding bird.

RECOGNITION
It is surprising that the Snipe has become the nominal species for its genus; in the early modern period, the ordinary *Gallinago* was the Woodcock, and our Snipe was then called *Gallinago minor* to distinguish it. This species was also commonly called *snipe* in Early Modern English, *neska* in Irish and *myresnipe* in Middle Scots.

The *Oxford English Dictionary* notes that the Early Modern English term *snite* is distinguished from the term *snipe* in a fifteenth-century manuscript and might refer to a different species. However, the early modern authors usually use one term or the other; where both terms are employed, they are always used together as a single reference (e.g. Sibbald 2020: 18 (II:3); Merrett 1666: 173), indicating that both were thought of as referring to the same animal.

In the nineteenth century, Great Snipe (*Gallinago media*) also overwintered in these islands (Shrubb 2013: 116). There do not seem to be any names for this species in the early modern texts, so if it was present in the early modern period some of the records on the map might actually refer to this species. Jack Snipe (*Lymnocryptes minimus*) was consistently distinguished as *Gallinago minima* in New Latin and *gid* or the *jack snipe* in Early Modern English.

DISTRIBUTION
The distribution of top-quality records for this species is statistically a poor fit with the known levels of recorder effort. It may have been locally distributed, locally abundant or of special local interest. All the top-quality records come from the South West and the Midlands of England, meaning the Snipe may have been more widely distributed or more abundant here than elsewhere. Habitat suitability modelling based on the top-quality, local records shows the species was associated with sites that have relatively mild winters and relatively warm summers, but this analysis is undermined by both the county-level records, which come from almost every region of Britain and Ireland, and the areas of optimum habitat identified in modern scholarship (Tapper 1999: 178–81).

The Snipe is one of the most commonly recorded species excavated from medieval sites and was particularly favoured on elite sites (Shrubb 2013: 18–19, 23–5, 127–8; Serjeantson 2006). It seems to have benefitted from clearance for agriculture (Williamson 2013: 10). It continued to be caught (and imported) for food in the early modern period, but it is not nearly as well recorded in our early modern texts. Populations were probably affected by drainage schemes and hunting pressure (Williamson 2013: 100–103). The Snipe seems mainly to have been thought of as a winter visitor rather than a summer breeder in early modern Britain and Ireland. Morton (1712: 428) in Northamptonshire and Borlase (1758: 245) in Cornwall are both aware of local populations of breeding birds, but state that the species is ordinarily seen only over winter. Taylor and Dale state that the population of Harwich and Dovercourt was exclusively made up of wintering birds:

> *Gallinago minor* Raii Ornith. 290 … The Snipe or *Snith*. This is found about Rivulets and running Springs, and without doubt is sometime Kitchen Provision here. This is a Bird of Passage being here in Winter only.
> (Taylor and Dale 1730: 400)

The Snipe declined due to the drainage of wetlands, the improvement of pastureland and overexploitation for sport and food. It seems to have been already declining in the lowlands by the end of the nineteenth century, when around 107,000 were sold each year as food in the Leadenhall and Newgate markets (Shrubb 2013: 127; O'Brien 2007c; Holloway 1996: 184–5; Tapper 1999: 178–81). Numbers recovered in the first half of the twentieth century, before declining again from the 1950s. Since then, the Snipe has continued to reduce its breeding range and has now been lost as a breeder from much of the lowlands of Britain and Ireland, due to habitat destruction, although it has increased in abundance in northern and western upland areas, especially in Highland Scotland.

Snipe records, 1519–1772. There are records from every region of Britain and Ireland except Connacht, North England and Wales. The record from Ulster is unreliable.

HERRING GULL

Larus argentatus

NATIVE STATUS Native

MODERN CONSERVATION STATUS	
World	Least Concern (Near Threatened in Europe)
UK	Red List
ROI	Amber List
Trend since 1772	Probably increased

The early modern recorders attested a few populations of Herring Gulls around Britain and Ireland, mostly around the coasts but once or twice further inland.

RECOGNITION

This species was called *Larus cinereus maximus* in New Latin and the *herring gull* and *great grey gull* by in Early Modern English. These names usually seem to be specific to the species, although Willughby also uses the term *great grey gull* to refer to the *wagel* (possibly our Iceland Gull, *Larus glaucoides*). These terms do, however, distinguish our species from *Larus cinereus minor*, the *sea-mall* (Common Gull, *Larus canus*), *Larus maximus ex albo & nigro varius*, the *great black and white gull* (Great Black-backed Gull, *Larus marinus*) and *Larus fuscus sive hibernus*, the *coddy-moddy* (Black-headed Gull, *Larus ridibundus*). All of these terms are well attested and serve to separate the species where they are used.

Unfortunately, most of the early modern sources do not distinguish records of the birds to species level. Most records simply refer to the species as a *sea-mew* or *gull*. These terms, along with the Irish *fuilinn* (not to be confused with the White-tailed Eagle *fuiller*), are not specific but have been included on our map here as diamonds to show the general locations where gulls were recorded in the early modern period.

DISTRIBUTION

There are not enough top-quality records to statistically analyse the distribution of this species. There are, however, generic gull records from the coasts of every region except Connacht, which is the worst-recorded region.

The Herring Gull is the best-recorded early modern gull. However, it may not have been the most numerous or the most culturally important. There is much better evidence for the exploitation of Black-headed Gull gulleries, some of which were inland (Gurney 1921: 184–91).

The Herring Gull seems generally to have been a coastal species in the early modern period. The pattern is actually not as strong as it might appear from our map, since some of the generic county-level records, as for example those from Yorkshire and Pembrokeshire, have been placed on the coast based on the assumption that they refer to coastal populations. But almost every more exactly located record does refer to the coast. There is one exception to this. A record from *The Natural History of Northampton-shire* (1712) suggests that Herring Gulls occasionally ranged inland even in the pre-industrial period:

> The Herring-Gull, and the Pewit or Black-Cap, Two other Birds of the Gull-Kind, I have also seen, tho' but very rarely, coast along our Rivers up toward the Sources of them.   (Morton 1712: 431)

Holloway (1996: 212–13) also found continued evidence of Herring Gulls occasionally being seen inland in the nineteenth century but noted that this was rare except in Highland Scotland.

The Herring Gull increased its abundance in the twentieth century. This may have been partly due to the end of egg-harvesting and perhaps direct persecution by gamekeepers (Sutcliffe 2021; Shrubb 2013: 207; Williamson 2013: 124), but the most important reason has been the colonisation of new habitats. Today it is still recorded on every coast of Britain and Ireland, but it has also now colonised the inland lowlands, and even nests in urban areas. This expansion seems to have been a development of the century from 1875 to 1975, and occurred when the Herring Gull (possibly following the Black-headed Gull) moved inland to landfill and sewage farm sites (Sutcliffe 2021; Williamson 2013: 179–80; Holloway 1996: 212–13; Parslow 1973: 95–6; Fitter 1959: 176–7).

Herring Gulls have been declining in abundance since the 1969–70 survey (Balmer et al. 2013). This has resulted in the abandonment of inland colonies in Lowland Scotland and the island of Ireland, and elsewhere reductions in coastal and island populations. The reasons for this include increased predation, decreased food availability from fishing and landfill, and the threat of botulism from landfill sites (Sutcliffe 2021; Balmer et al. 2013). To some extent, this represents a return to the pre-twentieth century status, but numbers are probably still higher than they were in the early modern period.

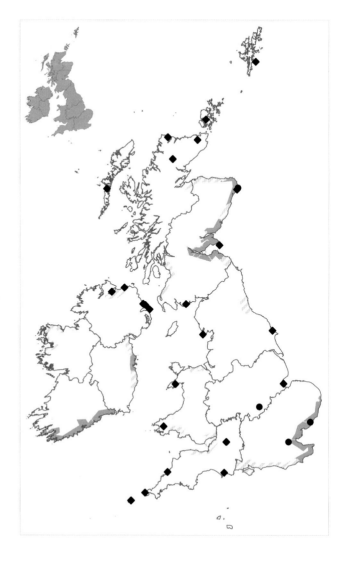

Herring Gull records, 1519–1772. There are records from every region of Britain and Ireland except Connacht, but only reliable records from Lowland Scotland, South England, Munster and Leinster.

GREAT AUK

Pinguinus impennis

NATIVE STATUS Native

MODERN CONSERVATION STATUS	
World	Extinct
UK	Extinct
ROI	Extinct
Trend since 1772	Certainly declined

The Great Auk was reliably recorded in three of our early modern sources – these are some of the very final records of the species from any country before it became globally extinct. All three records relate to a population breeding on St Kilda to the west of the Hebrides of Scotland. There are also unreliable records from the Farne Islands and the Isle of Man.

RECOGNITION

The species is most often known as the *gairfowl* in the surviving early modern sources, but there is some evidence that this name was half-forgotten. Robert Sibbald, for instance, lists the *gare* under 'Chapter VII: Those birds we have which are of an uncertain class' (Sibbald 2020: 22 (II:3)).

DISTRIBUTION

There are not enough top-quality records to statistically analyse the distribution of this species.

In the early modern sources, the Great Auk was only regularly recorded on isolated St Kilda, far from mainland Scotland. There are also unreliable records from the Isle of Man and the Farne Islands. Archaeological remains suggest that the Great Auk was once more common in the Isle of Man and also elsewhere especially in the Outer Hebrides, as well as in the Northern Isles, Caithness, Oronsay and the Scilly Isles and less commonly in Sutherland, as well as the Isle of May, Ailsa Craig and Cos. Antrim, Donegal, Clare and Waterford. However, the Great Auk seems to have already declined and have been lost from these areas before the early modern period (Yalden and Albarella 2009: 165; Fuller et al. 2007; Serjeantson 2001; D'Arcy 1999: 108; Bourne 1993; Baxter and Rintoul 1953: 681–2). In Orkney the species seems to have been rare from the beginning of the first century CE. Our early modern references include some of the final field descriptions of the bird:

> The Sea-Fowls are, first, *Gairfowl*, being the stateliest, as well as the Largest of all the Fowls here, and above the Size of a *Solan* Goose, of a Black Colour, Red about the Eyes, a large White Spot under each Eye, a long broad Bill; stands stately, his whole Body erected, his Wings short, he Flyeth not at all, lays his Egg upon the bare Rock, which, if taken away, he lays no more for that Year; he is *Palmipes*, or Whole-Footed, and has the Hatching Spot upon his Breast, *i.e.* a bare Spot from which the Feathers have fallen off with the heat in Hatching; his Egg is twice as big as that of a *Solan* Goose, and is variously spotted, Black, Green, and Dark; he comes without regard to any Wind, appears the first of *May*, and goes away about the middle of *June*.
>
> (Martin 1698: 48–9)

The Great Auk went extinct globally less than a century after our last source was written. The extinction of the species was initially driven by unsustainable hunting especially for food, feathers and fuel by sailors and fishers (Shrubb 2013: 49, 136; D'Arcy 1999: 110–12). The bird was especially targeted because it could not fly, and therefore could not nest on inaccessible cliffs or escape danger and did not develop a fear of humans (Serjeantson 2001; Bourne 1993; Ritchie 1920: 142). As this species became increasingly rare, specimens and eggs were increasingly sought for inclusion in private

Great Auk records, 1519–1772. There is a reliable record from St Kilda off the west of Scotland, and unreliable records from the Isle of Man and the Farne Islands.

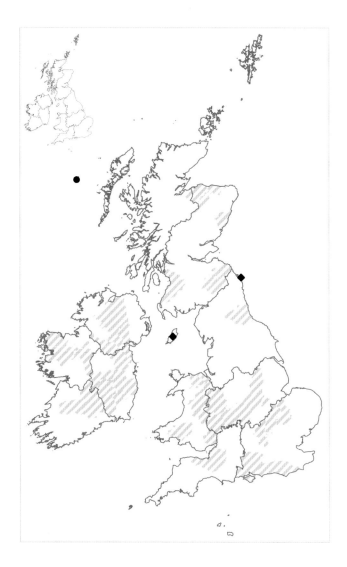

and commercial museums, which probably led to its final extinction (Serjeantson 2001; Lambert 1998). The extinction was shocking to naturalists writing in the nineteenth century and was intensely researched (Urry 2021; Lambert 1998). It was last recorded off Orkney in 1812. The final British record is from St Kilda in 1840, and the last record worldwide was from Eldey off Iceland in 1844 (Fuller et al. 2007; Lambert 1998; Holloway 1996: 436–7).

ATLANTIC PUFFIN

Fratercula arctica

NATIVE STATUS Native

MODERN CONSERVATION STATUS	
World	Vulnerable
UK	Red List
ROI	Red List
Trend since 1772	Uncertain

The Puffin was regularly referred to by early modern recorders around the coastlines and islands of Ireland and north and west Britain, including in some areas where it was later lost after the introduction of Brown Rats.

RECOGNITION

In the early modern period, this species was most commonly referred to by the Early Modern English name *coulterneb*, which was specific. It was also routinely called by the name *puffin*, or the Latin *puphinus*, which was also commonly used at the time for the Manx Shearwater. Here is an example from John Caius (1570):

> The Puphinus or Pupinus
>
> This is one of our sea birds. It has a body the size and shape of the little goose which the Greeks call βοσκάδα. It has red, webbed feet, positioned more to the back than the other birds with webbed feet, except the Pygoscelis penguins. The beak hangs much thinner in width and grows longer in length. There are four divided red sections at the top and two parts carved in the bottom which are of a pale orange colour. The area which is between these and the head is blueish and it is the shape of a moon, seen ten days after it is new.   (trans. from Venn 1912: 52)

In this instance, the description of the beak here could only apply to the Atlantic Puffin, not the Manx Shearwater. However, other references certainly refer to the latter species (as described on p. 116). I have included ambiguous records on the map for the Manx Shearwater.

In New Latin this species was called the *Anas arctica* of [Carolus] Clusius. This term is specific to our species and has now been adapted into our current Latin species name *Fratercula arctica*. The usual Scottish Gaelic term was *bouger* (modern Gaelic *buthaigre*), although Martin Martin (1703: 227) also attests *albanich* (literally 'the Scotsman') as being used by fishers around Ailsa Craig. K'Eogh (1739: 74) provides the Irish term *canoga*. Lhuyd in his additions to Camden's Britannia provides the Welsh *pâl* and Willughby & Ray (1678: 325) provide the south Welsh term *helegug*. The species also had several other local and national names in Early Modern English including *pope*, *bottle-nose*, *sea-parrot*, *mullet* and *golden head*. Wallis (1769: 341) adds *tommy-noddy* which seems to be cognate with the Orkney term *tominories* (Brand 1701: 180).

DISTRIBUTION

The distribution of top-quality records for this species is not statistically different from the known level of recorder effort. The gaps on the map may just reflect decreased survey effort in some regions. It may have been widespread.

Puffins were commonly consumed by humans in the early modern period and were even exported for consumption in other countries during Lent (Shrubb 2013: 24). The early modern distribution map of the Puffin agrees very well with the modern distribution map, and there may have been only relatively minor changes in distribution over the last few centuries – although the population is likely to be smaller now than it was in the early modern period. In terms of distribution, there are also a few locations given in the early modern texts where Puffins no longer occur; the populations have been lost. These include islands where the accidental introduction of rats has since exterminated the populations known from the early modern period, including Ramsey Island off the coast of Pembrokeshire and perhaps Raasay off Skye. The populations on the Calf of Man and Ailsa Craig were also lost due to rats but have now begun to recover (Harris and Wanless 2011: 45–57).

Puffins were perhaps the most targeted seabird in the nineteenth century, with 89,600 killed on St Kilda in 1876 alone (Shrubb 2013: 134–7). This suppressed numbers in some areas, but the scale of exploitation seems to have been sustainable on other islands. A more serious threat has been climate change, which has caused reduction in sandeels in the North Sea. This has led to widespread declines since the beginning of the twentieth century, although there was a brief recovery in the colder climate of the 1960s and 1970s (Balmer et al. 2013; Harris and Wanless 2007; Holloway 1996: 236–7; Ford 1982: 126–7).

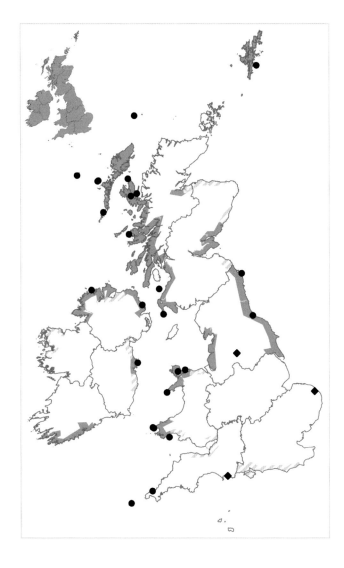

Puffin records, 1519–1772. There are coastal records from every region of Britain and Ireland except Connacht and the Midlands of England. The record from South England is unreliable.

EAGLE OWL
Bubo bubo

NATIVE STATUS Usually thought to be non-native

MODERN CONSERVATION STATUS	
World	Least Concern
UK	Not Present
ROI	Not Present
Trend since 1772	No change

There is a small amount of evidence that the Eagle Owl may have been present as a wild species in Scotland during the early modern period.

RECOGNITION
There is no established name for this species during this period, but a few accounts do appear to describe it. The account by James Wallace refers to a *stock-owl* living in Orkney. The *Dictionary of the Scots Language* gives a few references to the use of this term in the nineteenth century to refer to the Eagle Owl, which stays close to the *stock* (trunk) of the tree during the day to camouflage itself.

> Sometime the *Stock-Owl* [Eagle owl?] and *Bittern* have been seen in this Country. Eagle and Earns as they call them, and Gledes or Kites are here in great plenty, and very hurtful to their young Store.
> (Wallace 1700: 46)

DISTRIBUTION
There are not enough top-quality records to statistically analyse the distribution of this species.

Wallace's description of the *stock-owl* is the only local record. The species is, however, also referred to by a few national recorders. Turner attests that the species was absent from England, but it seems to be included as a wild species by Sibbald in Scotland (under *Bubo maximus*) and possibly by O'Sullivan in Ireland (under *bubo*). The species was mentioned more commonly, mainly in Shetland and occasionally on some of the other islands and in mainland Scotland in the nineteenth century (Baxter and Rintoul 1953: 267–8). It may have been a common vagrant. The Eagle Owl seems to have been frequently kept by falconers in the medieval Norse world (Haley-Halinski 2021: 120–4), and assuming this practice continued into the early modern period some of the records may have been of escaped birds.

Eagle Owl records, 1519–1772.
There is one possible
record from Orkney.

PEREGRINE

Falco peregrinus

NATIVE STATUS Native

MODERN CONSERVATION STATUS	
World	Least Concern
UK	Green List
ROI	Amber List
Trend since 1772	Probably increased

The early modern recorders provided a collection of uncertain references to Peregrines around the northern and western coasts of Britain, Ulster and Munster, but never further inland.

RECOGNITION

The most reliable name for this species is *peregrine* in Early Modern English. It was also sometimes referred to more ambiguously as the *falcon-* or *tercell-gentle*, which was a falconer's term borrowed from medieval French. This also provided a New Latin name *Falco gentilis*, which was later adapted as the scientific nomenclature for another bird, now *Accipiter gentilis* (Goshawk). It is perhaps not surprising that this term has been reallocated, as even in the early modern period it was ambiguous. Willughby tentatively suggested *falcon-gentle* might have referred either to the Peregrine or a closely related, otherwise unknown, species (Ray 1678: 79). The only way to be confident about which species is intended is using this name is where a text records the Goshawk separately under a different name.

The map also includes generic references to *falcons*. These are indicated by diamonds on Anglesey and the Northern Isles and Hebrides. These records are most likely to have referred to the Peregrine (Baxter and Rintoul 1953: 284) which was, at the time, the nominal species of *Falco*. The other native *Falco* species the Merlin (*Falco columbarius*), the Hobby (*Falco subbuteo*) and the Kestrel (*Falco tinnunculus*) were not and are not commonly called falcons at all by naturalists. However, the falcons were not always distinguished (Yalden and Albarella 2009: 135–6), especially by recorders without binoculars or much field knowledge, and some of the records may have been of other species.

DISTRIBUTION

There are not enough top-quality records to statistically analyse the distribution of this species.

It is worth noting, though, that all of the surviving records of wild Peregrines refer to coastal populations, if we include the ambiguous records of *falcons* there is also a clear bias toward the Northern Isles and Hebrides of Scotland. The very impressive number of uncertain records from Orkney are provided in James Wallace senior's *Description of the Isles of Orkney*:

> Hawks and Falcons have their Nests in several places of these Islands, as in the Noup, Swendal and Rapnes in Westra at High berrie and Aith-head in Waes at Braebrake, Furcarsdale and Rackwick in Hoy, at Halcro, Greenhead and Hocksa in South Ronaldsha, at Bellibrake and Quendal in Rousa; at Rousum-head and Lamb head in Stronsa, in the Calf of Eda, at Gatnip Gultak, Mul head in Deirness, Copinsha, Black Craig of Stromnes, Yeskrabie, Birsa, Marwick and

Costahead in the Mainland. The Kings Falconer comes every year, and takes the young, who has twenty pound Sterling in Salarie, and a Hen or Dog out of every House in the Countrey, except some Houses that are priviledged. (Wallace 1693: 17)

The identification of the species here as 'hawks and falcons' is not especially reliable, but the fact that the royal falconer picks the birds up each year suggests that they are likely to have been Peregrines rather than, for example, Sparrowhawks.

It is likely that wild Peregrines occurred inland in Britain and Ireland in the distant past, but the medieval archaeological evidence is compromised by the keeping of Peregrines as falconry birds. Unlike Sparrowhawk nests, no Peregrine nests are mentioned in the Domesday survey (Yalden and Albarella 2009: 67; Yalden 1987), so the Peregrine is likely to have been rarer than the Goshawk. If our records can be taken to suggest a coastal distribution, the Peregrine may have been rarer in the early modern period even than it was in the first half of the twentieth century. Based on the distribution of the surviving records, it seems most probable that the early modern population was depleted despite Ratcliffe's (1993: 63–4) conservative assumption that the Peregrine never declined

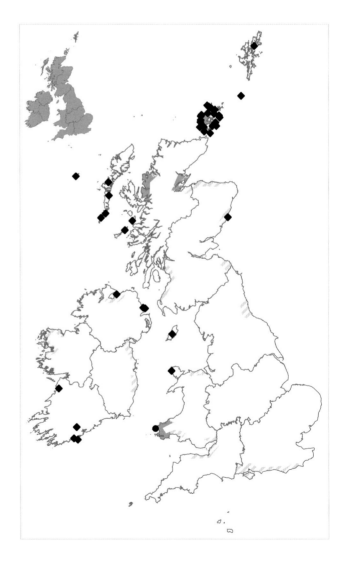

Peregrine records, 1519–1772. There are possible coastal records from Highland and Lowland Scotland, Ulster and Munster and Wales, but the only reliable records are from Wales, and Highland Scotland including Orkney, where there is a cluster of records.

until 1800. Populations were probably suppressed by capture for falconry (Shrubb 2013: 29–31, 208) and (later) extermination for pest control.

From the end of the early modern period, interest in falconry waned and if it is correct that populations had been suppressed by collection for falconry, then they would have had a chance to recover. The species also seems to have benefitted from the Industrial Revolution, since increased quarrying created additional inland rockfaces for nesting (King 2021; Williamson 2013: 81). By the end of the nineteenth century, Holloway (1996: 130–1) suggests Peregrines were widespread not just around most of the coasts of Britain and Ireland, but also inland in Scotland, north England and Wales. The degree to which Peregrine populations were resilient to the intense persecution for pest control carried out in the nineteenth and early twentieth centuries is debated (King 2021; Shrubb 2013: 189, 191; Hardey 2007; Lovegrove 2007: 76; Parslow 1973: 60). However, if it is correct that the distribution of the Peregrine was mainly coastal in the early modern period, it is possible that the species may even have been recovering during the early nineteenth century. Peregrine populations were, however, unambiguously suppressed in Britain in the middle of the twentieth century, first by the Destruction of Peregrine Falcons Order (1940) which mandated culling of Peregrines to help homing Pigeons during the Second World War, and second by the use of DDT and hydrocarbon insecticides (King 2021; Hardey 2007), which killed Peregrines and reduced their breeding success rate.

The decline of the Peregrine has now once again been overcome with the banning of many of most dangerous pesticides and the reduction (although not the end) of persecution for pest control. The Peregrine has increased its range by 200% since 1968–72 (Balmer et al. 2013). Abundance now seems to be at an all-time high, due in part to the colonisation of inland areas of Britain and Ireland, especially lowland areas where prior to the second half of the twentieth century Peregrines were not found but including also urban environments with high towers for nesting and large populations of Pigeons for food (King 2021; Lovegrove 2007: 294). The Peregrine may still be declining in some parts of Scotland due to persecution and habitat loss (Hardey 2007).

GREAT GREY SHRIKE

Lanius excubitor

NATIVE STATUS Native

MODERN CONSERVATION STATUS	
World	Least Concern
UK	Extinct as breeder?
ROI	Does not breed
Trend since 1772	Uncertain

This small vole-hunting bird was the most commonly recorded species of shrike, and an unreliable string of records refer to it across much of England.

RECOGNITION

Two *Lanius* species regularly winter in Britain, the Great Grey Shrike and Red-backed Shrike (*Lanius collurio*). Our species, *Lanius excubitor* seems to have always been a rare winter visitor, but the Red-backed Shrike was formerly widespread as a breeder across England and Wales as recently as the end of the nineteenth century (Holloway 1996: 374–5). This latter species declined through the twentieth century due to changes in land-use and reduction in the large invertebrates it feeds on, and is now a very rare breeder (Macdonald 2019: 77–8; Balmer et al. 2013). Surprisingly though, the Great Grey Shrike is far more commonly recorded than the Red-backed Shrike in the early modern sources.

In the early modern period, the Great Grey Shrike seems to be reliably referred to as *Lanius cinereus major* in New Latin or the *greater [ash-coloured] butcher bird* and possibly *wereangel* in Early Modern English. This usually makes it easy to distinguish from the Red-backed Shrike, which Willughby & Ray call *Lanius minor rubescens*. Morton (1712: 424) distinguishes another species, the *lesser ash-colour'd butcher-bird* (probably meaning the Red-backed Shrike again). *Lanius cinereus major* is twice shortened to just *Lanius cinereus*, which given Morton's term seems an untrustworthy name.

Other names were also widely used but less specific like *shrike*, *shriek*, *murdering-bird* and *butcher*. These terms seem to be generic and have been excluded from our map.

DISTRIBUTION

There are not enough top-quality records to statistically analyse the distribution of this species.

All of the records of this species come from the North, South and Midlands of England, but most of the records are uncertain either in terms of location or in terms of species. The lack of records from Scotland is surprising. Sibbald (2020: 15 (II:3)) provides a national record, but this is uncertain since he describes the bird as a kind of *kite*. The species was later recorded as a regular or occasional winter bird in some parts of Scotland in the second half of the nineteenth century (Baxter and Rintoul 1953: 140). There are no records from Ireland.

Great Grey Shrike records, 1519–1772. There are reliable records from North England and the Midlands, and there is an unreliable record from South England.

The Great Grey Shrike is rarely recorded in the archaeological record, and Yalden (2009: 227) suggested it is likely to have been only an uncommon winter visitor since the medieval period. It remains only a wintering species today. There is, however, one unreliable record of a breeding population from early modern England:

> The *Lanius cinereus major*, the greater Ash-colour'd Shrieke or Butcher-Bird: a very uncommon Bird, except it be in the Mountainous Parts of the North; And yet tho a Mountainous Bird, it breeds sometimes in Northamptonshire, and particularly in Whittlewood Forest, where 'tis called the Night-Jarr. Now and then one of them is seen at Winwick, in Eltinton Grounds, and in Stoke-Albany Park, particularly in the Month of August.   (Morton 1712: 424)

This record is complicated by the use of the term *nightjar* which usually referred to the species we still call the Nightjar (*Caprimulgus europaeus*), making this record unreliable. These locations are marked with diamonds.

Shrikes were occasionally taken as food or as bait for falcon-catching in mainland Europe (Shrubb 2013: 160), but it is not clear if numbers were high enough for this to occur in early modern Britain and Ireland.

The Great Grey Shrike's winter distribution has also changed and some of the sites recorded by the early modern sources no longer have populations.

Today, this shrike is a winter visitor found uncommonly but widely across Britain, especially in the South and South West of England and in Wales (Balmer et al. 2013). It is not known to breed anywhere in Britain, and it does not winter in Ireland. It is not clear if the species has changed its distribution since the early modern period.

MAGPIE

Pica pica

NATIVE STATUS Native to Britain

MODERN CONSERVATION STATUS	
World	Least Concern
UK	Green List
ROI	Green List
Trend since 1772	Certainly increased

The Magpie seems to have been usually overlooked by the early modern local recorders, but there are important records from Ireland which provide evidence of when the species was first recorded here.

RECOGNITION
The bird was called *pica* in New Latin and *pie* or *magpie* in Early Modern English. The only possible confusion was with the so-called *sea-pie* or *Pica marina* (Oystercatcher), but the two species are always clearly distinguished in our sources.

DISTRIBUTION
The distribution of top-quality records for this species is statistically a poor fit with the known levels of recorder effort. It may have been locally distributed, locally abundant or of special local interest. Habitat suitability modelling based on the sites where the Magpie was recorded in the early modern period suggests that the species may have had specific requirements. It is only reliably recorded on coastal sites which today have dry winters and cool summers. But in this case there also seems to have been a bias in the results due to an increased interest in the Magpie in Munster and Leinster (as a novelty) compared to elsewhere.

For the most part, the local recorders seem to have ignored the Magpie if it was present. The bird may have been occasionally caught for pest control or kept as a pet as it was in the medieval period and the nineteenth century (Shrubb 2013: 17, 175; Lovegrove 1990: 148–50; Gurney 1921: 65), but it is rarely mentioned in the early modern natural history sources. However, the Magpie does appear to have attracted particular attention and records when it began to colonise Ireland in the early modern period. For instance:

> 4. Of the Pie kind.
>
> 18. Pica varia Caudata, Will. p.87. The Magpie or Pianet.
>
> This bird was not known in Ireland 50 years ago, but are now very common. It destroys small birds. They build their nests with great cunning, fencing them with sharp thorns, having only one hole for themselves to pass in and out. (Smith 1750b: 325)

Magpie records, 1519–1772. There are records from South England, Lowland Scotland, Munster and Leinster.

This idea of the Magpie as a novel species in Ireland is corroborated, since less than a century before in 1669, Lorenzo Magalotti, a traveller in the retinue of Cosmo III, Grand Duke of Tuscany, had also visited Co. Cork and declared that Magpies were at that time absent from the province of Munster (and Ireland in general) (Magalotti 1821: 103). Robert Payne, who settled in Munster in the sixteenth century also confirmed that Magpies were absent from Munster and Ireland then in *A Brief Description of Ireland* (Payne 1841: 14). Another source attested the Magpie was still rare in Ireland in 1684 (MacLysaght 1939: 416–17).

There are earlier attestations of the Magpie in Ireland. Yalden (2009: 227) lists five Magpies in the archaeological record. O'Sullivan (ed. O'Sullivan 2009: 112–13) in 1626 referred to the *pica/hurraca* (Modern Spanish *urraca* = *Pica pica*) as an Irish bird. However, the description here of a black bird with white speckles which can mimic human speech suggests confusion with the Starling (*Sturnus vulgaris*).

The record of Magpie from the Isle of Mull is deeply uncertain. It is from a piece of folklore about the meaning of Magpie calls and was collected by Edward Lhuyd from John Beaton of Mull (ed.

Campbell and Thomson 1963: 67). It is included here based on the assumption that the folklore was in active transmission on Mull and therefore that Magpies were regularly heard there. In support of these assumptions, Lhuyd does comment on the absence of some other species from Mull. However, it should be noted that Beaton may have learned the folklore from a much older medieval manuscript, and that, even if the folklore had currency, the description 'calles much about a house' could refer to a captive bird. Magpies are now not normally found on Mull (Balmer et al. 2013).

Today, the Magpie is a common and widespread bird found in every lowland area of Britain and Ireland except the Northern and Western Isles of Scotland. However, the population has not been stable throughout the modern period. The Magpie is one of the species most intensely persecuted by gamekeepers to the extent that in Highland Scotland and East Anglia in particular population levels were significantly suppressed after the end of the early modern period, and the bird was extinct in some counties by the end of the nineteenth century (Williamson 2013: 148; Lovegrove 2007: 148–50; Holloway 1996: 378–9; Tapper 1992: 88–9).

The Magpie started to recover following the end of intense pest control during the First World War, and a reduction in the use of poison from 1911 (Pritchard 2021q; Lovegrove 2007: 148–50; Young 2007a; Holloway 1996: 378–9; Parslow 1973). In the 1940s, the bird began to colonise more urban areas and visit gardens more often. Numbers were affected by the loss of hedgerows in the 1950s, but since 1970 the population has risen significantly and the Magpie has now started to expand beyond the central lowlands of Scotland (Balmer et al. 2013; Young 2007a). Since around the turn of the millennium it has increased its abundance in Scotland and Ireland, but also started to decline in western England and Wales (Pritchard 2021q).

CHOUGH

Pyrrhocorax pyrrhocorax

NATIVE STATUS Native

MODERN CONSERVATION STATUS	
World	Least Concern
UK	Green List
ROI	Amber List
Trend since 1772	Uncertain

The Chough was especially associated with Cornwall in the early modern sources, but it was also recorded elsewhere on the coasts of South England, Lowland Scotland, Wales and on every coast of the island of Ireland.

RECOGNITION

The New Latin terms *coracias* and *pyrrhocorax* are specific to this species in Britain and Ireland. The most usual English term used was *Cornish chough* which was specific to species level, but it is also occasionally referred to more generically as a *chough* or a *daw* (along with *Corvus monedula*, the Jackdaw) or a *kae* (along with the rest of the crow family) (Yalden and Albarella 2009: 160; Lovegrove 2007: 150). In these cases, the species can be identified by context or description. The map for this species does not include any uncertain records. None of the early modern sources attests the legend of Arthur as a chough, which seems to have been popularised from the nineteenth century (Ramsay 2012).

DISTRIBUTION

The distribution of top-quality records for this species is statistically a poor fit with the known levels of recorder effort. It may have been locally distributed, locally abundant or of special local interest. Habitat suitability modelling based on the sites where the Chough was recorded in the early modern period suggests that the species may have had specific requirements. It is only recorded on coastal sites that today have dry winters and relatively warm summers. But in this case, the biggest reason is likely to have been the cultural idea of the Chough as a belonging to Cornwall.

The Chough was most associated with Cornwall during the early modern period. Most sources mentioning it refer to it within Cornwall. Many of these records are vague Cornwall-level references and so are not separately mapped here, but even the local records that do exist such as at Padstow show the species was more widely found in the South West than it is today. The records provided by the sources also suggest that the species had a greater range outside of the South West in the early modern period. The Chough was twice recorded on the south coast as far east as Dover (Chamberlayne 1683: 21–2). The records from County Down, Lough Erne and Mochrum in Galloway also fall beyond the ordinary present range of the species:

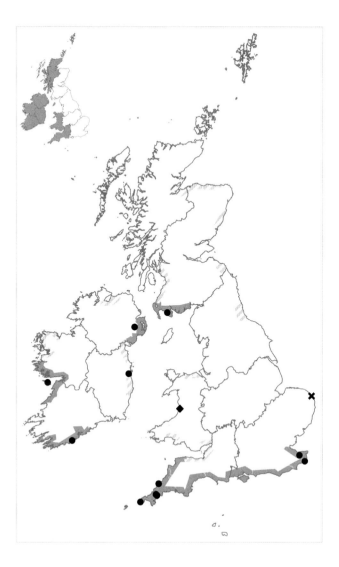

Chough records, 1519–1772. There are records from every region of Ireland, as well as South and South West England, Wales and Lowland Scotland.

near the sea [in Galloway there are] several sorts of wild-geese, wild-ducks, ateales, small teales, sea-maws, gormaws, and an other fowl, which I know not the name of; it is about the bigness of a pigeon; it is black, and hath a red bill. I have seen it haunting about the Kirk of Mochrum.

(ed. Symson 1823 [1684]: 79)

Despite the inclusion of the Chough in the Grain Acts of the sixteenth century, there is little evidence it was regularly targeted for pest control until the nineteenth century (Lovegrove 2007: 151). However, the early modern population may have already been reduced from the maximum distribution, since records from the Iron Age suggest the Chough was formerly found in the Hebrides and Northern Isles of Scotland (Yalden and Albarella 2009: 228; McKay 2007). The lack of records from Highland Scotland and the east coast of Lowland Scotland may be an oversight, since some early nineteenth-century recorders indicate that it was common there (Baxter and Rintoul 1953: 32–3).

The Chough seems to have started the modern period with a stable population. It further benefitted from the creation of additional habitat around quarries and mines across Britain during the Industrial Revolution and is still found inland on abandoned sites in North Wales, although numbers in these areas have been declining for the past few decades (Cross et al. 2021; Williamson 2013: 81; Baxter and Rintoul 1953: 32–3). On the coast, numbers are now stable. However, the Chough declined significantly during the nineteenth century (Holloway 1996: 380–1). It went extinct in the South of England (but not the South West) around 1830–60 and in the North of England, Lowland Scotland and (briefly) Anglesey around 1850–90. It is sometimes argued that cold weather might have led to its extinction in the nineteenth century. Choughs are dependent on environmentally sensitive tipulid (cranefly) larvae for their food, and their survival seems to be better with warm summers and mild winters (Reid et al. 2008; McKay 2007), but the Chough was widely distributed throughout the height of the Little Ice Age in the seventeenth century, so climatic deterioration is not a convincing argument for the nineteenth-century reduction by itself. Hunting of eggs and specimens for natural history collections and use as pets seems to have had an impact, as do the agricultural changes. The key reason for decline may have been hunting for pest control by gamekeepers (Lovegrove 2007: 151–2; McKay 2007; Holloway 1996: 380–1). The Chough finally went extinct in Cornwall in the first half of the twentieth century but recolonised at the start of the twenty-first century (Balmer et al. 2013), and there is now a strong breeding population there. The bird survived best in Wales and Ireland, where there have been continuous populations from the early modern period to the present day.

HOODED CROW

Corvus cornix

NATIVE STATUS Native

MODERN CONSERVATION STATUS	
World	Not Evaluated
UK	Green List
ROI	Green List
Trend since 1772	Uncertain

The Hooded Crow was recorded as a breeder by a few early modern sources in the Northern Isles and Ireland, but during the Little Ice Age it was more widespread during the winter, with records from much further south and west in Britain than is currently seen, leading to the term *Royston crow*.

RECOGNITION

Despite the nineteenth-century confusion between the Hooded Crow and Carrion Crow (*Corvus corone*) (Holloway 1996: 388), they were usually distinguished in natural histories from early modern Britain and Ireland. This species was most commonly called the *Royston crow* in Early Modern English or the *Cornix cinerea* in New Latin. It was also occasionally called other names like the *gray crow*, *scald crow*, and *pied crow* and *bran lwyd*. These all appear to be specific to Hooded Crow. John Ray (1713a: 39 (I)) adds that the bird was sometimes called *the sea crow*, but this term ordinarily seems to apply to the Cormorant so I have not included any references using that name on this map. Uncertain references to the *cornix* could apply to this species, but they could also apply to, for example, the Jackdaw, so are also not included.

There are a few generic records referring to *crows* which seem most likely to refer to this species based on location, since they are from areas where Carrion Crows have not historically been common. These have been identified with diamonds because they are uncertain and could for example refer to Rook (*Corvus frugilegus*) instead (Lovegrove 2007: 158–9). The Scottish Gaelic term *feannog* is taken in the same way.

DISTRIBUTION

The distribution of top-quality records for this species is not statistically different from the known level of recorder effort. The gaps on the map may just reflect decreased survey effort in some regions. It may have been widespread.

The records have a seasonal divide. There is evidence that the Hooded Crow nested in the Northern Isles and in Ireland (in this case Co. Cork). However, in our sources the Hooded Crow is most often associated with inland areas of South England, especially Royston in Hertfordshire. It was also recorded in Cambridgeshire, Norfolk, Wiltshire and Cornwall. In these locations, the Hooded Crow is described as a winter visitor. I have distinguished the two kinds of records by only shading the locations where breeding was mentioned in the records.

> In Summer time (saith Aldrovandus) it lives in high Mountains, where it also builds: In the Winter (compelled as is likely by the cold) it descends into the Plains. On the Heaths about Newmarket, Royston, and elsewhere in Cambridge-shire, it is frequently seen in Winter time.
>
> (Ray 1678: 124)

In early modern Britain, the Hooded Crow was regularly targeted for pest control along with the Carrion Crow and other corvids following the pest control acts of fifteenth-century Scotland and sixteenth-century England and Wales (Lovegrove 2007: 159–60; Baxter and Rintoul 1953: 4). The control does not seem to have been enough to restrict populations. The Hooded Crow and other corvids have continued to be targeted for pest control in the modern period, especially by gamekeepers and sheep farmers. This has suppressed some regional populations, but in itself does not seem to have had a significant impact on the overall population or led to any regional extinctions (Lovegrove 2007: 160–1; Young 2007b). Hooded Crows now commonly breed in Ireland, the Highlands of Scotland and the Isle of Man, and are found on the east coast of England in winter (Balmer et al. 2013). The species is currently increasing in abundance in Ireland but has significantly declined in Scotland.

Intriguingly, however, the Hooded Crow has reduced its distribution since the early modern period. Most obviously, the so-called *Royston crow* is no longer commonly seen at any time of year around Royston (or indeed in Newmarket or Wiltshire, where it was recorded as a visitor by our early modern sources). This is a modern development, but it came about within a century after the end of the early modern period (Pritchard 2021r; Holloway 1996: 386–9). The reason for this may be related to climate change. The Hooded Crow has a mutually exclusive range with the Carrion Crow, except for a hybridisation zone through Scotland, where both species are seen. The hybridisation zone has

Hooded Crow records, 1519–1772.
There are reliable records from
Munster, Lowland Scotland, South,
South West and North England, but
most of these are wintering records.

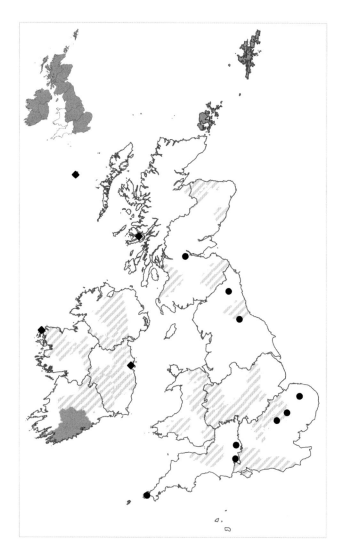

moved northwards during the warm years of the twentieth and twenty-first centuries, but faster in the lowlands than in the uplands (Young 2007b; Holloway 1996: 388; Cook 1975). It now almost follows the border between Highland and Lowland Scotland, and Hooded Crows are increasingly considered to be highland birds. During the Little Ice Age of the early modern period, it seems probable that the hybridisation zone would have been further south, and so the records from Royston and other areas might represent the species' former winter distribution due to this colder climate. If this is correct, the term *Royston crow* may itself be an artifact of the Little Ice Age.

RAVEN

Corvus corax

NATIVE STATUS Native

MODERN CONSERVATION STATUS	
World	Least Concern
UK	Green List
ROI	Green List
Trend since 1772	Uncertain

The Raven was occasionally recorded by the early modern sources, especially in cities where it was begrudgingly respected for its role in keeping the streets clean.

RECOGNITION

Corvus corax was the only species called *raven* in Early Modern English, so usually it is possible to be confident about what species early modern recorders are referring to. Generic Latin references to *corvus* usually also refer to the raven but these are less trustworthy. K'Eogh (1739: 76) provides the Irish *fiagh-duf*. The Scots term was *corbei*. The Welsh name is *cigfran* (literally 'meat-crow'), not to be confused with the *ydfran* ('grain-crow', meaning the Rook) or the *cogfran* ('chattering-crow', meaning the Jackdaw).

DISTRIBUTION

The distribution of top-quality records for this species is not statistically different from the known level of recorder effort. The gaps on the map may just reflect decreased survey effort in some regions. It may have been widespread.

At the beginning of the early modern period, the raven appears to have been widespread. It is recorded occasionally across Britain and Ireland, including in the east of Britain where it remains rare today. Along with the Red Kite, and smaller corvids, it especially prospered in urban areas where it took advantage of the poor refuse disposal and scavenged in the streets (O'Connor 2017). It was believed at the time that mould and rot could create disease-causing miasmas, and the Raven appears to have been respected (although probably never legally protected) as the provider of a valuable sanitation service (Raye 2021b). It was found not just in the city of London but also in other urban areas:

> Coruus maior Rauens in good plentie about the citty [of Norwich] wch makes so few Kites to bee seen hereabout. they build in woods very early & lay egges in februarie.
>
> (ed. Browne 1902 [1662–68]: 27)

The Raven was targeted for persecution under the Grain Acts of fifteenth-century Scotland and sixteenth-century England, but interestingly, the English Act of 1566 adds that the Raven was not to be targeted within two miles of any city or town (ed. Anon 1819, sec. 8 Eliz I c.15). The result of this was that in the early modern period the Raven seems to have been left alone in lowland England and Wales but was intensively targeted for persecution in the uplands (Lovegrove 2007: 164–5).

The Raven was lost from London almost immediately after the end of the early modern period, possibly due to improvements in sanitation standards. Increased persecution through this period by gamekeepers and shepherds and the conversion of common land pasture to arable, together with collection of birds and eggs for natural history collections and as pets, meant the Raven was quickly lost from the Midlands, Cheshire and Lancashire as well as London in the eighteenth

Raven records, 1519–1772. There are reliable records from every region of Britain and Ireland except South West England, Ulster and Connacht.

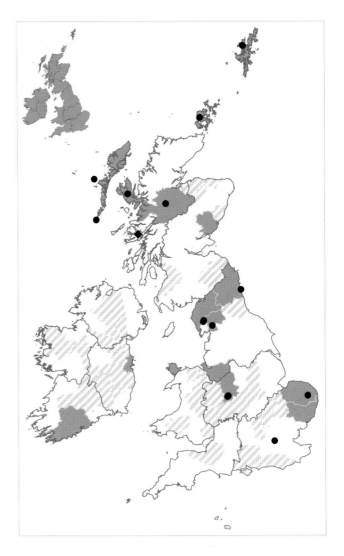

century. In Lowland Scotland it was notably declining by the end of the eighteenth century. By the end of the nineteenth century the Raven was extinct in the few inland counties of Ulster where it had previously been found. It was also gone from the Midlands and part of South England, and was not common anywhere in the lowlands (Raye 2021b; O'Connor 2013: 114; Shrubb 2013: 174; Ratcliffe 1997; Holloway 1996: 390–1; Thomas 1991: 274; Lovegrove 1990: 167–8; Baxter and Rintoul 1953: 1–2). It remained common only in upland areas of Ireland, Scotland and Wales. This was the context when the bird was introduced to the Tower of London as a Victorian tourist attraction (Sax 2007).

Following the abrupt reduction in gamekeeper persecution during the First World War, and the reduction in the use of pesticides (although Ravens do not seem to have been significantly impacted by DDT), the Raven has significantly recovered, and has spread from the uplands of northern and western Britain, and the coasts of Ireland across into the lowlands (Balmer et al. 2013; Mearns 2007). In Ireland the current range might represent an all-time widest distribution for the bird, since it is not mentioned in any of the (poorly recorded) inland counties of the early modern period, and it seems to have been rarer in the interior than on the coasts even in the nineteenth century (Holloway

1996: 390–1). In Scotland, the species declined slightly from the middle of the twentieth century due to land-use changes (especially afforestation and increased use of poison), but has now recovered (Mearns 2007).

SKYLARK
Alauda arvensis

NATIVE STATUS Native

MODERN CONSERVATION STATUS	
World	Least Concern
UK	Red List
ROI	Amber List
Trend since 1772	Uncertain

The Skylark was a bird of the open countryside. Prior to the enclosure of the open-field systems it was especially commonly recorded in the Midlands and the South of England.

RECOGNITION
This species is normally referred to in Early Modern English as the *lark*, *skylark* or *field-lark*. All of these terms appear to be specific to the Skylark; in English, *Lullula arborea* was always distinguished specifically as the *woodlark* and Skylark was called the *tit-lark*. In New Latin our species was called *Alauda vulgaris* to distinguish it from others, as Sibbald explains:

> The lark is distinguished from other birds by its heel or by the very long rear toe talon, as also by its singing, which occupies it in the air during flight.
>
> *Alauda vulgaris*; the Common Lark. It has a very sweet and splendid meat.
>
> *Alauda arborea*; the Wood-Lark.
>
> *Alauda pratorum*; the Tit-Lark.
>
> *Alauda cristata*.　　　　　　　　　　　　　　　　　(Sibbald 2020 [1684]: 17 (II:3))

Occasionally this species is also called *calandra* or *chalandrius* in Latin; these terms are also used for other species (especially *Melanocorypha calandra*, the Calandra Lark of the Mediterranean), but they usually occur in our texts alongside more specific terms for *Alauda arvensis*. The Irish term was (and continues to be) *fuiseog* (O'Sullivan 2009: 132).

DISTRIBUTION
The distribution of top-quality records for this species is statistically a poor fit with the known levels of recorder effort. It may have been locally distributed, locally abundant or of special local interest. In particular, there are more records than we would expect from the Midlands and South of England. The Skylark seems to have been especially adapted to more open countryside (Williamson 2013: 6–10, 41–2; Lovegrove 2007: 40). This is surprising because in the nineteenth century the Skylark was particularly abundant in the Downs, which fall to the south of the distribution indicated for the early modern period (Shrubb 2013: 162; Holloway 1996: 278–9). The Downs are poorly recorded in the early modern period, so records from this area are not likely to have survived in any case, but prior to the enclosure of fields in the Midlands, the population there may well have been just as high (Shrubb 2013: 163).

The Skylark was regularly eaten and is the passerine most commonly mentioned in the early modern market price lists and household (kitchen) accounts collected by Shrubb (2013: 35, 161–2). It was caught in nets and traps, and by falconers (usually with Sparrowhawks). It has been consistently found on archaeological sites from every period since the spread of agriculture in the Neolithic (Yalden and Albarella 2009: 71, 138). The bird was occasionally also seen as a pest on arable fields (Lovegrove 2007: 256).

At the end of the early modern period, changes in agriculture with the introduction of a turnip and clover crop rather than a fallow year may have benefited Skylarks (Williamson 2013: 108–9). The Skylark began to be hunted more intensively in the nineteenth century, with hundreds of thousands sold in the London markets each year, and many more exported to Paris (Shrubb 2013: 159, 161–3; Holloway 1996: 278–9). The bird was also often taken as a pet for its pretty song. It continued to be commonly eaten in London into the 1940s (Shrubb 2013: 21). Hunting seems to have suppressed some populations, especially in Highland Scotland, but the Skylark continued to be found in every county of Britain and Ireland at the end of the nineteenth century (Holloway 1996: 278–9).

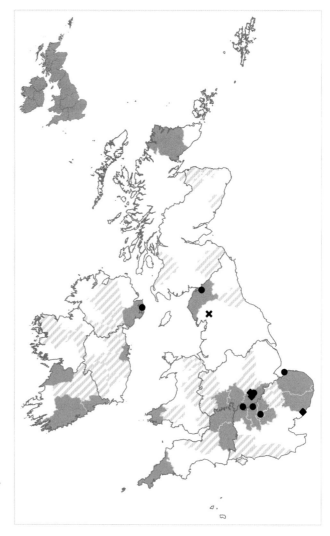

Skylark records, 1519–1772. There are reliable records from every region of Britain and Ireland except Connacht and Lowland Scotland.

Today, the Skylark continues to be very widespread outside of the uplands and urban areas. However, it is declining rapidly as a breeder in England and the Republic of Ireland, due to the switch to autumn ploughing and sowing (which results in vegetation that is too tall during nesting season, and less overwinter stubble) (Spence 2021; Balmer et al. 2013; Dougall 2007).

RING OUZEL

Turdus torquatus

NATIVE STATUS Native

MODERN CONSERVATION STATUS	
World	Least Concern
UK	Red List
ROI	Red List
Trend since 1772	Probably increased

The Ring Ouzel was occasionally recorded by the early modern sources across Britain and Ireland. The species appears to have been more widespread in Ireland and in the Midlands of England in the early modern period than it is today, but the lack of early modern records from Scotland is surprising.

RECOGNITION

This species is commonly recorded by the English name *ring-ouzel*, the Welsh names *mwyalchen* and *drynlhuan y graig*, the Irish *lon y leac*, and the New Latin names *Merula torquata* and *Merula saxatilis* (the two were considered to be different species by the early modern recorders, but actually both seem to be pre-Linnaean synonyms for the Ring Ouzel) (Roos 2015: 185 (n.49)).

The term *ouzel* by itself was generic in the early modern sources and often referred to, for example, the Blackbird (*Turdus merula*). More seriously, nineteenth- and twentieth-century translations of New Latin sources often wrongly translated *turdus* (a generic word for one of the thrushes) as *ouzel*, which sometimes leads to incorrect identifications of old records.

DISTRIBUTION

The distribution of top-quality records for this species is statistically a poor fit with the known levels of recorder effort. It may have been locally distributed, locally abundant or of special local interest. There are more records than we would expect from Leinster and Wales and none from Scotland, the South or South West of England, possibly indicating a local distribution. Habitat suitability modelling based on the sites where the Ring Ouzel was recorded in the early modern period suggests that the species may have had specific requirements. The lack of records from early modern Scotland suggests the species might have been unsuited to the colder and wetter winters. Yet Scotland is where the species is most numerous at present, and it was well recorded there in the nineteenth century (Holloway 1996: 316–17). The lack of early modern records from Scotland is therefore hard to explain. Based on how well recorded each region is, we would normally have expected 4 of the 10 records of this species to come from Scotland. However, no early modern populations are recorded, although Sibbald (2020: 17 (II:3)) notes it as present nationally. Baxter & Rintoul (1953: 192) suggest that it was included in Gordon's *Genealogical History of the Earldom of Sutherland*, but presumably this is a misunderstanding of the record of the 'blackbird or ossil' (Gordon 1813: 3) (almost certainly referring to a single species, probably *Turdus merula*). A more probable explanation is that the species was culturally of less interest to Scottish recorders, or that it was called by a name which has been overlooked in our texts.

In England, this Ring Ouzel is rarely mentioned by key sources; most of the locations are given by national authors writing bird handbooks. However, its distribution as recorded here does seem different to the distribution it has today. In Wales the distribution around Eryri matches the core range of the modern population (Smith 2021). In England the distribution recorded is more southerly – today it is mainly found as a summer breeder in the North rather than the Midlands. It was also recorded in the Midlands in the nineteenth century (Holloway 1996: 316–17), but has since been lost England south of the Peak District, perhaps due to the warmer summers brought on by climate change (Rollie 2007; Williamson 1975).

In Ireland although the Ring Ouzel is now very rare and mainly seen only in the far south-west and north-west, it was uncommonly seen across most counties in the nineteenth century (Holloway 1996: 316–17). The early modern records are therefore likely to be reliable.

Ring Ouzel records, 1519–1772.
There are reliable records from the
North and Midlands of England,
Ulster, Leinster and Munster in
Ireland as well as from Wales.

Merula Saxatilis, the Rock Ouzel … It is common on the high Mountains of Merionethshire, and Caernarvonshire in Wales, where it is called Muyalch y kraig, i.e. the Ouzel of the Rock, as it is in Irish, Lon y leac, in the same Sense. It is to be found on the Mountains of Mourne and Carlingford, and in Plenty on Mount-Leinster, in the County of Carlow.

Merula torquata, the Ring-Ouzel or Amzel … we conjecture that it is not of a Kind different from the immediately beforementioned Bird; but that they are Male and Female of the same Species, their Size, Shape, Places of Habitation, and Manner of feeding exactly agreeing.

(Harris and Smith 1744: 226–7)

The Ring Ouzel appears to have had a stable distribution through the nineteenth century, and was uncommon but widespread across all regions except South England, including populations in the Midlands (Holloway 1996: 316–17). It started to decline very quickly in the twentieth century, partially due to the loss of heather habitat in the uplands caused by afforestation, and perhaps due in part to an increase in warm summers brought on by climate change (Smith 2021; Balmer et al. 2013; Rollie 2007).

The Ring Ouzel still nests across much of the uplands of Britain, and in a few locations on the northern and southern coasts of the island of Ireland. It has continued to decline significantly since the 1960s, and has been lost from Exmoor, central Wales, and the western range of the southern uplands in Scotland (Balmer et al. 2013). It continues to be occasionally recorded on the coasts in winter.

SONG THRUSH

Turdus philomelos

NATIVE STATUS Native

MODERN CONSERVATION STATUS	
World	Least Concern
UK	Amber List
ROI	Green List
Trend since 1772	Uncertain

The Song Thrush was recorded by a few early modern sources around Britain and Ireland.

RECOGNITION

This species was most often called the *thrush* or *mavis* in Early Modern English, and occasionally the *throstle* or *thrissel*. Shrubb (2013: 168) has suggested that the term *thrush* was generic in the early modern period, and that it might have applied most commonly to the Blackbird (*Turdus merula*). This might be true in the household accounts, but in the more scholarly natural histories of Britain and Ireland consulted for our maps, the *Turdus* species were consistently distinguished in Early Modern English: the *throstle* or *thrush* is clearly distinguished from the *black-bird*, *redwing*, *fieldfare*, *ring-ouzel* or *missel-bird*.

The *Turdus* genus.

Under this name both the blackbird and starling are included: It is easily distinguished because it is medium in size, between a pigeon and a lark. It has a beak of mediocre length and thickness, moderately bent downwards. Its bone is yellow on the inside. It has a long tail. It lives from mixed berries and insects. It sings. Those we have are:

Song Thrush records, 1519–1772.
There are reliable records from every
region of Britain and Ireland except
Connacht and Lowland Scotland.

Turdus viscivorus; the Shreitch. [the Mistle Thrush]
The one called simply *Turdus* or *Viscivorus minor.* [the Song Thrush]
The splendid taste of its meat is recommended.   (Sibbald 2020 [1684]: 17 (II:3))

K'Eogh (1739: 88) attests the Irish terms *smo-logh* and *smologh-ceol* (literally 'song thrush', possibly a calque on the English term). In New Latin it is referred to as *Turdus,* or *Viscivorus minor,* to differentiate it from *Viscivorus major* (which is our Mistle Thrush, *Turdus viscivorus*). The term *Turdus* alone is not enough to distinguish this species and has not been included on this map – so that, for instance, when Lesley stated in the sixteenth century that the *Turdus* was barely found in Scotland this was translated by Lesley's translator Dalrymple as referring to the *feldifare* (*Turdus pilaris*) (Lesley 1675: 24, 1888: 40).

DISTRIBUTION

There are not enough top-quality records to statistically analyse the distribution of this species, but there are county-level records from almost every region of early modern Britain and Ireland, suggesting that the bird might have been widespread.

Like the other species of thrush, the Song Thrush was regularly caught for food. Sometimes it was kept and fattened up before being sold, and was used especially in pies (Shrubb 2013: 168–71). This does not seem to have affected population levels. The Song Thrush had a distribution approximately matching its modern range at the end of the nineteenth century. In fact, there is little evidence of any significant change in distribution between the early modern period and the present day, with the exception that there are no historical records of breeding in Shetland until the twentieth century (Holloway 1996: 320–1). Apart from occasional population declines after bad winters, its population seems to have been stable, despite the demand for the thrush as food increasing through to the twentieth century (Shrubb 2013: 14, 159, 168–71). Today, the Song Thrush is found all around Britain and Ireland and is especially common in the lowlands. The bird did decline as a breeder in these islands in the 1970s and 1980s, perhaps due to agricultural changes, especially increased drainage, and also increases in predation (Balmer et al. 2013). The population in Britain has now begun to recover, but in Ireland the decline continues.

WHEATEAR
Oenanthe oenanthe

NATIVE STATUS Native

MODERN CONSERVATION STATUS	
World	Least Concern
UK	Amber List
ROI	Amber List
Trend since 1772	Probably increased

The Wheatear was included by a handful of recorders around Britain and Ireland especially in south-east England.

RECOGNITION
This species was most reliably referred to as the *wheat-ear* in Early Modern English. According to the *Oxford English Dictionary* this name ultimately derives from **whiteeres* (white-arse), but by the early modern period it was generally called *wheat-ear*, and the etymology may have been forgotten – although the name *white-tail* was still occasionally used (e.g. Ray 1678: 233). The name *wheat-ear* was unique to the species. It was also called the *smich, smatch, chock, chac* and, most confusingly, the *stone-chaker* in Early Modern English; all of these latter terms have to be carefully distinguished from the bird that was referred to as the *stone-chatter* or *stone-smich* (*Saxicola torquata*, the Stonechat).

The New Latin term *oenanthe*, when used by itself, referred to this species, but when modified it could also refer to others – so that the Whinchat was called *Oenanthe secunda*, and the Stonechat was called *Oenanthe tertia*. A few authors also give the New Latin term *caeruleo* which could lead to confusion with (*Parus*) *caeruleus*, which referred in early modern Britain and Ireland to the Blue Tit.

DISTRIBUTION
The distribution of top-quality records for this species is statistically a poor fit with the known levels of recorder effort. It may have been locally distributed, locally abundant or of special local interest. There are far more records from the South of England than we would expect, suggesting the Wheatear might have been more widely distributed or more abundant in this region. Habitat suitability modelling

Wheatear records, 1519–1772.
There are reliable records from the
South, South West and Midlands
of England, and the Northern
and Western Isles of Scotland.

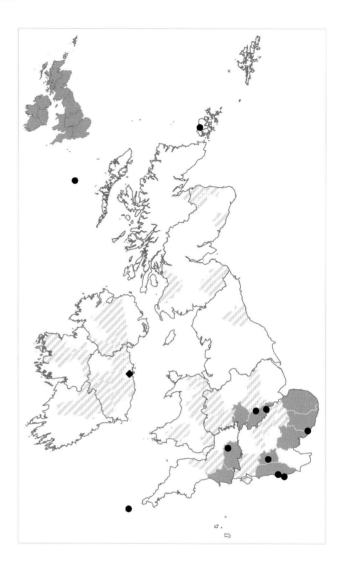

based on the sites where the Wheatear was recorded in the early modern period suggests that the species may have had specific requirements. It is only recorded on lowland sites which today have relatively mild, dry winters.

The distribution recorded by the early modern sources does not match with the modern distribution of this species at all. Today, the species mainly breeds in northern and western Britain and northern, north-eastern and western Ireland; apart from passage migrants, it is now rarely seen in South England. If these records have not been misidentified, this implies that the species' distribution has changed drastically over the last few centuries. In the past, this was explained due to the warmer climate of the first half of the twentieth century (Williamson 1975). However, it is also possible that it has followed the Rabbit (Raye 2021c). Wheatears are dependent on Rabbits in the lowland portions of their range (Kämpfer and Fartmann 2019), and often nest in abandoned warrens (as the early modern recorders noted). In early modern Britain and Ireland, though, Rabbits were almost exclusively found on the coasts and islands. The exception is the South of England, where Rabbit warrens could be encountered inland, as we have shown, and where Wheatears nested.

The Fallow Smich or Wheat Ear: *Oenanthe sive vitiflora*, Aldrovandan. There are not only great Numbers of them upon the Downs in Sussex, as Mr. Willughby observes; but in Northamptonshire too, and that not only in the Clay-land Fields as about Oxendon, but in those of a lighter Soil, as about Oundle, where some call them Clod-Hoppers. In our Red-land they nest in forsaken Coney-burrows, and in the Clefts or Intervals of the Keal and Quarry-stone.

(Morton 1712: 427)

The Wheatear's population in South England was lost in the second half of the twentieth century, due to habitat loss caused by the outbreak of myxomatosis which killed Rabbits in the lowlands, and decreased densities of sheep in the uplands (Balmer et al. 2013).

CROSSBILL

Loxia curvirostra

NATIVE STATUS Native

MODERN CONSERVATION STATUS	
World	Least Concern
UK	Green List
ROI	Green List
Trend since 1772	Uncertain

The Crossbill was recorded by sources as ordinarily present in East Anglia, and there are records of irruptions elsewhere in south Britain and Ireland.

RECOGNITION

There are three species of Crossbill that commonly nest in these islands. The Common Crossbill, the Parrot Crossbill (*Loxia pytyopsittacus*) and the Scottish Crossbill (*Loxia scotica*), the latter a species endemic to Scotland. These three are almost impossible to distinguish in the field or by genetic testing but differ in habitat, diet and call, and generally do not interbreed (Yalden and Albarella 2009: 41–2; Summers 2007). The early modern sources did not distinguish the different species of crossbill; however, the Parrot Crossbill seems to have only colonised in the modern period, and the Scottish Crossbill is confined to Scottish pine forests, so since there are no local records from Scotland, all the early modern records on our map are likely to refer to the Common Crossbill. This species was called *cross-bill* and *shell-* or *sheld-apple* in Early Modern English and *loxia* or *curvirostra* in New Latin.

DISTRIBUTION

The distribution of top-quality records for this species is not statistically different from the known level of recorder effort. The gaps on the map may just reflect decreased survey effort in some regions. It may have been widespread.

The early modern sources record the Crossbill breeding only in East Anglia, including Norfolk and Essex. It is unclear why the Scottish Crossbill went unrecorded. The Scottish Crossbill's habitat is pine, and during the early modern period pine was relatively common, albeit mainly found on poorer soils in Highland Scotland, which was a well-recorded region. Pinewoods were used as commons for grazing livestock and collecting wood for tools, construction, fuel and light (candles) (Smout 2009), so were regularly visited by humans. The national record by Sibbald (2020: 18 (II:3)) uses the terms mentioned above, so the Crossbill was presumably not called by a different name or unrecognised in Scotland.

Crossbills regularly go through population irruptions where large flocks migrate, sometimes to distant areas where they are not normally seen. These seem to have been thought especially worthy of recording by our key sources, particularly when they occurred in Ireland, and form most of our records, but they do not represent the normal breeding range of the species. These records have been shown on the map as diamonds:

> They are indeed sometimes, but very rarely, found in the Western Parts of England about Worcester, driven thither (as they were in the Winter of 1707 into the County of Down) by some accidental Winds and Weather from the colder Climates of Europe, perhaps from the North of Scotland, (where we are told by Dr. Sibbald that they are found) rather than from Norway or Denmark, as some have though. They are not Natives of this County [Down]; yet many of them were seen at Waringstown in Winter 1707.   (Harris and Smith 1744: 229)

The distribution of the Crossbill in Britain and Ireland significantly expanded in the nineteenth century. Irruptions have allowed it to quickly exploit new environments. It possibly became established in Ireland for the first time following an irruption in 1838, and bred in several counties by the end of the nineteenth century (Holloway 1996: 412–13). The species actually declined in Ireland

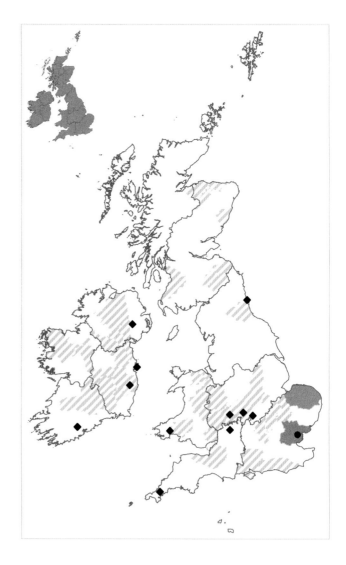

Crossbill records, 1519–1772. There are records from every region of Britain south of Scotland, and from every region of Ireland except Connacht, but most of these records are of irruptions. A permanent population seems to have lived in South England.

in the twentieth century and was lost as a breeder in the middle of the century (Balmer et al. 2013; Parslow 1973: 176–7), before recovering in the second half of the century after further irruptions. It now breeds in all four regions.

The Crossbill has also significantly expanded its breeding range in Britain since 1968–72 (Balmer et al. 2013). It breeds in coniferous forests, and benefited significantly from the forestry plantations created after the Second World War (Pritchard 2021s; Williamson 2013: 154). As well as Ireland, the Crossbill colonised Wales in the twentieth century and has now significantly increased its range across Scotland and England.

FRESHWATER LAMPREYS

River Lamprey (*Lampetra fluviatilis*) and Brook Lamprey (*Lampetra planeri*)

NATIVE STATUS Native

MODERN CONSERVATION STATUS	
World	Least Concern
UK	Not Evaluated
ROI	Least Concern
Trend since 1772	Probably increased

The two freshwater lamprey species were not distinguished in the early modern sources. They are described as living across much of England and parts of Ireland and Wales.

RECOGNITION

These species were most commonly called *lamprey* in Early Modern English and *lampetra* in Latin. O'Sullivan (ed. 2009: 174) attests the Irish *luimpraea*, and Morris gives the Welsh terms *llyswen bendoll*, *lleprog* and *lamprai* (Morris 1747). When used alone, these terms are not specific and could equally refer to Sea Lamprey (*Petromyzon marinus*) which also enters river systems. Records using these terms only are shown as diamonds on our map because of this uncertainty.

The reliable names used for the freshwater lamprey were *lampern* in Early Modern English (a term occasionally still used to distinguish the species in modern times (Wheeler 1979: 64–5)) and *Lampetra parva fluviatilis* (or some close variant) in New Latin. Other regional and specific terms are also attested in Early Modern English like *suckstone*, *stone-grig* and *pride*. These all appear to distinguish the freshwater lampreys from the Sea Lamprey. However, our authors almost never distinguish between the River Lamprey and the Brook Lamprey, so these terms can only be said to refer to the freshwater lamprey species.

DISTRIBUTION

For the obligate freshwater riverine species, records at river-level have been shown on the map by shading every county with similar topography that river runs through.

The distribution of top-quality records for this species is statistically a poor fit with the known levels of recorder effort. It may have been locally distributed, locally abundant or of special local interest. There are no reliable records of the species from Scotland or Ulster. Habitat suitability modelling based on the sites where the freshwater lampreys were recorded in the early modern period suggests that these species may have had specific requirements. They are only recorded on lowland sites which today have relatively mild winters and warm summers. An equally widespread distribution for this species would also not fit with the river-level records from England and Wales.

Thinking about the freshwater lampreys in Scotland, Sibbald (1701) includes what he calls *Lampetra fluviatilis* in his unpublished *Account of the fishes and other aquatick animals taken in the Firth of Forth*, but unlike for most of the species in that text, the record is not specific about location and

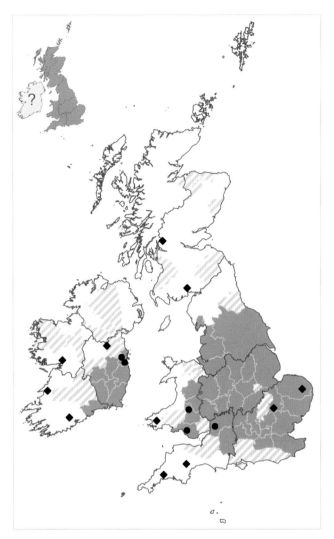

Lamprey records, 1519–1772. There are reliable records from every region of Britain south of Scotland and from Leinster and part of Munster in Ireland. There are also unreliable records from Connacht and Lowland Scotland.

seems to indicate that the species is found somewhere in Scotland. He also includes only what he calls the *sea-lamprey* in his *History, Ancient and Modern, of the Sheriffdoms of Fife and Kinross* (ed. Sibbald 1803), without any records of freshwater lampreys.

Many freshwater species were absent from Scotland in this period, because the rivers of Scotland were not connected to the rivers of continental Europe during the last glacial period like the rivers of the south east of England were (Maitland 1977 1987; Wheeler 1974). However, the freshwater lampreys are not likely to have been among these because the River Lamprey lives in the sea as an adult and easily colonises new areas, and the Brook Lamprey has a high tolerance for cold conditions, so might have survived the glacial period in Scotland. Both species were also mentioned immediately after our period in records from the *Old Statistical Account of Scotland* (Maitland 1977). The most likely explanation for our evidence is that the freshwater lampreys were less common in Scotland or of less cultural interest.

The River Lampreys that live in Loch Lomond do not migrate into marine environments, unlike other River Lampreys. They are smaller, darker and have a larger head, eye and sucker, and presumably might represent a population that was isolated during the last glacial period (Maitland 2007: 71–7; Wheeler 1974). There is one very tentative record which might refer to this population in one of our texts:

> About fourteen miles hence is a meare or lake called Loemund, in Perth, wherein are the flitting islands, which move (my host, Mr. Fleemeing, affirmed he hath seen it): it is most rough in calm weather; the fish are without fins. (Brereton 1844 [1634–35]: 113)

The identification of the species here is on the basis that few other freshwater fishes lack fins.

Lampreys were widely eaten in the medieval and early modern periods, and one London ordinance established a close season in 1630 (Locker 2018: 47–8, 67–70; Mac Laughlin 2010: 72; Wheeler 1979: 64–5). From the seventeenth century, they also increasingly came to be used as bait for marine fishing (Almeida et al. 2021; Wheeler and Jones 1989: 62).

The freshwater lamprey species are likely to have benefitted in the modern period from the creation of the canal network across Britain (Wheeler 1974). At the same time, however, they were being increasingly caught for use as bait. From the 1780s, this fishery amounted to several hundred thousand River Lampreys caught each year for use in the North Sea fisheries (Almeida et al. 2021). This level of fishing may have been unsustainable: by the end of the nineteenth century the lamprey fishery had collapsed, probably due to overexploitation as well as increased pollution.

The freshwater lampreys are now widespread and recorded in every region of Britain and Ireland (Henderson 2014: 4–5; Davies et al. 2004: 46–7; Igoe et al. 2004).

STURGEONS

European Sea Sturgeon (*Acipenser sturio*) and Atlantic Sturgeon (*Acipenser oxyrinchus*)

NATIVE STATUS Native in Britain, Extinct in Ireland

MODERN CONSERVATION STATUS	
World	Critically Endangered
UK	Not Evaluated
ROI	Extinct
Trend since 1772	Certainly declined

There were reliable records of sturgeons from the coasts of Britain and Ireland, and records from the rivers further inland in parts of North, South and the Midlands of England.

RECOGNITION

As well as European Sea Sturgeon (*Acipenser sturio*), there was a second species of sturgeon present in Europe in the early modern period. This was the Atlantic Sturgeon (*Acipenser oxyrinchus*), formerly thought of as the American Sturgeon. This fish seems to have colonised Europe (including Britain and therefore likely also Ireland) by sea in the medieval period and went extinct in these islands after the end of the early modern period (Locker 2018: 53–4; Ludwig et al. 2009).

The names used for the sturgeons in our sources are most commonly the *sturgeon* in Early Modern English, *ystursion* in Welsh and *sturio* or *acipenser* in Latin. These terms all appear to be generic – they probably referred to either sturgeon species. Willughby and Ray do distinguish a second species of sturgeon, although it is not clear if they are truly distinguishing the two species, or just two different specimens of a single species:

> The [sturgeon] fish which is described by the learned Willughby was 13 inches long, and with its dorsal fins 8.5 inches from the end of the beak, the nose half an inch, the eyes 1 7/8ths of an inch, the mouth 2 inches, the first section of the fins 3 and a half inches, the second section 8 inches, the vent 8 3/8ths of an inch, the anal fins 9 inches.
>
> In addition, we note another sturgeon either a species or merely a sex, different from the description, with a much shorter beak, a larger and broader head, and a colour less black and more grey-blue.
> (Willughby & Ray 1686: 239–40)

Occasionally, the sturgeon is also described as a *royal fish* – or *piscis regis* in Latin or *bradan ri* in Irish – referring to the royal right to all sturgeons caught in British waters (Maitland 2007: 89). However, since this term was also used for cetaceans, where it is not employed alongside more reliable names described above such records are given as diamonds on the map to show that they are unreliable.

DISTRIBUTION

The distribution of top-quality records for this species is not statistically different from the known level of recorder effort. The gaps on the map may just reflect decreased survey effort in some regions. It may have been widespread.

Sturgeon records, 1519–1772.
There are records from around the
coasts of Britain, except mainland
Highland Scotland. The record from
Wales is unreliable. There are also
records from Leinster and Munster
in Ireland. There are inland
records from the South, North
and Midlands of England.

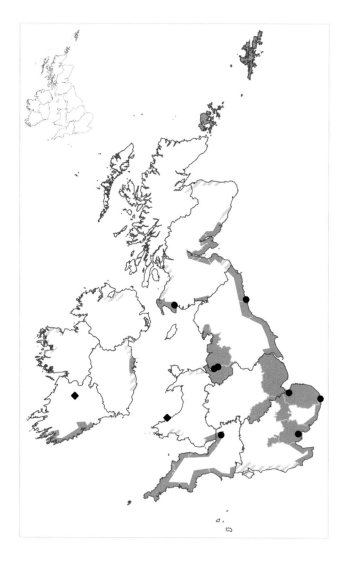

Sturgeon seem to have been more common in the early modern period than they are today. The early modern sources provide 21 reliable local, river- or county-level records from early modern Britain and Ireland. Yet sturgeon populations seem to have already declined by the early modern period due to pollution and blocked waterways (Locker 2010). The early modern sources make it clear that the species was rare rather than common in the areas where they described it, and today sturgeon are far more common on the coasts, and only enter the lower parts of rivers (Maitland 2007: 91). Therefore, to avoid overestimating the range of the species, unlike the other freshwater fishes, the distribution has been mapped conservatively, so that records of sturgeon in rivers have been assumed to apply only to the estuary unless a location is mentioned further upriver – the fish is likely to have occasionally been found inland in some of the counties where only the coasts have been shaded. Nonetheless, the sturgeon appears to have been occasional rather than common even in the range shown here.

> The Sturgeon, a Fish, that for its Magnitude, and the Rarity of meeting with it in an Inland
> County as this [Northamptonshire] is, may justly challenge the First Place. I have certain
> Information of Three Sturgeons that have been taken in the River Nyne above Peterborough.
> One at Allerton-Mill, Seven Feet, Nine Inches and Half in Length 123 l[b]. in Weight, on April 7th
> 1682. Another nigh Castor-Mill, Eight Feet Two Inches long, Three Feet Three Inches Deep, and

152l[b]. Weight, on May 20th 1698. Draughts of these Two in thei[r] just. proportions are still to be seen in Mr. St. John's House at Longthorpe. And a Third more lately, Octob. 23rd 1704, nigh Allerton-Mill, that was Five Feet Nine Inches in Length. 'Tis reported, that there once was a Sturgeon taken at Wadenho-Mill, another nigh Denford in the same River.

(Morton 1712: 419–20)

Today, apart from isolated populations, the European Sea Sturgeon is only known to breed in the Gironde river system in France (Handford 2004). It has been suggested that the fish formerly bred in the River Thames and the River Severn. There are early modern records from these watercourses, and elsewhere (Smyth 1885: 319; Griffiths and Binnell 1758: 209–11), as the map shows. It is not clear whether the sturgeon bred in the shaded areas of the map of early modern Britain and Ireland or just visited – visiting sturgeon may have been more common before the decline of the species.

The sturgeon was commonly eaten in elite settings in medieval and early modern England (Locker 2018: 47–8, 67–70). There was increasing demand for sturgeon and its roe in the nineteenth century, and great quantities of both were imported to London (Locker 2010). Overfishing, together with pollution, habitat deterioration and river blockage are likely to have been the causes of the catastrophic decline in the sturgeons (Handford 2004). The Atlantic Sturgeon is now extinct around Britain, while the European Sturgeon is critically endangered and exceptionally rare even as a visitor. The *Database & Atlas of Freshwater Fishes* contains records of only 13 individuals from the whole of the twentieth century (Handford 2004). Catches of this fish now usually make the news (Maitland 2007: 88). This represents a significant decline since the early modern period, when the sturgeon was widely recorded, albeit rare, in the seas and rivers around Britain.

BREAMS

Common Bream (*Abramis brama*) and Silver Bream (*Blicca bjoerkna*)

NATIVE STATUS Native to Britain, not Ireland

MODERN CONSERVATION STATUS	
World	Least Concern
UK	Not Evaluated
ROI	Not Evaluated
Trend since 1772	Probably increased

The freshwater bream species were recorded in the early modern sources across much of lowland England, Ireland and – surprisingly – at one site in Scotland.

RECOGNITION

This map shows records of what was most reliably called *Cyprinus latus* in New Latin. In Early Modern English this species was referred to as the *bream*, which is unfortunate because this term was also used for a few marine species, sometimes distinguished as *sea-bream* in our sources (e.g. our Red Porgy *Pagrus pagrus* and Ray's Bream *Brama brama*). Our map shows inland records, records in freshwater habitats and references from lists of freshwater species only, so it is unlikely that there is any confusion with the marine species.

To add to the possible uncertainty, though, the early modern authors do not distinguish Common Bream from Silver Bream, so the records from England might refer to either species. The most likely records to be confused are those from the South and Midlands of England. The records from elsewhere are more likely to refer to Common Bream only, since Silver Bream is thought to have been found only in the Midlands and Yorkshire during the early modern period (Peirson 2004).

DISTRIBUTION

For the obligate freshwater riverine species, records at river-level have been shown on the map by shading every county with similar topography which that river runs through.

The distribution of top-quality records for these species are statistically a poor fit with the known levels of recorder effort. They may have been locally distributed, locally abundant or of special local interest. These species seem to have been regionally absent or less abundant in some areas. Most obviously the bream was absent from parts of Connacht, but also, given the number of records from lowland England, also perhaps Highland Scotland and South West England. Habitat suitability modelling based on the sites where bream were recorded in the early modern period suggests that these species may have had specific requirements. They were not recorded in lowland areas with the coldest winter or summer temperatures. This fits with the ecology of the fishes as specialists of lowland, high-nutrient, slow-flowing rivers and lakes (Lyons 2004; Peirson 2004).

Both Common Bream and Silver Bream are obligate freshwater species and could not have survived the icy conditions of the last glacial period (McCormick 1999; Wheeler 1974). At the end of the last glacial period, only south-east England is believed to have been connected to the rivers of Europe for long enough to permit recolonisation, meaning that returning obligate freshwater species may have been unable to reach Scotland or Ireland. This makes some sense of the early modern distribution. Based on this model, and the lack of references in the *Old Statistical Account of Scotland*, Maitland (1977) suggests that both species were only introduced to Scotland in the nineteenth century. There is one early modern record for Scotland from Castle Loch in Lowland Scotland, made in about 1683 by a Dr [George?] Arch[i]bald as a contribution to Robert Sibbald's *Scotia Illustrata*:

> In that Castle Loch at Lochmaben are various Fishes, besides these two formerly mentioned;
> Pikes, Green backs, Breams, Vetches, Pearches with some others.
>
> (Archibald in Mitchell 1908: 188)

This list of species would have made Castle Loch attractive to anglers in the early modern period, as it is today. Sibbald clearly trusted the record, since he listed the bream as a Scottish species in his *Scotia Illustrata* (2020: 25 (II:3)). If this record is reliable, it is not clear how or when the bream came to be established on the site. It may have been introduced for anglers. The bream was already commonly being transported and kept in domestic ponds for sport/food by the early modern period

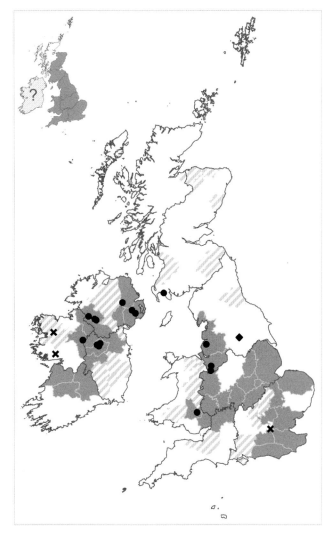

Bream records, 1519–1772. There are records of wild populations in every region of Britain south of Scotland and every region of Ireland.
There is a record of a possibly captive population in Lowland Scotland.

(Hoffmann 1996), as for example the fish noted at Temple Newsam House by William Brereton in 1635 (ed. 1844: 73) – this record is shown as a diamond on our map. Breams were not mentioned later in the *Old Statistical Survey* of Lochmaben published in 1793, but were mentioned in the *New Statistical Survey* published in 1845, which led Maitland to suggest they were introduced to Castle Loch between the two surveys (Maitland 1977; Sinclair 1793: 236–7). However, given that the *Old Statistical Survey* only identifies three of the 'fifteen or sixteen' species known to be present on site in this time, this does not necessarily mean that breams were absent in the eighteenth century; the record from the seventeenth century suggests the fish may have been there but simply not recorded.

The early modern distribution in Ireland is also of interest. For the same reason as with Scotland, it is often thought that no obligate freshwater species are native to Ireland (King et al. 2011; Maitland 2004; McCormick 1999; Griffiths 1997). Yet the number of records from early modern Ireland implies that the Common Bream was already well established here by the early modern period. The first record from the Irish texts in the atlas database is from the first quarter of the seventeenth century (Gainsford 1618: 146) and it is mentioned in seven additional texts before the end of the century. The distribution of records from the early modern period almost exactly matches the distribution of records from the middle of the twentieth century in Ireland, before the fish was introduced to additional rivers by the Inland Fisheries Trust (Kennedy and Fitzmaurice 1968). To explain this, it should be pointed out that bream was in some places the first fish widely adopted for use in ponds during the medieval period, soon to be replaced by Pike, Tench and Roach, then Common Carp (Locker 2018: 55). Bream populations may have been imported to Ireland before the other species because of this.

Today the freshwater bream species are relatively widespread in lowland England and Ireland. The Silver Bream is confined mainly to eastern England, but the Common Bream is more widespread, although still rarely recorded in Scotland, the North and South West of England, Wales and Leinster (Lyons 2004; Peirson 2004). The early modern map matches the areas where the Common Bream is now most widespread. Both species seem to be stable.

Both bream species may have been able to colonise additional waterways due to creation of the canal network in the nineteenth century (Wheeler 1974). Common Bream have also been introduced to additional sites in South Wales, south-west Scotland, Devon, Cornwall and the Channel Islands in the modern period, as ornamental fish and for anglers (Lyons 2004).

CARP

Common Carp (*Cyprinus carpio*) and Crucian Carp (*Carassius carassius*)

NATIVE STATUS Not Native (naturalised)

MODERN CONSERVATION STATUS	
World	Least Concern (*C. carpio* is Vulnerable)
UK	Not Evaluated
ROI	Not Evaluated
Trend since 1772	Probably increased

Carp populations were recorded by our early modern sources in the south of both Britain and Ireland but not further north.

Carp species records, 1519–1772. There are records from the South, South West and Midlands of England, and part of North England. There are also records from Munster and Leinster in Ireland. Some records show captive populations only, and the counties they are in have not been shaded.

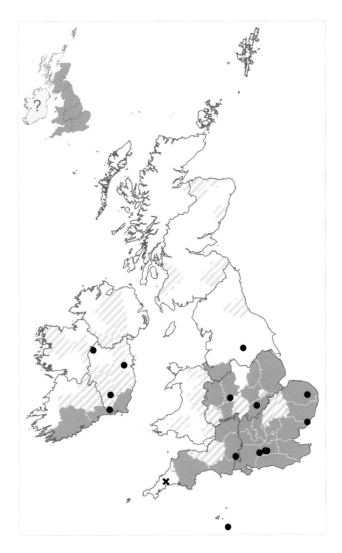

RECOGNITION

This map shows populations of what were called *carp* in Early Modern English and *cyprinus* or rarely *carpio* in New Latin; this must be distinguished from *Cyprinus latus*, now the Bream. Unfortunately, the early modern sources do not distinguish between different kinds of carp. At least two species are known to have been present in the early modern period, the Crucian Carp and the Common Carp. The Grass Carp (*Ctenopharyngodon idella*) was only introduced comparatively recently and so is not intended here, but the other two species cannot be easily distinguished in the early modern texts, except that populations in riverine environments are likely to be Common Carp (Bolton 2004a,b). This means that some records cannot be identified to either species:

> [I]n the drowned Coal-pit-open-works S[outh] W[est] of Wednesbury [in Staffordshire], into which Pike, Carp, Tench, Perch, &c. being put for breed, they not only lived, but grew and thrived to as large a magnitude as perhaps they would have done any where else, and were to the palate as gratefull. (Plot 1686: 241–2)

DISTRIBUTION

For the obligate freshwater riverine species, records at river-level have been shown on the map by shading every county with similar topography which that river runs through.

The distribution of top-quality records for these species is statistically a poor fit with the known levels of recorder effort. They may have been locally distributed, locally abundant or of special local interest. There are far more records than expected from South England. For these species only counties with populations which seem to be freely living in natural environments have been shaded. The pattern of these records is especially clear in Scotland, where several sources were deeply interested in recording fish populations, but no carp were recorded. Habitat suitability modelling based on the sites where the carp species were recorded in the early modern period suggests that they may have had specific requirements. They are only recorded on sites which today have relatively warm summers and relatively mild, dry winters.

The references collected from the early modern sources represent some of the first records of both species in these islands. Both seem to have been introduced (Jeffries et al. 2017; Williamson 2013: 16; Locker 2010; Lever 2009: 244; Maitland 2007: 132–6). The Crucian Carp has usually been considered in the past to be native to south-east England, but genetic analysis suggests this species was actually introduced to Britain in the fifteenth century. In the early modern period, it was especially prized for keeping in ponds in cooler temperatures because it does not need the same high temperatures to breed as the Common Carp (Svanberg and Cios 2014). Later it seems to have been introduced to additional sites, especially as a sport fish for anglers. The Common Carp is native to south-eastern Europe and was, curiously, transported across the rest of the continent in wet grasses, rushes or straw and introduced to additional waterbodies throughout the medieval period (Locker 2018: 59–61; Hoffmann 1994; Bond 1988). The earliest historical records of the fish in Britain are from the fourteenth century. It was most probably transported from the Low Countries to Britain as part of the commercial fishery and fishpond businesses at, for example, the Stews in Southwark in London (Locker 2010; Currie 1991). It was then transported to Ireland by 1626 at the latest, but possibly much sooner after it was introduced to Britain (Brazier et al. 2012). The Common Carp was especially valued as a food. By the end of the sixteenth century, it had become the most important species for household ponds and stews because it matures quickly and can tolerate water with very little oxygen, but by the end of the period it was also kept for its aesthetic value.

The core modern range of the two carp species in Britain is approximately the same as the early modern range recorded in our sources, but the species are now also found on sites in Cornwall, Wales, North England and Lowland Scotland, outside of this core range (Bolton 2004a,b). In Ireland, due to introductions from 1950 onwards, the Common Carp (but not the Crucian Carp) is now present in Ulster and Leinster as well as Munster and there are a few sites known from the east of Connacht (Brazier et al. 2012; Lever 2009: 243–4). This change includes introductions by anglers in the nineteenth and twentieth centuries, possibly aided by the temperature increases between the Little Ice Age in the early modern period and global warming in the modern period (Maitland 1977, 1987, 2007: 130–1, 136). Both Common and Crucian Carp require water temperatures in excess of 18°C (Crucian Carp 20°C) to breed. This limits the breeding of all carp populations in these islands, but especially some Scottish and Irish populations (Locker 2010; Lever 2009: 241–4), and the latitudinal line at which rivers reliably reach this temperature during summer in Britain and Ireland is likely to have been further south in the early modern period. The Common Carp also seems to have benefitted from the creation of the canal system in the modern period (Wheeler 1974). The distribution of the Common Carp seems to be relatively stable, although it likely only maintains its population levels in some areas due to artificial restocking. The Crucian Carp is threatened due to competition with the Common Carp, hybridisation with Common Carp and Goldfish and the introduction of the Asian Tapeworm (Bolton 2004a).

ROACH
Rutilus rutilus

NATIVE STATUS Native to Britain, not Ireland

MODERN CONSERVATION STATUS	
World	Least Concern
UK	Not Evaluated
ROI	Not Evaluated
Trend since 1772	Uncertain

The Roach was recorded by early modern authors mainly in England. It was not recorded in Scotland.

RECOGNITION
This species is generally called the *roach* in Early Modern English. It was also referred to occasionally in New Latin as the *rutilus*, *rubellus* and once the *aulecula*. All of these terms were also used for the Rudd in Irish English. Since the Roach is not currently believed to have been present in early modern Ireland, all records using the terms from Ireland have been included on the map for the Rudd and are not included here. It is possible that some of the records using the terms from Britain were also intended to refer to the Rudd.

From a conservative standpoint, the only reliable records on our map would be from Norfolk and from the Rivers Thames and Trent, where the sources confirm both *roach* and *rudd* to be present. In practice though, the confusion in terminology seems to have been mainly confined to Ireland. It is analogous to the confusion about the Irish *weasel*. As we saw in 'Polecat' (p. 54), since the Least Weasel is absent from Ireland, the Stoat is often referred to as a weasel. Similarly, since the Roach was traditionally absent from Ireland, the Rudd was referred to as a *roach*.

DISTRIBUTION
For the wholly aquatic riverine species, records at river-level have been shown on the map by shading every county with similar topography which that river runs through.

Since the records of the Roach in Britain and Ireland have been identified based on location, it would be a circular argument to statistically analyse the species' former distribution between the two islands.

Wheeler (1974) places the Roach in the group of obligate freshwater fishes that cannot tolerate ice, and which colonised Britain from eastern England via the land bridge. The species appears to have been able to distribute relatively well, and Roach remains were, for instance, part of the archaeological record of medieval York (O'Connor 1988: 114).

Roaches were commonly eaten in the past, as they are today. Because, like Common Carp, they can live in water with little oxygen, they were popularly kept in stews (fish holding pools) for later consumption in the medieval and early modern periods (Locker 2010 2018: 47–8, 49–50, 55, 67–70; Hoffmann 1996), and escapes from captivity might have aided their distribution across south Britain. They are also one of the most commonly excavated fishes from medieval sites, and ordinances were passed to protect them by the Corporation of London as early as the seventeenth century (Wheeler 1979: 66–7).

In Scotland, it is worth noting that there are no early modern records of either Roach or Rudd. The modern populations here may have been introduced later either purposefully for sport, or accidentally as bait used to catch Pike (Locker 2010). However, Maitland points out that the species is listed in Loch Lomond (and possibly elsewhere in south-western Scotland) in the *Old Statistical Account of Scotland* in 1796 (Maitland 1977, 2007: 163, 253; Sinclair 1796). This suggests either that a population in south-western Scotland was overlooked in the early modern sources, or that the species was introduced there at the end of the eighteenth century. Either way, through the nineteenth and twentieth centuries the fish was introduced to additional sites on the east coast of Scotland. It also benefitted from the creation of the canal network from the end of the eighteenth century onwards, which linked together various lowland waterways and allowed the Roach to increase its distribution (Wheeler 1974). The species has also now been introduced to Ireland and became widespread in the 1960s and 1970s (Carter 2004a).

Today the Roach is widely recorded across lowland England, Wales, Scotland and Ireland, but less well recorded in upland North England, Highland Scotland, inland Wales and Cornwall (Carter 2004a). It is resistant to pollution and was one of the few river species that did not significantly decline in the eighteenth and nineteenth centuries (Williamson 2013: 161). Its population seems to be stable.

Roach records, 1519–1772.
There are records from every region
of Britain south of Scotland.

RUDD

Scardinius erythrophthalmus

NATIVE STATUS Native to Britain, not Ireland

MODERN CONSERVATION STATUS	
World	Least Concern
UK	Not Evaluated
ROI	Not Evaluated
Trend since 1772	Uncertain

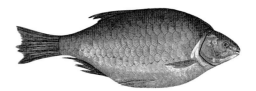

Our sources recorded the Rudd in Ireland and the east of England, but not in Scotland or Wales.

RECOGNITION

This species was most reliably called the *rudd* or *finscale* in Early Modern English. Yet all of the references provided on this map for Ireland instead refer to the species as the *roach*. The species we now call Roach is not known to have been present in Ireland in the early modern period, and therefore all references to the Roach in Ireland have been identified instead as the Rudd. It is very possible that in addition to these, some of the records of *roach* from Britain might also refer to the Rudd, but no British records referring to the *roach* have been included on this map. The abundance of records from Ireland compared to Britain might thus be due to identification bias. However, assuming that the Roach was not present in Ireland in the early modern period, every record on this map should be reliable.

The Early Modern English term *cock-roche* (=our cockroach, *Blattodea* spp.) is rarely mentioned by the sources, and records of these species are easily distinguished from records of the Roach because they are not found in water.

DISTRIBUTION

For the obligate freshwater riverine species, records at river-level have been shown on the map by shading every county with similar topography which that river runs through.

Because the records of the Rudd from Ireland have been identified partly based on location, it would be a circular argument to analyse the distribution of the species based on these same records. In Britain there are not enough records to infer the range, although the lack of records from Scotland, Wales and South West England might agree with the argument (Maitland 2007: 169; Burrough 2004) that the native range of the Rudd in Britain is limited to east England; the fish has been recently spread by anglers outside of this area. Maitland's search of historical sources found no evidence of the Rudd in Scotland before its introduction in 1912 (Maitland 1977).

The Rudd is usually thought to have been introduced to Ireland (King et al. 2011; Moriarty and Fitzmaurice 2000; McCormick 1999), although it is in an anomalous group of freshwater fishes (along with the Gudgeon, Loach and Perch) for which there is not a clear introduction date (Kennedy and Fitzmaurice 1974). As with the Pike and the Common Bream, it seems to have been already widely distributed in Ireland by the early modern period. The earliest Irish record is from 1682 (Dowdall in Gillespie and Moran 1991: 210), and the Rudd was recorded in Ireland by four more texts before the end of the century. The fish needs temperatures of at least 15°C to breed (Maitland 2007: 170), which would seem to preclude it surviving through the glacial period in Ireland even on the south coast below the extent of the ice sheets.

Rudd records, 1519–1772.
There are records from every region
of Ireland and from the South,
North and Midlands of England.

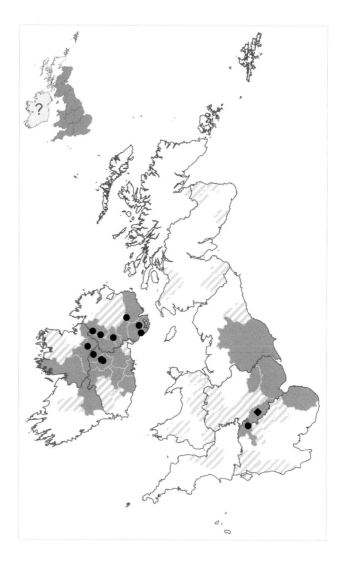

Today the Rudd is recorded mainly in the east of England, and in the north and south of Wales, Somerset, northern England, Dumfries & Galloway (Burrough 2004). For the most part the fish has been introduced to these additional sites by anglers for sport, but presumably it might have also benefitted from the creation of the canal system at the end of the eighteenth century (Wheeler 1974). It is now recorded widely in Ireland. The species is threatened by habitat deterioration and eutrophication of lakes. In Ireland, the Rudd has been declining due to the introduction of the Roach (Maitland and Campbell 1992).

TENCH
Tinca tinca

NATIVE STATUS Native to Britain, not Ireland

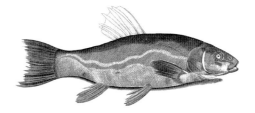

MODERN CONSERVATION STATUS	
World	Least Concern
UK	Not Evaluated
ROI	Not Evaluated
Trend since 1772	Probably increased

The Tench was recorded in our early modern sources across much of England, and Munster and Leinster in Ireland. There are no early modern records from Wales or Scotland.

RECOGNITION
This map shows records identified as *tench* in Early Modern English and *tinca* in New Latin. These terms are generally specific to the Tench. Rutty (1772: 368) and Thomas (ed. Dodsley 1775: 9) do distinguish a *sea tench* (possibly Black Seabream, *Spondyliosoma cantharus*), but all the records included on this map appear to be to in the context of freshwater habitats, as in this list of fish in the River Thames:

> How remarkably good its Salmon! what fine large Flounders, Smelts, Shads, Trout, Graylin, Perch, Carp, Tench, Barbell, Chub, Roach, Dace, Gudgeon, Pike, and other Fish, as Eel, Lampreys, Bleak, Ruffe, &c. (too many to mention) are there caught above London-Bridge; nay and oftentimes Sturgeon.   (Griffiths and Binnell 1758: 20–21)

DISTRIBUTION
For the obligate freshwater riverine species, records at river-level have been shown on the map by shading every county with similar topography which that river runs through.

The distribution of top-quality records for this species is statistically a poor fit with the known levels of recorder effort. It may have been locally distributed, locally abundant or of special local interest. There are more records than expected from the North and Midlands of England and from Leinster, but none from Scotland. Habitat suitability modelling based on the sites where the Tench was recorded in the early modern period suggests that the species may have had specific requirements. It is only recorded on sites that today have relatively mild, dry winters and warm summers, although in this case the distribution might be more simply explained by a lack of opportunity to colonise other areas. Maitland's search of the historical record suggests the first introduction of the Tench to Scotland was in the nineteenth century (Maitland 1977).

The Tench also seems to have been introduced to Ireland. It may have been introduced via medieval and early modern fishponds along with the Common Carp and Common Bream (Maitland and Campbell 1992; Kennedy and Fitzmaurice 1970). The first record in the early modern Irish sources is from 1682 (Dowdall in Gillespie and Moran 1991: 210), but this is the only record in the database from the seventeenth century. The species continued to have a patchy distribution here into the twentieth century.

The Tench was routinely eaten in the past and it was also valued for its medicinal properties (Maitland 2007: 173–7). It was especially valued in historical ponds because, like the Common Carp and Roach,

Tench records, 1519–1772. There are records from every region of England, as well as from Munster and Leinster.

it can live in water with very low oxygen levels and can be transported out of water as long as it is kept in damp conditions (Locker 2010; Maitland 2007: 175; Hoffmann 1996).

The modern distribution of the Tench in these islands is similar to that recorded in the early modern sources. The core of the range in Britain is in England south of Yorkshire (Maitland 2007: 174–5; Hindes 2004). In Ireland the core range in 1955 was the River Shannon, its tributaries and associated loughs (Kennedy and Fitzmaurice 1970). Additional populations along the coasts of north England, and in Wales, Lowland Scotland and across much of Ireland have been created in the modern period mainly due to stocking by anglers (for sport, not food), although this species also seems to have benefitted from the creation of the canal network from the end of the eighteenth century (Maitland 1977; Wheeler 1974). Tench need very warm summer temperatures to breed and are likely preserved in some parts of these islands only due to regular restocking (Maitland 2007: 175). The population of the Tench seems stable.

PIKE

Esox lucius

NATIVE STATUS Native to Britain, Possibly Native to Ireland

MODERN CONSERVATION STATUS	
World	Least Concern
UK	Not Evaluated
ROI	Not Evaluated
Trend since 1772	No change

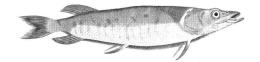

Early modern recorders observed Pike commonly through much of Britain and Ireland, but with few records from Connacht, Highland Scotland, West Wales and Cornwall, and none from any of the smaller islands.

RECOGNITION

This species was the only one referred to as *pike* in Early Modern English, *pyck* or *gedd* in Middle Scots and *lucius* in Latin. This makes identification straightforward (although the similar-looking term *merlucius* refers to the Hake).

Occasionally, other local Early Modern English terms like *pickerel* (exclusively used for *E. lucius*) and *jacke* (refers to several species) are also employed, but these are normally accompanied by *pike* or *lucius* for clarity. K'Eogh (1739: 73) attests the Irish terms *galiesk* and *luse*. Morris attests the Welsh term *penhwyad*, but this appears to also sometimes have referred to the Perch.

DISTRIBUTION

The distribution of top-quality records for this species is statistically a poor fit with the known levels of recorder effort. It may have been locally distributed, locally abundant or of special local interest. Habitat suitability modelling based on the sites where the Pike was recorded in the early modern period suggests that the species would have been well adapted to conditions across Britain and Ireland, but it clearly was absent from some areas. Most obviously, there are only absence records from Cornwall and Galway, and there are fewer records from north of the central belt of Scotland. The numbers of top-quality records in the south of England and Wales are also below what we would expect given the number of records of this species and the overall distribution of records; it may have been less common here, even though still widespread. This is surprising because the Pike is in the group of freshwater fish species that colonised Britain by river after the last glacial period. It was originally thought to have been native only to south-east England (see 'Breams', p. 208), and to have only been introduced more widely during the medieval period, but this actually does not fit with the archaeological record: remains of Pike have been excavated from Mesolithic Lake Flixton, Yorkshire; by the River Witham in Lincolnshire from both the Bronze Age and the Iron Age; and from Roman Carlisle, York, and once medieval York (Locker 2018: 10–13 25–6; O'Connor 1988: 114). The species' very wide distribution in Britain likely attests to its dispersal ability more than its medieval popularity.

The Pike is one of the ten best-recorded species in the early modern sources. It was one of the most commonly eaten freshwater fishes in the medieval and early modern periods (Locker 2018: 36–41, 47–8, 67–70; Harland et al. 2016). Its early modern distribution pattern fits relatively well with the known modern distribution, except that the Lake District seems to have been a well-known early modern spot for the species whereas now Pike are more rarely found there, and South England is now a stronghold for the species. The Pike was popularly kept in ponds in the medieval and early modern period because it can live in water with low oxygen levels and be transported alive outside of water as long as it kept damp, although it will eat smaller fish in captivity (Locker 2010 2018: 54–9; Maitland 2007: 190; Hoffmann 1996). It has been spread further in modern times by the creation of the canal network and by hydroelectric diversion schemes, as well as being intentionally introduced to further sites by anglers (Maitland 1977 2007: 185–6; Wheeler 1974).

The records from Wales are interesting because in modern times Pike are much less widespread in the country than they are elsewhere in the British Isles. It is possible that the population here has actually reduced its range over the last 300 years.

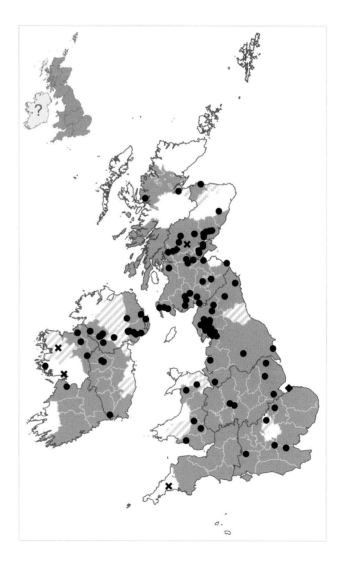

Pike records, 1519–1772.
There are records from every region of Britain and Ireland.

In Scotland, Pike are recorded by the early modern sources as having a stronghold in Dumfries & Galloway, but being generally distributed as far north as Perth, with a northern band across Moray and Ross-shire. This essentially matches the known range of the fish in modern Scotland, although there have been additional introductions outside of this area over the last few decades (Maitland 2007: 185–6; Harding 2004).

The Pike is well recorded on the early modern island of Ireland, especially in Ulster. Like every other obligate freshwater fish, the Pike is usually assumed not to be native because although there was a river connecting the south-east of England with Europe, there was no river connecting Ireland with Britain or Europe after the end of the last glacial period (King et al. 2011; McCormick 1999; Griffiths 1997; Wheeler 1974). In this case, it is often suggested that the species was introduced in the early modern period for food (following Went 1957). This is partially based on Gerald of Wales' attestation that the Pike was absent from Ireland in the twelfth century (O'Meara 1982: 38), and partially based on O'Flaherty's word that it was not yet widespread in parts of Connacht in the seventeenth century:

> The water streames [of West Connacht], besides lampreys, roches, and the like of no value, breed salmons (where is recourse to the sea), eels, and divers sorts of trouts. There was never a pike or bream as yet engendered in all this countrey, nor in the adjacent parts of Mayo or Galway counteys. (ed. O'Flaherty 1846 [1684]: 11)

However, this introduction scenario does not fit easily with the evidence provided by the early modern sources, for two reasons. First, the Pike was clearly already widespread and well known in parts of Ireland by the early modern period. The first record from the early modern sources is from the first quarter of the seventeenth century (Gainsford 1618: 146), and the Pike's Irish distribution is recorded in 14 additional sources before the end of the seventeenth century. Second, and less convincingly, the early modern records come especially from Ulster. This is not a good fit with, for example, the Common Carp (p. 210) and Tench (p. 217), also believed to be introduced in the late medieval period for food, where most of the records come from Munster and Leinster. This might suggest a different colonisation route. The reason that this is a less convincing objection is that the pattern of records does fit with the Rudd, which is usually also thought to have been introduced (King et al. 2011; McCormick 1999; Griffiths 1997). It is also true that the Pike is much hardier at lower temperatures than Common Carp or Tench, as described previously. But the early modern introduction theory, at least, is also challenged by the archaeological evidence, since Pike are included in the thirteenth–fourteenth-century assemblage at Trim Castle (Hamilton-Dyer 2007). Unless we suggest these Pike were imported for food, this evidence would push back the latest possible introduction of the Pike into the medieval period. Further, genetic analysis suggests multiple subpopulations exist among Irish Pike, one of which seems to have colonised Ireland shortly after the end of the last glacial period (Pedreschi and Mariani 2015; Pedreschi et al. 2014). This new scenario would fit with the early modern evidence if we allow that the Pike was not able to naturally colonise some parts of Ireland, and has therefore only been introduced or been able to colonise some watercourses in the modern period, just like in Scotland (Maitland 1977 2007: 185–6; Wheeler 1974). If the Pike is in fact accepted to be native, this opens up the possibility that other freshwater species which were widespread in the early modern period might also have colonised Ireland naturally by the same route.

SMELT

Osmerus eperlanus

NATIVE STATUS Native

MODERN CONSERVATION STATUS	
World	Least Concern
UK	Not Evaluated
ROI	Least Concern
Trend since 1772	Certainly declined

The Smelt seems to have been far more common in the early modern period than it is today. It was recorded by naturalists in every region of Britain and in Co. Dublin and Co. Cork in Ireland.

RECOGNITION

This species was commonly referred to as the *smelt* in Early Modern English and *eperlanus* or *viola* (from the smell of violets) in New Latin (Maitland 2007: 193). These terms are specific to this fish. It was also called the *spirling* or *spurling* in Middle Scots. This term is specific to our species but is also given as *sparling*. When spelled with the letter <a> this is a generic term also used for the Herring and the Sprat. Records using this term are only included on this map where it appears alongside the reliable term *smelt* or where the description or context makes the intended species clear.

Morris attests the Welsh term *brwyniad*, but this also seems to be generic (Morris 1747). Archibald provides the term *russ-fish* in Middle Scots, which seems to be specific:

> Russ-fish, so called from their smell and colour, being sea-green coloured and smelling like a Bundle of green bushes. These seem spurlings.   (Archibald [1683] in Mitchell 1908: 190–1)

DISTRIBUTION

Because the Smelt lives most of its life in estuaries, and only comes a short distance upriver to breed (Atkinson 2004), for this species, references to rivers have been shown by shading the county or counties that contain the estuary. Inland counties are only shaded where fish are specifically said to occur in them, as for example in Northamptonshire. The shading is based on county boundaries, not necessarily on how far the Smelt swam inland.

As a partially saltwater species, the Smelt appears to have been able to colonise Britain and Ireland by sea, so was not reliant on land-bridges (Maitland 1987). Up until the 1920s there was also at least one freshwater population which did not migrate at Rostherne Mere in Cheshire (Atkinson 2004), but the only population recorded from Cheshire in the early modern sources was in the River Dee in Chester.

The distribution of top-quality records for this species is not statistically different from the known level of recorder effort. The gaps on the map may just reflect decreased survey effort in some regions. It may have been widespread in the estuaries of Britain and Ireland.

Smelt records, 1519–1772.
There are records from every
region of Britain and from
Munster and Leinster in Ireland.

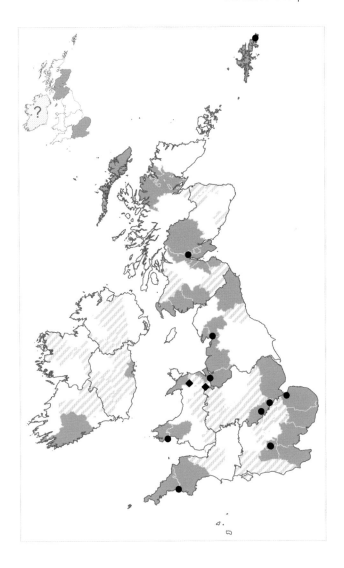

According to the archaeological record, Smelt were eaten in medieval England from the ninth century onwards (Locker 2018: 53; Harland et al. 2016; Sykes 2001). This continued in the early modern period, when demand for Smelt made this the most important fishery on the River Thames (Wheeler 1979: 48–50). In the seventeenth and eighteenth centuries, the Thames fishery was controlled with close seasons, presumably to prevent overexploitation. This level of management was successful into the nineteenth century, when the fishery appears to have become unsustainable. By the end of the nineteenth century, populations on the Thames had collapsed (Wheeler 1979: 49–50). It has since been lost from many other rivers due to pollution and poor river management (Maitland 2007: 199–200; Atkinson 2004).

Based on the early modern evidence, it is clear that the species has dramatically declined. It is found regularly on the shores of Munster and Leinster in Ireland (Henderson 2014: 78), and is known to breed in Lowland Scotland and in the South and Midlands of England (Atkinson 2004). There are no modern records from the South West of England, from Wales, or from the Scottish Highlands. Although the Smelt is recovering in some areas, it is still vulnerable and seems to be considerably depleted compared to the early modern distribution.

COMMON WHITEFISH (POWAN, GWYNIAD, SCHELLY)

Coregonus lavaretus

NATIVE STATUS Native to Britain, not Ireland

MODERN CONSERVATION STATUS	
World	Least Concern
UK	Subspecies are Critically Endangered
ROI	Not Present
Trend since 1772	Probably no change

The British freshwater whitefishes were recorded in early modern Britain and Ireland within their present ranges: the powan in Loch Lomond, the gwyniad in Llyn Tegid (Bala Lake) and the schelly in the Lake District.

RECOGNITION

The British freshwater Whitefish subspecies (but not the Irish Pollan) are considered together here, since the latest research appears to indicate that they represent divergent populations of a single species (Crotti et al. 2020).

In the early modern period, all three populations of Freshwater Whitefish were considered part of the New Latin group *albula*, with inconsistent species names. The vernacular names commonly used for each population today were already in use by the early modern period so that we find early modern recorders mentioning the Middle Scots *powan*, Welsh *gwyniad* (plural *gwyniaid*) and English *schelley*. Maitland (2007: 210) points out that the term *schelley* has also occasionally been used for Chub (*Squalius cephalus*), but luckily all records of *schelley* from our sources also add the New Latin *albula* or Welsh *gwyniad* which makes the meaning clear.

The Early Modern English term *whitefish* was not specific. It could occasionally refer to our species but more often referred to the fish-and-chip-shop whitefish species, included in the 'Whitefishes' section of this atlas. However, all the records included on the map are secure because they can all be distinguished by whether they describe freshwater or marine fish.

DISTRIBUTION

There are not enough top-quality records to statistically analyse the distribution of this species.

In other Northern European populations, the Freshwater Whitefish is migratory between fresh water and the sea, but in Britain all populations currently live their whole lives in lakes (Winfield 2004a). These populations seem to have been isolated in place at the end of the last glacial period (Wheeler 1974). They were found in the same waters as today in the early modern period, when it was common knowledge that, for instance, the Llyn Tegid populations never moved into the River Dee (called

Common Whitefish records, 1519–1772. There are records from Wales, North England and Lowland Scotland.

Afon Dyfrdwy on the south side of the lake), despite the fact that the river runs through Llyn Tegid (e.g. Prise 1663: 5).

There is, however, one uncertain reference of the species migrating in the early modern period. This is Sibbald's record of the *Albula nobilis* of von Schoneveld. This term originally referred to *Coregonus hiemalis*, the now-extinct Lake Geneva Whitefish (Willughby & Ray 1686: 185). Sibbald used this term to refer to a freshwater fish that migrates to the Firth of Forth (ed. Sibbald 1803: 125). He was not using this term to refer to one of the fish-and-chip whitefish species, because he lists it under the 'trout kind' (immediately after *Salmo salar*) rather than under the 'cod kind', where the fish-and-chip-shop whitefishes are included. The most likely fish he could have intended is *Coregonus lavaretus*, but if so, this is likely to be a mistake – since there do not seem to be any other records of *Coregonus lavaretus* populations continuing to migrate between rivers and the sea from Britain after the end of the last glacial period.

Apart from some recent introductions to other lakes, as for example Blea Water, Carron Valley Reservoir and Loch Sloy, the Freshwater Whitefish is still found almost exclusively in the locations that are indicated on the early modern map (Maitland 2007: 215; Winfield 2004a). Because there are so few locations where this fish remains in Britain, it is vulnerable. The Ruffe also extensively predates Whitefish eggs and can drive populations to extinction. Worryingly, the Ruffe has become established along with the Whitefish in both Llyn Tegid and Loch Lomond (Maitland 2007: 215; Winfield 2004a).

IRISH POLLAN

Coregonus autumnalis pollan

NATIVE STATUS Native to Ireland, not Britain

MODERN CONSERVATION STATUS	
World	Endangered
UK	Not Present
ROI	Vulnerable
Trend since 1772	Uncertain

The Irish Pollan was recorded in Lough Neagh in the early modern period.

RECOGNITION
The Pollan was a freshwater fish found in Britain and elsewhere in Europe. It is similar to the Common Whitefish (*Coregonus laveratus*) and Vendace (*Coregonus vandesius*), but actually represents a subspecies of the Arctic Cisco.

The Pollan is difficult to identify in the early modern texts because it does not seem to have a standard name. The term *pollan*, which is now used for the Irish Pollan, was used in the early modern texts analysed for this volume exclusively to refer to the shads. This distinction has caused confusion for previous scholars. To clarify, the records of *pollans* in Harris and Smith's *The antient and present state of the county of Down*, Henry's *Topographical Description of the Coast of County Antrim and North Down*, Nevill's *Some observations upon Lough-Neagh in Ireland* and *The Journeys of Samuel Molyneux* are untrustworthy. At the very least, these texts show considerable confusion with the *Alosa* spp., to the extent that they cannot be relied on as records of the Pollan, but it is most likely that all these records referring to *pollans* actually intended to refer to shads. The only reliable early modern reference from our texts calls the species the *freshwater whiting*.

DISTRIBUTION
This species is only reliably recorded in Lough Neagh, where the text specifically distinguishes the species from a shad:

> [The Shad]
>
> But this Lake [Lough Neagh] affords a Fish, which is uncommon in other Parts, called by Aldrovandus the Alosa, by Rondoletius, Gesner, and others the Clupea and the Thrissa, by the English the Shad, or the Mother of Herrings, and by the Irish the Pollan, or Fresh Water Herring … This Fish was for a time supposed to be a peculiar Inhabitant of Lough-Neagh, but time has corrected that Error, and it is now known that Lough-Earn, in the County of Fermanagh, has the same sort of Fish, though not in so great Plenty; and there are also some of them in the River Severn.

[The Irish Pollan]

There is another Species of Fish in this Lough [Neagh], for any thing we can learn, peculiar to it, called the fresh Water Whiting, in shape exactly resembling the Sea-Whiting; but less in size, almost entirely White, and a very ordinary soft insipid Food.   (Harris and Smith 1744: 237–8)

I have added headings here to distinguish the two species. Even though the term *pollan* is used in the first part of the quotation, the species listed here is clearly the Shad (*Alosa* spp.) rather than the Irish Pollan. Only the second part of the quotation is a record of the Irish Pollan. The record here says the Pollan was living only in Lough Neagh; the Lough Erne record in this text refers to the Shad.

There are now known to be five populations of Pollan on the island of Ireland, in Lower Lough Erne, three loughs on the River Shannon (Lough Derg, Lough Ree and Lough Allen) as well as Lough Neagh. All but one of these populations appear to have gone unnoticed in the early modern period. This may have been because the population at Lough Neagh was the strongest, as it is today. In most loughs the Pollan makes up less than 1% of the fish population, but in Lough Neagh it is around 25% (Harrison et al. 2012). Lough Neagh is also the only lough where commercial fishing is possible.

Pollan records, 1519–1772. There is a single record from Lough Neagh.

The Pollan is the only landlocked population of *Coregonus autumnalis*, and is endemic to Ireland. This species seems to be a survivor from the glacial period that adapted to warming temperatures by stopping its migration (Harrison et al. 2012; Ferguson 2004; Rosell et al. 2004). It was presumably present in Britain during the last glacial period as well as Ireland but must have gone extinct in Britain as well as in the rest of Europe. This means that it is not considered native to Britain. Modern Irish summers are already nearly too hot for the Pollan (Rosell et al. 2004), so the species may have prospered during the Little Ice Age of the early modern period.

The Pollan is exclusively encountered in five loughs in Ireland, with other populations of *Coregonus autumnalis* found in Siberia, Alaska and Canada. This leaves it vulnerable to habitat loss. The Pollan is also threatened by the introduction of the Roach and the Zebra Mussel, as well as eutrophication and climate change (Harrison et al. 2012; King et al. 2011; Rosell et al. 2004; Harrod et al. 2001).

SALMON
Salmo salar

NATIVE STATUS Native

MODERN CONSERVATION STATUS	
World	Least Concern
UK	Not Evaluated
ROI	Vulnerable
Trend since 1772	Certainly declined

Almost every recorder based in early modern Ireland, mainland Scotland, Wales and upland England recorded the Salmon, usually on multiple sites. Salmon was also still widely recorded in lowland England at this time, but not as regularly.

RECOGNITION
The species is called by the Early Modern English term *salmon*, or an orthographic variation in almost every early modern text, which makes it easy to recognise. Occasionally, other terms like *grey trout* or *graulin* are used, but these are in addition not instead of *salmon*. The Irish term was *bradane*.

Morris provides at least 13 different terms for various ages and types of Salmon in Welsh:

> This Fish hath different Names in different Counties of Wales according to its age & Sex. Some of the antients called it *Maran*, others *Eog*, and in Cardigan they still called the Large Salmon they take in the Sea *Pysgod Eog*, which they pronounce Euog. That with a Crooked bill or beak they call *Cammog*, which they say in the male, whose Feminine is called *Chwiwell … Hwyfell. Adfwlch & Penllwyd* are other different kinds of Salmon. (Morris 1747, fol. 294v)

DISTRIBUTION
For the obligate freshwater riverine species, records at river-level have been shown on the map by shading every county with similar topography which that river runs through.

The distribution of top-quality records for this species is statistically a poor fit with the known levels of recorder effort. The number of records from Highland and Lowland Scotland and Connacht is higher than expected given the number of records and the general distribution pattern, whereas the number of site-level records from England, especially the South and the Midlands of England, is much

Salmon records, 1519–1772.
There are records from every region
of Britain and Ireland, including the
Hebrides and the Isle of Man, with
contested records from Orkney.

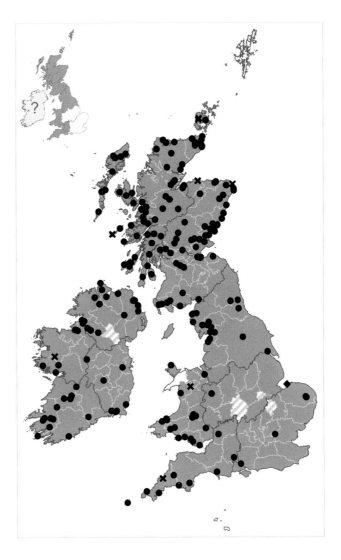

lower than we would expect. But habitat suitability modelling based on the sites where the Salmon was recorded in the early modern period suggests that the species would have been well adapted to conditions across Britain and Ireland. And since there are still records from every region, just more records from some regions than others, this suggests the bias in records reflects an abundance pattern: Salmon was likely more common in the north and west of Britain and in Connacht than it was in the south and east of Britain and Ireland.

The Salmon was often eaten in the medieval and early modern periods, especially in the eighteenth century (Locker 2018: 67–70; Harland et al. 2016; Mac Laughlin 2010: 72, 163–4). Apart from the Eel, Salmon seems to have been the only widely recorded freshwater fish on Irish medieval and post-medieval sites, and was commonly exported (Hamilton-Dyer 2007; Butlin 1991). The right to catch Salmon at particular locations could be jealously guarded, and many people were prosecuted for catching salmon illegally in sixteenth- and seventeenth-century Ireland (Mac Laughlin 2010: 206–7). A close season was first imposed for Salmon in England in the year 1030, and Salmon were regularly caught in fish weirs around this time (Locker 2010). This continued into the early modern period, with special ordinances by the Corporation of London dictating the close season and net sizes in the Thames. By this point, though, Salmon populations were already being affected by the increasing numbers of water mills, which may have excluded the species from rivers of the South and Midlands of England (Lenders et al. 2016). It has even been hypothesised that the Europe-wide decline in populations of freshwater fishes like salmon around 1000 CE led to the development of the marine fisheries (the 'fish event horizon') which fed people in the second half of the medieval period and beyond (Locker 2018; Barrett et al. 2004).

Salmon declined further shortly after the end of the early modern period. The causes for this decline were probably initially river obstruction, overfishing and pollution. For the most part, these factors have continued into the modern period with, for example, the increasing importance of hydroelectric dams, although legislation has reduced pollution in Britain and Ireland (King et al. 2011; Locker 2010; Maitland 2007: 226, 230–1). The loss of the fishery on the River Clyde, just after the end of the early modern period, due to overexploitation was much commented on at the time (Maitland 2007: 225). The loss of the Salmon from the River Thames in the nineteenth century also attracted a great deal of interest. Although Salmon are unlikely to have ever been especially numerous in this lowland river, they were certainly present and commonly exploited in the early modern period (Wheeler 1979: 51–60). The loss of Salmon in the Thames was probably caused by a combination of pollution, overexploitation, the construction of weirs and other habitat disturbances.

The Salmon was the best-recorded species in the early modern sources. Just like today, the highest density of Salmon sightings in the early modern period comes from mainland Scotland, North England and South Wales. Salmon are now rare in the rivers of South England and the Midlands, and the species appears to be in long-term decline (Henderson 2014: 80–1; Shelley 2004).

BROWN TROUT

Salmo trutta

NATIVE STATUS Native

MODERN CONSERVATION STATUS	
World	Least Concern
UK	Not Evaluated
ROI	Least Concern
Trend since 1772	No change

The early modern sources recorded Trout populations in almost every county of Britain and Ireland.

RECOGNITION

The Brown Trout is not closely related to the non-native Rainbow Trout (*Oncorhynchus mykiss*) which has been introduced by anglers for sport.

The English term *trout*, the Irish term *breach* and the New Latin term *trutta* when used without any adjectives were usually specific to Trout, but these terms with extra adjectives could refer to other species, e.g. *grey trout* (Salmon). However, Trout can be very diverse in appearance so a bewildering number of additional terms can refer to the same species, e.g. *cean-dubh* for the Loch Levan Trout, and *shoat* in early modern Devon and Cornwall, although this was later also used as a term for the Grayling. In early modern sources from Wales, the term *suin* (sewin) was and still is commonly used to refer to the Sea Trout, which is a form of Brown Trout that migrates to the sea as an adult. In contrast, the river trout appears (according to Morris) to have been termed the *brithyll*, and the species was split into various varieties (e.g. *brithyll du*, *brithyll brych*) (Morris 1747), but other species need to be carefully distinguished, e.g. *brithyll mair* (Three-bearded Rockling, *Gaidropsarus vulgaris*) *brithyll y dom* (Three-spined Stickleback, *Gasterosteus aculeatus*). The term in Ulster Irish for the species was *budogh*.

DISTRIBUTION

For the obligate freshwater riverine species, records at river-level have been shown on the map by shading every county with similar topography which that river runs through.

The distribution of top-quality records for this species is statistically a poor fit with the known levels of recorder effort. It may have been locally distributed, locally abundant or of special local interest. The number of site-level records from Munster, the South West of England, the Midlands and Lowland Scotland is lower than we would expect, and the number of records from Ulster, Connacht, South England and Wales is higher than expected. However, habitat suitability modelling based on the sites

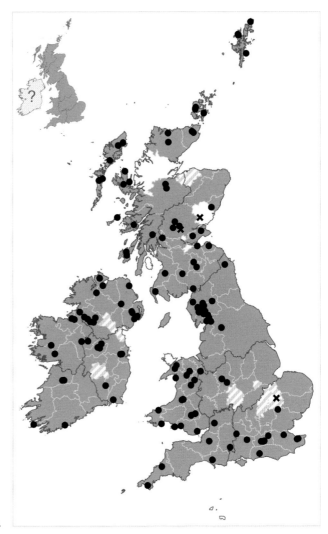

Trout records, 1519–1772. There are records from every region of Britain and Ireland, including the Hebrides, the Northern Isles and the Isle of Man.

where the Trout was recorded in the early modern period suggests that the species would have been well adapted to conditions across Britain and Ireland, and it is widely attested in every region, so the bias in the recording pattern might be more easily explained by a local abundance. Because Sea Trout migrate between marine areas and rivers, there seems to have been no problem for the Trout in colonising every part of Britain and Ireland after the end of the last glacial period (Wheeler 1974).

The Trout is one of the ten best-recorded species in the early modern sources. It was commonly eaten in some areas in the medieval and (less commonly) the early modern periods (Harland et al. 2016) and was caught by anglers for sport – most of the site-level records refer to fisheries and angling spots:

> From Cromlin river to the river of Casla [Co. Galway] are two miles. This river falls into Casla haven and has salmon and trout-fishing on it yearly. These white trouts are called Lihain, and come on to the rivers from the sea as salmons doe. (ed. O'Flaherty 1846 [1684]: 63)

Surprisingly, this interest in the Trout in historical records does not agree with the archaeological record: in Ireland Trout seem to have been rarely found on archaeological sites (Hamilton-Dyer 2007), although this might reflect a sampling bias.

The Trout is sensitive to pollution because it needs high oxygen levels in the water (Locker 2010). This led to declines in some populations, initially during the medieval and early modern periods, due to trades like mining, fulling and tanning. These declines increased in severity during the Industrial Revolution and following the increase of sewage and industrial pollution (Paris et al. 2015; Coates 1998: 62; Wheeler 1979: 69–70).

Today, the Trout is still found in rivers and lakes across all of Britain and Ireland, including the smaller islands. Rivers and standing waters in Britain and Ireland are regularly stocked and restocked with Trout for anglers, and this may have introduced the fish to some additional areas (particularly in Highland Scotland) or helped its recovery in areas where it was lost following declines during the Industrial Revolution (Paris et al. 2015; Maitland 2007: 241; Wheeler 1979: 67–70). Yet it was clearly already widespread in the early modern period.

The Trout appears to have very slightly expanded its range in England since the 1970s, presumably because of improved pollution control (Davidson 2004). In Scotland there have been declines particularly in Sea Trout on the west coast, perhaps in part due to genetic degradation caused by restocking (Maitland 2007: 241). In Ireland, despite declines and some local extinctions, Trout remain widespread, and populations are relatively stable (King et al. 2011).

ARCTIC CHARR
Salvelinus alpinus

NATIVE STATUS Native

MODERN CONSERVATION STATUS	
World	Least Concern
UK	Not Evaluated
ROI	Least Concern
Trend since 1772	No change

The Arctic Charr was recorded by our early modern naturalists in the upland areas of the north and west of Britain and Ireland (especially in lakes, lochs, loughs and llyns).

RECOGNITION
The Arctic Charr populations of Britain and Ireland are geographically and genetically distinct and are sometimes separated into multiple species. For the purposes of this volume, they are considered together.

In the early modern period this species was called the *char, red charr or gilt-charre* in Early Modern English, the *torgoch* in Welsh, the *gelletroch*, *redwaimb* and *sysbinge* in Middle Scots, and the *Umbla minor* in New Latin.

In the past, the Scottish and Welsh fish were thought to be of a different species to the English and Irish ones, but now they are considered simply to be separate populations of Arctic Charr. These terms are all generally specific to our species, but *umbla* must be distinguished from the *umbra* (Grayling). The Brook Charr (*Salvelinus fontinalis*) was not introduced until after the end of the early modern period and so is not included on this map.

DISTRIBUTION
The distribution of top-quality records for this species is statistically a poor fit with the known levels of recorder effort. It may have been locally distributed, locally abundant or of special local interest. Having said that, the populations of Charr now known in the west of Ireland other than Co. Donegal could have been easily overlooked in the early modern sources, since Connacht was so poorly recorded.

Habitat suitability modelling based on the sites where the Charr was recorded in the early modern period suggests that the species may have had specific requirements. It is only recorded on sites which today have cool summers. Most of the references come from upland lakes in the north and west of Britain and Ireland. There is also at least one early modern reference from a river system – River Brathay which connects to Windermere (Leigh 1700: 142). All records are shown as dots rather than by shading entire river systems or regions.

The species was occasionally eaten in the early modern period, and there seem to have been some local fisheries (using nets), as for example in Lough Melvin, Co. Donegal:

> on the northwest shore [of Lough Melvin] which is in the Co. Donnegall, they used before the revolution to take, in great plenty, a fish like a large red trout, with red fins and very beautyfull, which was then call'd charr. That it is a fish not to be taken with angling, but only with nets; And that since that time having neither boats nor nets they have taken none, tho' sometimes they see some of those red fin'd fish near the shore.   (Henry 1740: 5 (folio 61r))

Charr records, 1519–1772.
There are records from Wales,
North England, Highland and
Lowland Scotland and Ulster.

Within Ireland, as well as Lough Melvin, there is also an early modern population of Charr recorded from Lough Eske and a possible record from Lough Erne.

Today the Arctic Charr is recorded most commonly in Scotland north of the central belt, and in the west of Ireland. It is also recorded in North England, south-west Scotland, north-west Wales and in Leinster (Maitland 2007: 245–7; Winfield 2004b). In these islands Arctic Charr are mostly confined to large waterbodies, where they have lived since the last glacial period. They are native to both Britain and Ireland (McCormick 1999). Some populations are known to have been lost (especially in North England and Wales), and the Charr has been introduced (usually accidentally) to a small number of additional still-water environments throughout Britain (Maitland 2007: 246). Because of its geographical isolation, the Charr is vulnerable to habitat changes and the introduction of invasive species. It is now declining in Britain and Ireland with some populations already lost (King et al. 2011; Maitland 2004). However, for the most part, the Charr does not appear to have changed its distribution since the early modern period.

GRAYLING

Thymallus thymallus

NATIVE STATUS Native to Britain, not Ireland

MODERN CONSERVATION STATUS	
World	Least Concern
UK	Not Evaluated
ROI	Not Present
Trend since 1772	Probably increased

The Grayling was recorded in the early modern sources especially in a few clusters around southern Britain.

RECOGNITION

This species was most often called the *grayling* in Early Modern English and the *umber*, *umbra* or *thymallus* in New Latin. These names are ordinarily specific enough to identify the species, although some care must be taken to differentiate the *graylord* (Saithe), the *Umbla minor* (Charr), and the *tunbridge grayling* (*Hipparchia semele*, the Grayling butterfly), a largely coastal insect which is only recorded once in our early modern sources. The species is especially well recorded in Wales and a few Welsh terms are also attested in our sources: *cawlidd*, *glas gangen*, *glas onnen*, and possibly *molfrith*. It is difficult to be clear how many of these were consistently specific to the Grayling – *cangan* at least is not specific since the term *glas gangen* used for this species needs to be distinguished from the term *coch gangen* used for *Squalius cephalus* (Chub).

DISTRIBUTION

For the obligate freshwater riverine species, records at river-level have been shown on the map by shading every county with similar topography which that river runs through.

The distribution of top-quality records for this species is statistically a poor fit with the known levels of recorder effort. It may have been locally distributed, locally abundant or of special local interest. In this case, the species seems to have been absent from most of these islands. Habitat suitability modelling based on the sites where the Grayling was recorded in the early modern period suggests that the species may have had specific requirements. It is only recorded on sites that today have relatively warm summers. However, in this case a more important reason for the limited distribution of the Grayling might have been lack of opportunity to expand.

The early modern distribution of this species is striking. The Grayling was recorded in three main areas centred on the River Wye, River Severn and the Brecon Beacons (including Llangorse Lake), the River Trent, its tributaries and the Peak District and the River Avon, its tributaries and the Downs. In other words, the Midlands, the Welsh Border counties, plus Gloucestershire, Wiltshire and Hampshire might be considered the Grayling's core native range. These records would suggest that previous authors were wrong to consider it not native to Wales. The species is also recorded in the archaeological record from Roman and medieval York (O'Connor 1988: 114). The Grayling was absent from the south east of England (as it generally remains today) because the habitat there is less suitable, perhaps due to muddy riverbeds, pollution and obstructions (Cove 2004).

There are also a few questionable records of the Grayling from elsewhere in Britain and Ireland. These include specifically marine or estuary records from Co. Dublin in Ireland and the Firths of Forth

Grayling records, 1519–1772.
There are reliable records from
Wales and the Midlands, South
and South West of England,
and unreliable records from
Lowland Scotland and Leinster.

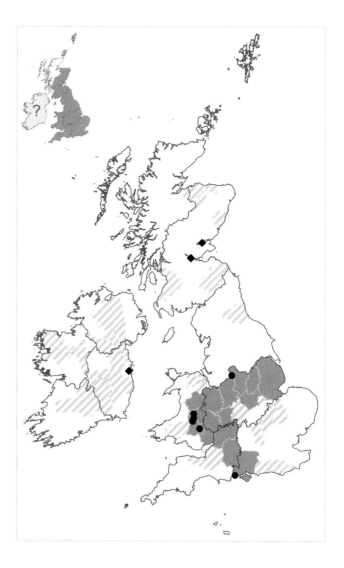

and Tay in Scotland. The Grayling does occur in brackish water elsewhere in Europe (Cove 2004), so these records are not inconceivable, but all Grayling in Britain and Ireland currently live in fresh water, and the fact that these sources do not mention freshwater populations as well suggests that these are mistaken records.

The Grayling has sometimes been thought of as vermin due to its competition with the Trout and Salmon. It also would not have been suitable for stocking as a food species in early modern ponds because it cannot tolerate low oxygen levels and is extremely sensitive to water quality (Locker 2010). However, it is now often valued by anglers as a catch in its own right and has been introduced to other areas of Britain (Maitland 2007: 259–61; Cove 2004). In addition to its early modern range, the fish is now also found in Lowland Scotland and the North of England where it has become widespread. The introduction to Scotland is especially well documented and it seems to have occurred in the nineteenth century. The Grayling continues to be threatened by pollution, overfishing, habitat destruction and physical barriers in rivers (Cove 2004).

PERCH
Perca fluviatilis

NATIVE STATUS Native to Britain not Ireland

MODERN CONSERVATION STATUS	
World	Least Concern
UK	Not Evaluated
ROI	Not Evaluated
Trend since 1772	Probably increased

The Perch was already commonly but patchily recorded across Britain (except Highland Scotland) and also recorded in Co. Cork, Ireland in our early modern sources.

RECOGNITION
This species is most reliably, but rarely called *Perca fluviatilis* in New Latin (which remains the taxonomic name today). This term appears to be specific to our species.

The more generic Latin term *perca* and the Early Modern English term *perch* are less reliable because they are occasionally used to refer to the European Bass (p. 286). To make matters worse, the term *bass* is also occasionally used in Early Modern English to refer to the Perch as well as the European Seabass. The two species also cannot always be distinguished based on location, because the Seabass regularly enters estuaries and even sometimes river systems. Luckily, it has been possible to distinguish almost all of our records based on context and description, or because the texts describe both species.

DISTRIBUTION
For the obligate freshwater riverine species, records at river-level have been shown on the map by shading every county with similar topography which that river runs through.

The distribution of top-quality records for this species is statistically a poor fit with the known levels of recorder effort. It may have been locally distributed, locally abundant or of special local interest. The number of records from Lowland Scotland and the North of England is higher than expected and the number in all four regions of Ireland and in Wales is lower than expected. The early modern recorders attest a continuous range from Lowland Scotland down through to the South of England (but not Cornwall, where there is an absence record) as well as from Co. Cork in Ireland.

The traditional scholarship on the history of the Perch in Britain (perhaps summarised best in Maitland and Campbell (1992: chap. 4)) is unsatisfactory. Based on Wheeler's (1974) belief that freshwater fish could not have distributed themselves from south-east England to Scotland after the last glacial period corroborated by Maitland's (1977) inability to find any Perch records from Scotland before 1790, it is usually thought that the species is only native to south-east England. The present-day wide distribution of the fish is believed to be the result of ecclesiastical centres (sometimes specifically medieval Cistercian monasteries) stocking the Perch for consumption during fasting periods.

There are two problems with the traditional narrative: (i) If the species was introduced to Scotland by medieval monasteries, it cannot have also been absent from Scotland between the medieval period and 1790. The former presence of this fish in Scotland is in fact supported by the records provided in the early modern texts analysed in this volume. The remains of Perch have also been found from archaeological excavations of Roman and medieval York (O'Connor 1988: 114). (ii) The idea that freshwater fishes are especially associated with monastery ponds and Lent is not supported by historical research, which finds that obligate freshwater species were high-status foods equally or perhaps even more associated with elite sites than religious ones, and were perhaps not widely fed to monks on ordinary fasting occasions, especially in the second half of the medieval period when fish were being introduced to new sites (Maccarinelli 2021; Locker 2018: chap. 4; Hamilton-Dyer 2007; Hoffmann 1996). Currie (1988) estimated that even a small monastery would need 21 acres of ponds, managed intensively, to feed each monk a fish a day, and that monasteries with 40 monks would need 90 acres of ponds, so that for most of the time during fasting, monks would likely have eaten vegetables or marine or migratory fishes. Nevertheless, it is true that the Perch was commonly kept as a pond fish (as a store of food) in the medieval and early modern period, and that it was one of the most frequently eaten

Perch records, 1519–1772. There are reliable records from every region of Britain south of Highland Scotland, and from Munster.

freshwater fishes in this period (Locker 2010, 2018: 36–41, 49–54, 56, 67–70). Although it needs high oxygen levels, the Perch can survive outside of the water in damp conditions and so is easy to transport, just like Common Carp and Pike (Maitland and Campbell 1992).

But we could just as easily make sense of the evidence by arguing that the Perch actually colonised most of Britain naturally along with the Pike, and that medieval stocking only established the Perch in new sites at a local level and created a small number of populations in Ireland. This would also better explain how the Perch found its way to the Mesolithic Lake Flixton in Yorkshire as well as Roman York (Locker 2018: 10–11 25–6). Yorkshire is not Scotland, but it is not clear why a fish that presumably spread from the south-east of England to the north of Yorkshire in 3,000 years, could not have naturally found its way to Lowland Scotland given another 9,000 years.

Today, the Perch in Britain has a very similar distribution to that shown by the early modern naturalists, with the exception of a few probable introductions to, for instance, Cornwall, Anglesey and the Isle of Man (Carter 2004b). It has also likely been introduced to more sites locally in the modern period by hydroelectric schemes as well as in canals (Maitland 1977; Wheeler 1974). However, in Ireland the distribution is different – the species is widespread today but was only reliably recorded in Co. Cork in the early modern period (the record from Co. Dublin explains the fish is 'scarce' (Rutty 1772: 368)). This is not the same as the situation in Scotland, where the Perch was already widely recorded in the early modern period. As with most of the other obligate freshwater species, it is most likely that the Perch is not native to Ireland but was widely introduced in the modern period by anglers (Moriarty and Fitzmaurice 2000; McCormick 1999; Wheeler 1974). The record from Co. Dublin, for example, strongly suggests an introduced population:

> *Perca fluviatilis*. The Perch.
>
> It is scarce here, and found only in ponds, and consequently not so much to be recommended; that found in pure rivers or in the Sea being excellent food, and allowed by Physicians to Valetudinarians, and even in fevers according to Mundius.
>
> This fish with its prickles, inflicts wounds very dangerous and difficult to cure. (Rutty 1772: 368)

LESSER SPOTTED DOGFISH/ SMALL-SPOTTED CATSHARK

Scyliorhinus canicula

NATIVE STATUS Native

MODERN CONSERVATION STATUS	
World	Least Concern
UK	Not Evaluated
ROI	Not Evaluated
Trend since 1772	Uncertain

There were a few records of the Lesser Spotted Dogfish dotted around Britain and Ireland, but far more of uncertain species of *dogfish*.

RECOGNITION

This species is reliably referred to as either the *Catulus major vulgaris* or the *Catulus minor* in New Latin. Although these two terms were coined to distinguish two types of dogfish, and Linnaeus (1758: 234–5) believed them to be different species (*Squalus catulus* and *Squalus canicula* Linnaeus), they have since been merged into our Lesser Spotted Dogfish (*Scyliorhinus canicula*). In theory these two terms can be distinguished from the *Catulus maximus*, which has now become modern *Scyliorhinus stellaris* (the Bull Huss/Nurse Shark) (Willughby & Ray 1686: 63), but in practice no early modern author attests this last term as a specifically British or Irish species.

The English terms *dogfish*, *rough hound* and *hound fish*, Irish *fiagach*, Late Cornish *morgay* and Welsh *morgi* all seem to be generic names which can apply to any similar species. Morris also attests at least three other names for the species in Welsh: *ci ysgarmer*, *ci brych* and *ci coeg*. On our map the reliable references using New Latin names are shown as dots and shading, and records referring to other names are shown as diamonds.

DISTRIBUTION

There are not enough top-quality records to statistically analyse the distribution of this species. However, based on the generic records (shown as diamonds on our map) the dogfishes/catsharks in general were widely distributed around early modern Britain and Ireland. The cartilage of sharks, skates and rays does not survive well in the archaeological record, but the species are occasionally found and seem to have been sometimes eaten in medieval and early modern Britain (Harland and Kirkwall 2016; Harland et al. 2016; Serjeantson and Woolgar 2006).

Today, the Lesser Spotted Dogfish is found commonly on all coasts of Britain and Ireland. It is regularly caught by anglers and sometimes eaten as 'rock salmon' (Henderson 2014: 18–19). It seems to have increased significantly over the last century, from being relatively rare to now being the most common elasmobranch species around Britain. This is probably due to trawl-fishing reducing the Thornback Ray population and warming ocean temperatures which benefit the dogfish (Sguotti et al. 2016; Rogers and Ellis 2000).

Dogfish records, 1519–1772. There are unreliable records scattered around the coasts of Britain and Ireland.

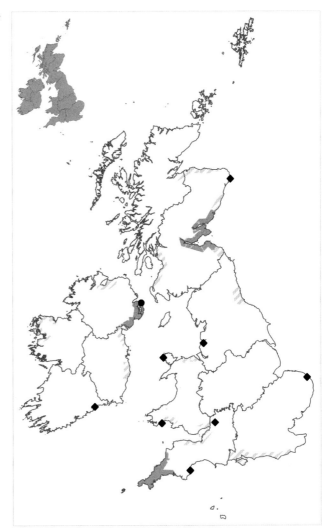

ANGELSHARK

Squatina squatina

NATIVE STATUS Native (locally extinct)

MODERN CONSERVATION STATUS	
World	Critically Endangered
UK	Not Evaluated
ROI	Critically Endangered
Trend since 1772	Uncertain

In the early modern period, the Angelshark was recorded around Britain and Ireland more widely than today. It was found most frequently in the southern parts of these islands.

RECOGNITION

The most reliable references of this species from the early modern period use the New Latin name *squatina*. The term *angel-fish* was also occasionally employed and is specific to this species. Morris (1747) attests the Welsh term *maelgi* (to be distinguished from *morgi*, meaning the sharks and rays generally). It is not clear if this term was ordinarily specific, but the context makes the meaning clear in this instance at least. The most common name for the species in Early Modern English was *monk-fish*; this name was also sometimes attached to another species which is usually still called the Monkfish today. In practice, though, this latter species was more commonly called the *frog-fish* or *Rana piscatrix* in the early modern period, so the references to this species can usually be reliably distinguished.

Occasionally, authors use the New Latin term *squatino-raia*. This was identified by Linnaeus (1758: 233) as *Rhinobatos rhinobatos*, the Guitarfish. The term was used in the same way by Willughby & Ray (1686: 79), who distinguished the *squatino-raia* from the *squatina* (meaning the Angelshark). However, the Guitarfish does not seem to have ever bred so far north as Britain, and the term appears to have been used in other British and Irish sources for the Angelshark – references using this term are identified on our map with diamonds.

DISTRIBUTION

The distribution of top-quality records for this species is statistically a poor fit with the known levels of recorder effort. It may have been locally distributed, locally abundant or of special local interest.

Few early modern authors describe the Angelshark. There are references from the Celtic Sea, Irish Sea, English Channel and North Sea. The records from early modern Cardigan Bay are particularly interesting because they indicate that a breeding population was present in the eighteenth century at the core of the modern Welsh breeding range (Moore and Hiddink 2022). The lack of records from other coastal recorders implies that the species may have been rare, but it probably had a wider distribution in the early modern period, as for example in the following description from *The History and Antiquities of Harwich and Dovercourt* in Essex. This record is also of interest because it suggests Angelsharks were regularly eaten in the past:

> Squatina … The Monk or Angel-fish. This is caught by the Fisherman on all Coasts, and is carried with the aforementioned flat Fish to supply the Markets; but this being frequently very large, is not sold whole, but cut into Pieces. There is a good figure of it in Willughby Tab. D. III.
>
> (Taylor and Dale 1730: 422)

Angelshark records, 1519–1772.
There are reliable records
from the coasts of South and
South West England, Munster
and Leinster and Wales.

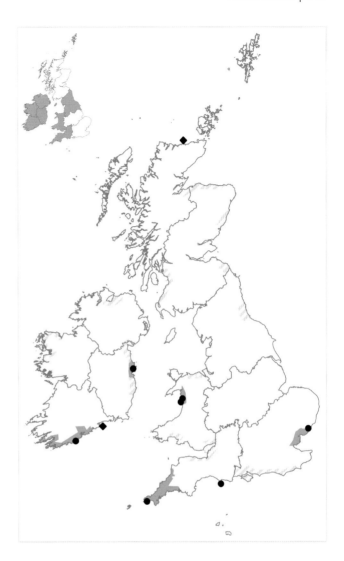

Today, the Angelshark is rarely encountered in the North Sea (or off the east coast of Britain generally) but occasionally found on other coasts, especially around Tralee Bay in Munster, Clew Bay in Connacht and Cardigan Bay north of Aberystwyth in Wales (Bom et al. 2020; Hiddink et al. 2019; Sguotti et al. 2016; Henderson 2014: 32–33). Like other elasmobranch species, the Angelshark is especially vulnerable to trawl-fishing and its decline may be particularly tied to the use of modern beam-trawlers (Bom et al. 2020; Hiddink et al. 2019).

COMMON SKATE COMPLEX

Dipturus batis sensu lato: Blue Skate (*Dipturus flossada*) and Flapper Skate (*Dipturus intermedius*)

NATIVE STATUS Native

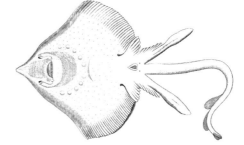

MODERN CONSERVATION STATUS	
World	Critically Endangered
UK	Not Evaluated
ROI	Not Evaluated
Trend since 1772	Certainly declined

The Common Skate complex seems to have been recorded around the coasts of Britain and Ireland.

RECOGNITION

The Common Skate has now been split into a complex of two species: the Flapper Skate, the larger species found in the North Sea, and the Blue Skate, the smaller species found in the Celtic Sea (Henderson 2014: 38–9).

The early modern nationalists used a range of names to refer to this species. Most usefully, the New Latin name *Raia laevis undulata* or *cinerea*, or *Raia laevis majoris* and the Middle Scots term *dinnen scate* seem to refer specifically to our Common Skate complex, both in the North Sea (in the Firth of Forth) and in the Celtic Sea (off Co. Cork).

Interestingly, there is another related term: *Raia laevis* or *Raia laevis vulgata* seems to have been employed to identify a species that was more common and smaller than *Raia laevis undulata*. This term is used to record species off the east coast of Ireland and off the north, east and south-west coasts of Scotland. It is possible that this term was mainly applied specifically to our Flapper Skate, although it may have applied to other smaller skate species as well, and therefore records of this species have been shown as diamonds and the distribution has not been shaded:

> 1. *Raiae laeves*, Skates and Flairs, dictae, which are frequent.
> 2. *Raiae asperae. Raia clavata* Rondeletii, Thornbacks.
> The lesser and younger of both kinds, are called Maids, or Maiden Ray.
> It is eaten both fresh and salted. For the latter purpose it is prepared by being pressed and salted a little, and alternatively dried in the sun and in the house for some months at Howth, Baldoyle and Skerries [Co. Dublin], and being well watered is sometimes eaten with parsnips, and by some deemed equal to Ling. The Liver is excellent. (Rutty 1772: 348)

This map also includes generic references to *skate*, *skait*, *skeat* and *scatt* in Early Modern English and *morcath* in Welsh. These cannot be identified to species level but probably referred to the Common Skate species complex, since that is likely to have been the most common species. These records are shown as diamonds and are not reliable.

DISTRIBUTION

There are not enough top-quality records to statistically analyse the distribution of this species. However, if the early modern records are trustworthy, they suggest it was found across Britain and Ireland in the early modern period.

The cartilage of sharks, skates and rays does not survive well in the archaeological record, but the species are occasionally found and seem to have been commonly eaten in medieval and probably early modern Britain and Ireland (Hamilton-Dyer 2007; Serjeantson and Woolgar 2006), as can be seen in the quotation above.

Over the course of the modern period, the Common Skate has changed from being a commercially viable species generally distributed around Britain and Ireland to a rarely seen Atlantic species (Henderson 2014: 39) only found in the extreme north of these islands. It was already depleted in the North Sea by the start of the twentieth century (Cardinale et al. 2015), and continued fishing pressure by anglers and bycatch in trawlers have made the species critically endangered (Fogarty 2017: 169–70; Sguotti et al. 2016; Rogers and Ellis 2000).

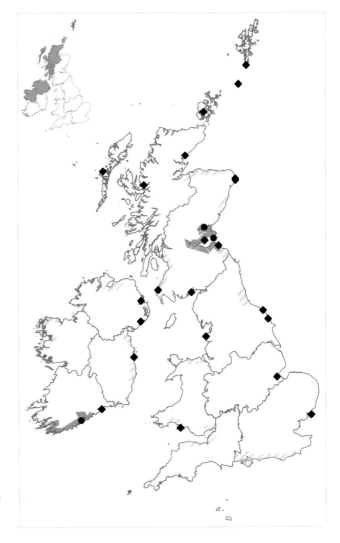

Common Skate records, 1519–1772. There are unreliable records scattered around the coasts of every region of Britain and Ireland except Connacht and South West England.

THORNBACK RAY

Raja clavata

NATIVE STATUS Native

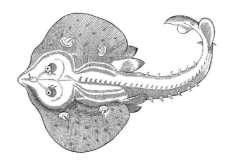

MODERN CONSERVATION STATUS	
World	Near Threatened
UK	Not Evaluated
ROI	Not Evaluated
Trend since 1772	Uncertain

The Thornback Ray was recorded widely in our early modern sources around the coasts of Britain and Ireland.

RECOGNITION

This species was normally referred to as the *thornback ray* or *thornback mayd* in Early Modern English. In New Latin it was called *Raia clavata*, which essentially remains the species' taxonomic name today.[7] Morris attests the Welsh terms *(mor)cath arw* and *cath brigog*, the latter of which needs to be carefully distinguished from *ci pigog* (meaning *Squalus acanthias*, our Spurdog) (Morris 1747). These terms are all specific to our species.

Raja clavata is by far the most commonly recorded species of ray. The other species seem to have more usually been recorded using only generic terms like *ray* or *mayd*. The Welsh generic terms *morcath* and *cath-fôr* seem to have applied to both the ray and skate species. These generic terms are not included on our map.

DISTRIBUTION

The distribution of top-quality records for this species is not statistically different from the known level of recorder effort. The gaps on the map may just reflect decreased survey effort in some regions. It may have been widespread.

The cartilage of sharks, skates and rays does not survive well in the archaeological record, but the species are occasionally found in archaeological excavations of human settlements and are likely therefore to have occasionally been eaten in medieval and early modern Britain and Ireland, perhaps especially in England (Harland et al. 2016; Hamilton-Dyer 2007; Serjeantson and Woolgar 2006):

> They are found in the like Places, as the Skate, and their Flesh is much of the same Taste, but is somewhat more hard of Digestion; the Liver is accounted a good Dainty among some: The Thornback is nourishing enough, and affords both a solid and durable Food. Some pretend, that it provokes Venery [acts as an aphrodisiac], and increases Seed; but its Flesh is hard and not easy of Digestion, causes Wind, and produces heavy and gross Humours, especially if eaten before you let it lie for some Time. They are to be chosen when they are plump, and as tender.
> (Griffiths and Binnell 1758: 220)

Today, the Thornback Ray is widespread and found on every coast of Britain and Ireland, although it is rarer on the North Sea coast than elsewhere (Wood 2018: 221). A century ago it was the most common

---

7 The modern name is actually *Raja clavata*, with a <j> instead of an <i> The letter <j> in Latin is sometimes pronounced as <i> or <y>, so there is no change in pronunciation. Linnaeus tended to substitute <j> for <i> when the letter was in the middle of the word (Stearn 1966: 50).

Thornback Ray records, 1519–1772. There are coastal records from every region of Britain and from Munster and Leinster in Ireland.

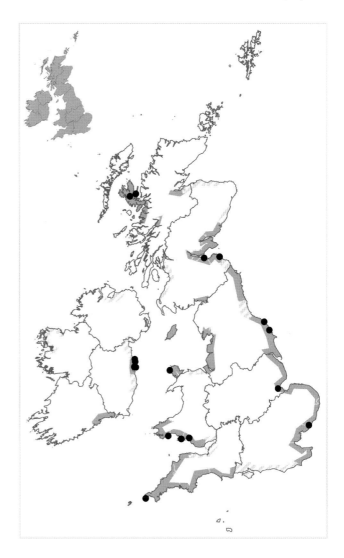

elasmobranch species, but it declined around 1970 (perhaps following the advent of mechanised beam-trawling), and although the population has stabilised, it is currently still suppressed (Sguotti et al. 2016; Rogers and Ellis 2000).

EEL

Anguilla anguilla

NATIVE STATUS Native

MODERN CONSERVATION STATUS	
World	Critically Endangered
UK	Not Evaluated
ROI	Critically Endangered
Trend since 1772	No change

The Eel was widely but unevenly recorded across Britain and Ireland by the early modern sources. Some sources record many fishing sites, some only that they are present in the local area.

RECOGNITION

The terms *eel*, *eele*, *ell* and other variant spellings in Early Modern English, *llyswen* in Welsh, *ascue* or *asequin* in Irish, and *anguilla* in Latin usually referred to this species in early modern Britain and Ireland. However, these terms were also modified (usually with a different adjective) for other species like the sandeel and the Conger Eel. For instance, Morris attests that in Welsh, *llyswen* is our species, *mor-llyswen* is the Conger Eel and *llyswen bendoll* is a lamprey (Morris 1747).

DISTRIBUTION

For the Eel, records at river-level have been shown on the map by shading every county with similar topography which that river runs through.

The distribution of top-quality records for this species is statistically a poor fit with the known levels of recorder effort. It may have been locally distributed, locally abundant or of special local interest. The Eel is a migratory species which lives part of its life in the sea and therefore would have had no problem in colonising Scotland and Ireland as well as southern Britain (McCormick 1999; Wheeler 1974). This might have left us to expect the species to be evenly distributed across these islands in the early modern period. However, there are fewer records than expected from the South, South West and Midlands of England and more than expected from Ulster, Connacht, Leinster and Wales, perhaps suggesting that Eels were more widespread or more abundant in these regions. But interestingly, habitat suitability modelling based on the sites where the Eel was recorded in the early modern period suggests that the species would have been well adapted to conditions across Britain and Ireland. The explanation for this might be the decline in English Eel populations, which is brilliantly explored by John Wyatt Greenlee (2020: 121–3, 311–12). As Greenlee explains, from at least the tenth century, some people in England paid their rent in Eels, and hundreds of thousands of Eels were recorded as being owed and paid each year. This number suddenly dropped in the fourteenth century and at the same time, legal restrictions were placed on methods of catching them. An earlier decline in the number of Salmon in England was probably caused by the obstruction of rivers with watermills, but this does not seem to have affected Eels, which are able to leave the water to bypass obstructions (Greenlee 2020; Lenders et al. 2016; Maitland 2007). Greenlee posits that the reason for the decline in Eels, and the continued suppression of populations through the early modern period, might instead

Eel records, 1519–1772. There are records from every region of Britain and Ireland, including the Hebrides, the Northern Isles, the Isle of Man and the Isles of Scilly.

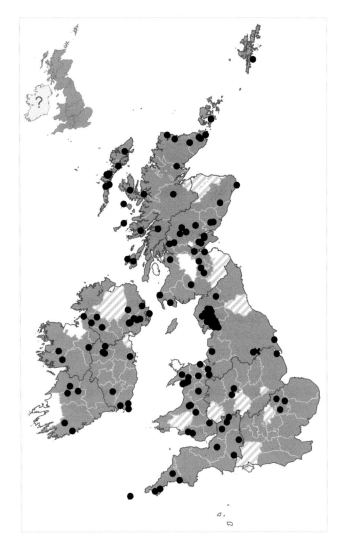

have been the low temperatures and increasing droughts and flooding of the Little Ice Age, together with the economic changes brought on by urbanisation and plague in the fourteenth century. This makes some sense. The importance of climate is supported by a consideration of the worldwide distribution of the species (Dekker 2003).

The idea that Eels declined due to the Little Ice Age would also explain why most early modern references from the Northern Isles also suggest Eels to have been especially rare there (e.g. Lesley 1675: 38). However, the accounts of Eels from the Northern Isles usually group them together along with other supposedly venomous species like the Toad, which were believed to be absent from several islands off Scotland (as explained on p. 305), so it is equally possible that there may be a cultural explanation for the lack of references. In contrast, Monteith's (1633) local account implies that there were a few locations where Eels could be found in Shetland in the early modern period – for example:

> [Bressay] hath in it eight Loughs, of no considerable length of Breadth, to wit, the Lough of Grein-sitter, the Lough of Brough, the Lough of Seateer, the Lough of Aith, the Lough of Gunielstay, two Loughs of Beoster, and the Lough of Kellabuster, all of which have small Brooks running from them, and are somewhat stored, with Common Trouts and Eels.
>
> (ed. Monteith 1711: 60)

Further, if the early modern suppression in Eel numbers was caused in large part by colder climate of the Little Ice Age, we might expect that the decline would affect the whole of Britain and Ireland equally, or even affect the northern and western parts of these islands more than the southern and eastern. This does not seem to have been the case. If the early modern evidence is trustworthy, the population in early modern England was suppressed more than the populations in Scotland, Wales or the island of Ireland. Localised declines might more plausibly be ascribed to overexploitation or pollution: hundreds of thousands, if not millions, of Eels were being caught in England before Eel rents began to be abandoned in the fourteenth century, and the passing of laws limiting capture suggests people at the time believed they were taking Eels at an unsustainable level. Harland et al. (2016) point out that the average size of Eel found at York peaks in the tenth century, and decreases from then, possibly suggesting that the fishery was impacting the number of mature Eels present in England. Perhaps the demand for Eels as a food was so great that it led to the medieval decline and continued to suppress populations into the early modern period.

It is also possible that the rivers of lowland England in the late medieval and early modern period had already reached such a poor state that pollution was starting to affect fish abundance (Hoffmann 1996). When the River Thames was at its most polluted, roughly 1850–1950, the pollution was so bad that the live-eel-sellers reportedly had to anchor at Gravesend and send daily shipments of Eels to the ships docked at Billingsgate, because if the boats full of Eels docked at Billingsgate all the Eels would die (Wheeler 1979: 38, 63; Buckland 1861: 268–71). Pollution was not this dire in the fourteenth century, but it may possibly have been sufficiently bad in Lowland England to be a factor in the Eel's decline and it could have continued to suppress populations into the early modern period.

The Eel continued to be eaten across Britain and Ireland through the early modern period. As mentioned above, to supplement the lack of Eels in England, huge shipments were brought from the Netherlands and other countries where there was an abundant supply. From the fifteenth century to the twentieth, Dutch eel-sellers imported live Eels and sold them from barges called schuyts on the River Thames outside of London (Greenlee 2020: chaps 7–8). Elsewhere in these islands, Eels were caught locally or received smoked or salted, and they continued to be commonly eaten, even by people who could not normally afford to obtain freshwater fishes (Locker 2018: 36–41, 47–54, 67–70; Harland et al. 2016; Mac Laughlin 2010: 74–5, 163–4; Hamilton-Dyer 2007).

The Thames and other rivers of lowland England have been cleaned up to some extent over the last 70 or so years, and there has been a partial recovery – most lowland watercourses in England are once again now visited by Eels, although certainly not in the numbers that were found in the medieval period. However, in the twenty-first century the Eel is now classed as critically endangered. The numbers of young 'glass' Eels currently arriving in British waters from their breeding grounds is only 1% of the numbers from the 1980s, and Eel populations are in steep decline across Europe (Greenlee 2020: 340–1; Henderson 2014: 60–1). There appear to be many factors involved in this, including (illegal) exploitation, pollution, habitat degradation, obstruction of rivers, climate change and increasing parasites and diseases.

CONGER EEL

Conger conger

NATIVE STATUS Native

MODERN CONSERVATION STATUS	
Europe	Least Concern
UK	Not Evaluated
ROI	Not Evaluated
Trend since 1772	Uncertain

The Conger Eel was regularly recorded around the coasts of early modern Britain and Ireland.

RECOGNITION

This species was usually called *conger eel* or *sea eel* in Early Modern English and also *conger* in Latin. O'Sullivan attests an odd sounding Irish name *congrio dergan* [little red conger?] (ed. 2009: 172), but I am not aware of any other attestations of this term. Sibbald (ed. Sibbald 1803: 121) provides the Middle Scots *heawe-eel*. Morris provides a few Welsh terms: *congren* and *c[l]yngyren* and *mor-lyswen* (Morris 1747). As long as the species is carefully distinguished from the Eel and sandeels it is generally straightforward to identify it based on the English and Latin terms.

DISTRIBUTION

The distribution of top-quality records for this species is not statistically different from the known level of recorder effort. The gaps on the map may just reflect decreased survey effort in some regions. It may have been widespread. The species has been occasionally found at archaeological excavations and seems to have been eaten in medieval and early modern Britain and Ireland (Locker 2018: 67–70; Hamilton-Dyer 2007; Kowaleski 2003).

The early modern sources occasionally refer to Conger Eels travelling up rivers; I would suggest this is most probably due to confusion in the early modern sources between the Conger Eel and freshwater Eel, as although the Conger Eel enters estuaries it does not tend to enter fresh water (Henderson 2014: 62). The best example of this is in a letter by John Parkhurst, later Bishop of Norwich, to Conrad Gessner:

> In English they are called a conger, a gongre, a conger ele and in England very small congers are elverz. They are caught in great quantities on dark nights in the River Severn by Gloucester and Tewkesbury. Indeed, they are gathered up in a certain kind of net and almost drawn from the water. Caught and boiled, rolled up and sold, they are also kept as titbits, and they are generally eaten with vinegar or mustard. Also, those of the smallest size ascend from the seas into the rivers. Their name seems to be based on their similarity to eels.
>
> And that's all about the little congers as I have received it from that best and most learned man, John Parkhurst from England.   (translated from Gessner 1554a: 292)

Based on the size and terminology, this source seems to be describing the harvest of what we call elvers or glass eels (meaning freshwater Eels) rather than of Conger Eels, and so the records have been included on the map for that species with the assumption that the use of the term *conger* was a mistake.

In the present day, the Conger Eel is much more commonly found in the Atlantic Ocean, Celtic Sea and Irish Sea coasts (the area classed by the International Council for the Exploration of the Sea as the Celtic Seas ecoregion) than in the North Sea (Wood 2018: 224; Henderson 2014: 62). The population around these islands appears to be stable (Rindorf et al. 2020).

Conger Eel records, 1519–1772.
There are coastal records
from every region of Britain
and Ireland except Ulster.

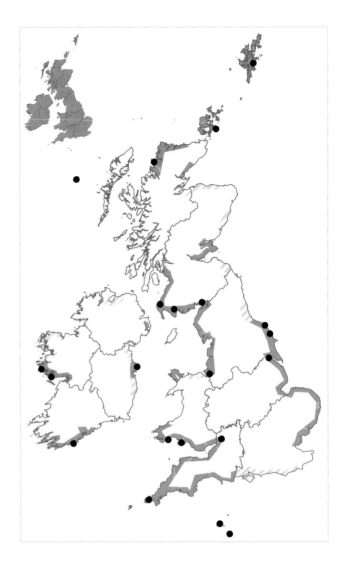

SHADS

Alosa spp.

NATIVE STATUS Native

MODERN CONSERVATION STATUS	
World	Least Concern
UK	Not Evaluated
ROI	Vulnerable
Trend since 1772	Certainly declined

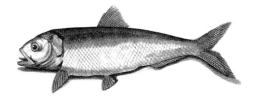

The shad was recorded by early modern sources around the coasts of Britain and Ireland and in a few rivers inland.

RECOGNITION

There are two species of shad in Britain and Ireland, the Twaite Shad and the rarer Allis Shad. There is also a subspecies which has adapted to live in fresh water only, the Killarney Shad. This last population is only found in Lough Leane in Killarney but seems not to have been noted in the early modern sources. Unfortunately, the early modern texts also do not distinguish between the Allis Shad and Twaite Shad. Sibbald does distinguish the Middle Scots terms *craig-herring* and *crue-herring*, but it is unclear if these two correspond to our modern species.

The terms used for both species include the Early Modern English *shad, schad, freshwater herring* and *mother of herring*. In Welsh it was called the *herling* or *herlyn*. In New Latin it was called *alosa* or *clupea*. Piers records the species under the Irish name *goaske* (modern Irish *gabhar*) (see Farran 1944). The two species need to be distinguished in texts from the Herring, which was also called *herring* and *clupea*, as well as *Trachurus trachurus* (the Atlantic Horse Mackerel) which was (and is) called the *scad*.

Confusingly, this species was also called the *pollan* (and once possibly *dolleyn*) in early modern Ireland, which is the term used today for the Pollan (p. 226). Both species were sometimes thought of as particular to Lough Neagh, but they can be distinguished by looking at other terms used and the context: the Pollan is thought to have been an obligate freshwater species in the early modern period, whereas the shad species migrate to the sea and are often compared to the Herring:

> The English call them fresh Water Herrings, for want of another Name; for Pollan is an Irish Name … I am of the Opinion that they are that sort of Fish that is caught in the Sea, or between the fresh and Salt Water, call'd Shads; and that the large ones come from the Sea, as the Salmon doth, and leave their Spawn in the Lough [Neagh] … And that which confirms me in my Opinion is, that at the Salmon Fishing at Colraine, they catch many of the large ones going up to the Lough.
> (Neville 1713)

DISTRIBUTION

Because this species spends most of its time at sea, only entering river systems to breed, for the map county-level records have been assumed to refer to the coast, and river-level records have been shown by shading only the county where the river enters the sea, unless otherwise specified in the texts.

The distribution of top-quality records for this species is statistically a poor fit with the known levels of recorder effort. It may have been locally distributed, locally abundant or of special local interest. There are fewer records than we would expect from Highland Scotland and North England and more than we would expect from Ulster, Munster and Wales, perhaps suggesting a regional distribution or abundance. Habitat suitability modelling based on the sites where the Shad was recorded in the early modern period suggests that the species may have had specific requirements. It is only recorded on lowland and coastal sites that today have relatively mild and dry winters.

It is worth noting that there are records from inland rivers and loughs including Lough Neagh, Lough Erne, Lough Iron, Lough Derravaragh, the River Trent, River Thames and River Severn. Shad enter river systems to breed each spring, so it is possible that these rivers carried breeding populations in the early modern period. These species have clearly declined over the course of the modern period and many breeding populations seem to have been lost due to overfishing, pollution and obstruction of rivers (Maitland 2007: 109; Aprahamian 2004; Wheeler 1979: 65–6, 83). The species seem to have been occasionally eaten in medieval and early modern England (Serjeantson and Woolgar 2006; Wheeler 1979: 65–6), but fishing for them is likely to have been sustainable in the early modern period.

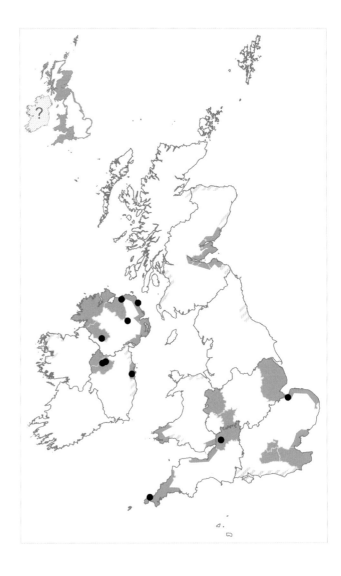

Shad records, 1519–1772.
There are coastal records from
Wales and Lowland Scotland,
and records further inland in the
South, South West and Midlands
of England, Ulster and Leinster.

Because the shads are migratory, and live most of their lives in the sea, they could have easily colonised Ireland after the end of the last glacial period (McCormick 1999; Griffiths 1997). Today, the shad species are found relatively rarely on the shores of Britain and Ireland and occasionally in river systems, especially (for the Twaite Shad) the River Severn and River Wye in Wales, the River Cree in Lowland Scotland, the Barrow and Suir in Leinster and the Blackwater in Munster. The Killarney Shad of course lives its whole life in Lough Leane (King et al. 2011; Maitland 2007: 112; Aprahamian 2004).

EUROPEAN PILCHARD/SARDINE

Sardina pilchardus

NATIVE STATUS Native

MODERN CONSERVATION STATUS	
World	Least Concern (Near Threatened in Europe)
UK	Not Evaluated
ROI	Not Evaluated
Trend since 1772	Uncertain

The Pilchard was recorded by the early modern sources in several places around Britain and Ireland, but was most frequently recorded in Cornwall, where Pilchard fishing was especially common.

RECOGNITION

This species is usually referred to as the *pilchard* in English. Tonkin provides *hernan* as the Late Cornish term (ed. Dunstanville 1811: 102), which needs to be distinguished from *hernan-gwidn* (literally the 'white pilchard' but meaning the Herring). In New Latin it was called *Harengus minor* to distinguish it from the nominal *Harengus* (= the Atlantic Herring). This makes it appear that the fish can be securely identified to species level in our sources. However, considering that in the modern period there is often confusion between the Pilchard and other similar species, it is plausible that some of the points on our map may be unreliable. Morris attests an English dialectal term *pinshers* (Morris 1747) which does not seem to be used elsewhere.

Some sources also refer to the *sardine/sardina*. Unlike today, during the early modern period this term was always used in coordination with other terms which show that it does not refer to the Pilchard (it is instead used to refer to the Herring, the Smelt and Sprat).

DISTRIBUTION

The distribution of top-quality records for this species is statistically a poor fit with the known levels of recorder effort. It may have been locally distributed, locally abundant or of special local interest. Habitat suitability modelling based on the sites where the Pilchard was recorded in the early modern period suggests that the species may have had specific requirements. It is not recorded in the coldest parts of Britain or Ireland. Most of the Pilchard locations provided on the map are from the south coast of Cornwall. This is partly due to a difference in local interest. The early modern sources relate that the Pilchard fishery was perhaps the most important industry in Cornwall in the period, and therefore there is much more information available about the local distribution of Pilchards from Cornwall compared to the rest of Britain and Ireland. However, the reason that there was a Pilchard fishery in Cornwall is that Pilchard is more common in Cornish waters than elsewhere in Britain and Ireland (Henderson 2014: 70), as Borlase recorded in the eighteenth century:

This fish comes from the north seas in immense shoals, and in the summer months, about the middle of July, reaches the islands of Scilly, and the Land's End of Cornwall; not driven by fish of the cetaceous kind (as some have thought), but shifting their situation as the season prompts, and their food allures them; thus by a tour to the warm southerly coasts of Britain, they strengthen and prepare themselves and their young ones to return to the great northern deeps, for the sake of spawning and securing themselves during the stormy season. The pilchard continues off and on in the south chanel, principally from Fawy harbour westward, and is taken sometimes in great numbers at Mevagissy, in the creeks of Falmouth and Helford harbours, in the creeks of St. Kevran, and in Mount's Bay; some pilchards are also taken in St. Ives Bay in the north chanel.   (Borlase 1758: 272)

Pilchard was also commonly caught elsewhere, as for example in early modern Ireland. Here, the fish is most commonly referred to on the south coast, and in the seventeenth and early eighteenth centuries several fisheries were established on the south and west coasts, but perhaps especially in Co. Cork (Went 1945). Pilchards were caught between November and May and fishing rights were jealously guarded by local communities (Mac Laughlin: 207–8). Munster's fishery was actually important to the international network of pirates operating around Europe and North Africa in the seventeenth

Pilchard records, 1519–1772. There are coastal records from the South and South West of England (where there is a cluster on the south coast), Lowland Scotland, Connacht, Munster and Wales.

century. Piracy provided off-season work for pilchard-fishers, and Pilchards provided food for the pirates who visited this area each summer (Mac Laughlin 2010: 150–5; Appleby 1990).

Today, the Pilchard remains much more abundant in the south of Britain and Ireland than in the north, and is especially rarely found in the North Sea, and far more abundant off Cornwall (Henderson 2014: 70–1). However, there has been a change: the Pilchard fishery off Munster began to collapse immediately after the end of the early modern period. The fishery off west Cornwall was more reliable for some time, and the Pilchard even temporarily replaced the Herring as the most common fish there during the warmer weather of the mid-twentieth century, but stocks eventually collapsed from their previous high in the 1960s and 1970s (Ford 1982: 140–1). There is an industry based around Pilchards ('sardines') today but it is unreliable due to fluctuating stocks (Fogarty 2017: 38–9; Henderson 2014: 70–1). The reason for the twentieth-century decline seems to have been overfishing and climate change (Southward et al. 1988; Ford 1982: 141).

HERRING

Clupea harengus

NATIVE STATUS Native

MODERN CONSERVATION STATUS	
World	Least Concern
UK	Not Evaluated
ROI	Not Evaluated
Trend since 1772	No change

The Herring was perhaps the most widespread fishery species recorded around early modern Britain and Ireland. It seems to have been found off every coast but was best recorded around western Scotland.

RECOGNITION

The term *herring* was used by almost all of the recorders, and this was generally specific to the Herring although it was rarely used for the Irish Pollan and Dace and needs to be distinguished from the term *herring gull*. The Irish term was *scadain* and the Scottish Gaelic was *skedan*. *Penwaig* and *ysgaden* were the Welsh terms, and Tonkin (ed Dunstanville 1811: 102n) gives *hernan-gwidn* as the Late Cornish. This last needs to be carefully distinguished from the generic *hernan*, which refers to the Pilchard.

DISTRIBUTION

The distribution of top-quality records for this species is statistically a poor fit with the known levels of recorder effort. It may have been locally distributed, locally abundant or of special local interest. But habitat suitability modelling based on the sites where the Herring was recorded in the early modern period suggests that the species would have been well adapted to conditions across Britain and Ireland. And in this case, the Herring is one of the ten best-recorded species in the early modern sources, and it is clear they could be found at times on every coast, so the pattern of records probably reflects regional abundance.

Herring (either 'white herring' preserved by salting, or 'red herring' preserved by both salting and smoking like a kipper) was perhaps the most commonly eaten fish in medieval Britain and Ireland,

Herring records, 1519–1772. There are coastal records from every region of Britain and Ireland including the Hebrides, the Northern Isles and a contested record from the Isle of Man.

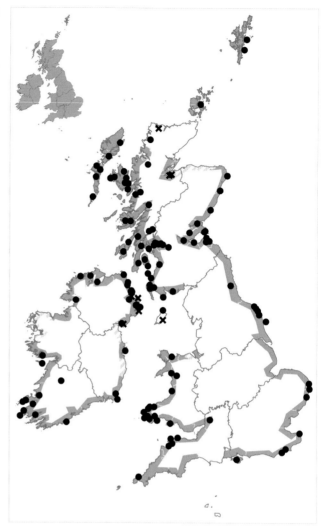

and one of the only fish easily attainable by ordinary people outside of urban areas and fishing communities at the beginning of the early modern period (Holm et al. 2019; Breen 2016; Locker 2016; Hamilton-Dyer 2007; Serjeantson and Woolgar 2006). British and Irish waters were very important internationally for Herring fisheries, and by the end of the thirteenth century, thousands of people from around Europe were fishing for Herring in the North Sea off the east coast of England (Holm et al. 2022; Kowaleski 2003). However, the fisheries off the east coast had already begun to fail in the fourteenth and fifteenth centuries, so fishing vessels began to ply elsewhere – by the early modern period often as far as Iceland and Newfoundland, and for Cod rather than Herring. The consumption of preserved fishes actually declined through the early modern period in Britain (although perhaps not in Ireland or elsewhere in Europe) and a wider range of fishes was eaten here by the end of the period (Holm et al. 2019; Locker 2016; Kowaleski 2003).

Herring is also one of the species thought to have been most influenced by the Little Ice Age (Ford 1982: 33, 140–1). These fish prefer temperatures of 3–13°C, and so at the end of the eighteenth century when water temperatures were consistently below this level, Herring are likely to have moved to the south of their distribution, away from Norway towards Britain (Lamb 2011: 190; Fagan 2000: chap. 3).

One primary source (quoted in 'Cod', p. 266) suggests that the Herring might have become rare around Shetland in this period. Yet there is some difficulty in interpreting this record in that local Herring fisheries are unreliable and often move or collapse. Our early modern records also mention failed fisheries off Anglesey, Co. Down, the Isle of Man, Sutherland and in the Moray Firth (Mitchell 1906: 192; Pococke 1887: 213–14; Sacheverell 1859: 14; Dodsley 1775: 7–8; Harris and Smith 1744: 241–2; Lesley 1675: 27–8).

Herring is still widely found throughout British and Irish waters today, so if anything, the range of this species is now more evenly spread than it was at the height of the Little Ice Age (Henderson 2014: 67). In the second half of the twentieth century, however, Herring fisheries around Britain and Ireland collapsed again due to overfishing (ICES 2021a). Although stocks have been recovering in the twenty-first century, the abundance of this fish during the early modern period may have been higher than it is today. There are few hints to quantify this, although Lewis Morris of Anglesey records a catch of over one million Herrings by 47 boats in one night off Aberystwyth in Ceredigion, Wales:

> 5th October 1745
>
> There were of Herrings taken in Aberystwyth, in Forty Seven Boats Two Thousand One Hundred & Sixty Mace in one night.
>
> Each mace is Five Hundred, and 126 to the Hundred.
>
> Then 126 x 9 = 630 in a mace
>
> And 630 x 2160 = 1,360,800
>
> Which is one million three-hundred & sixty thousand, eight hundred herrings.
>
> (Morris 1747: 311)

HAKE

Merluccius merluccius

NATIVE STATUS Native

MODERN CONSERVATION STATUS	
Europe	Least Concern
UK	Not Evaluated
ROI	Not Evaluated
Trend since 1772	Uncertain

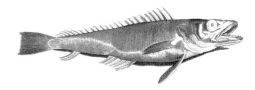

The Hake was recorded commonly in the early modern sources around Ireland and Britain south of Highland Scotland.

RECOGNITION

This species was usually referred to as the *hake*, and less often the *poor-jack* (once the *poor-john*) in English. In Welsh it was called *cegddu*, and O'Sullivan (2009: 172–3) attests the Irish *coilm oir*. In New Latin it was called the *Asellus primus* or *merlucius*. These terms all appear to be specific to our Hake. However, care needs to be taken to distinguish the other kinds of fish called *Asellus*, including, for instance, *Asellus longus* (Ling) and *Asellus major* (Cod).

DISTRIBUTION

The distribution of top-quality records for this species is statistically a poor fit with the known levels of recorder effort. It may have been locally distributed, locally abundant or of special local interest. There are more top-quality records than we would expect from Munster, Wales and the South West of England and less from North England or Scotland. Habitat suitability modelling based on the sites where the Hake was recorded in the early modern period suggests that the species may have had specific requirements. It is only recorded on sites that today have relatively warm summers and winters. This bias in the natural history sources is actually supported in the medieval and early modern export records and by archaeological evidence which shows that the species was most commonly eaten in Ireland and the South West of England in the past (Breen 2016; Mac Laughlin 2010: 98–100; Hamilton-Dyer 2007; Serjeantson and Woolgar 2006; Butlin 1991), and was less frequently consumed in other parts of England. To some extent this reflects the modern distribution of the species. Hake are much more common in the deeper waters off South West England and Ireland than they are around South England, the Midlands, or North England (Henderson 2014: 90; Kowaleski 2003). Yet today Hake are just as common around Scotland, particularly Highland Scotland, as they are around England, and the small number of records of the Hake from Scotland is surprising.

The species remains relatively rare off the coasts of Essex, where it was recorded as an occasional catch in 1730 in the *History and antiquities of Harwich and Dovercourt*:

> *Marlucius vulgaris*, Bellon[ius] De Aquat[icus] 124 … The Hake. This is sometimes caught here [off Harwich and Dovercourt, Essex]. When salted and dried, it is call'd Poor Jack.
>
> (Taylor and Dale 1730: 429)

Hake records, 1519–1772. There are coastal records from every region of Britain and Ireland except Highland Scotland, South England and Munster.

The Hake declined in the twentieth century, and at one point was essentially absent from the North Sea. Stocks have now recovered due to more sustainable quotas (ICES 2022a; Baudron and Fernandes 2015).

BURBOT
Lota lota

NATIVE STATUS Native to Britain, not Ireland

MODERN CONSERVATION STATUS	
World	Least Concern
UK	Extinct
ROI	Not Present
Trend since 1772	Certainly declined

The now extinct Burbot was recorded in early modern rivers across eastern England.

RECOGNITION

This species was reliably called the *burbot* or *bird-bolt* in Early Modern English and *Mustela fluviatilis* in New Latin. These terms are reliable where they appear but must be carefully distinguished from similar species: many of the mustelids (i.e. Weasel, Stoat, Polecat, Ferret) were also given New Latin names starting with *Mustela* in the early modern period, but the highest chance of confusion is with Three-bearded Rockling (*Gaidropsarus vulgaris*), a marine fish which was called *Mustela vulgaris* and *Mustela marina* in New Latin and also the *bourbee* in Early Modern English.

The Burbot was also regularly called the *eelpout* in Early Modern English, but this term is also used for *Zoarces viviparus* (still called the Viviparous Eelpout) (Worthington et al. 2010). The map only shows records using the term *eelpout* where it is used alongside one of the more reliable names, or it is present in a freshwater context. It is also once called the *coney-fish* (Merret 1666: 190).

DISTRIBUTION

For the obligate freshwater riverine species, records at river-level have been shown on the map by shading every county with similar topography which that river runs through.

The distribution of top-quality records for this species is statistically a poor fit with the known levels of recorder effort. It may have been locally distributed, locally abundant or of special local interest. If the species was widely distributed across Britain and Ireland, we would have expected records from Lowland and Highland Scotland, Ulster, Wales and the South West of England at least, and probably no records from the Midlands instead of five.

The early modern records of the Burbot are confined to an area of eastern England from Norwich to the River Derwent. This agrees approximately with previous studies, which suggests a maximum distribution split into four main areas: (i) the River Trent Catchment and River Ancholme, (ii) rivers draining into the Wash, (iii) Yorkshire rivers, (iv) eastward-flowing rivers of Norfolk and Suffolk (Worthington, 2010). It was presumably unable to disperse further after colonising this area at the end of the last ice age (Locker 2010; Everard 2021: 32). The British population seems to have been distinct from other Burbot populations, but was most closely related to the western European clade (Everard 2021: 25–7).

> Burbolts being a Fish not frequently met with in the Southern Rivers of England, are often found in this County [Yorkshire], especially in slow Rivers and standing Waters, as in the River Foss in York, and also in Derwent; but in no Place more frequent, than in the Fen Ditches of the Levels, about four miles from Doncaster.
> (Richardson 1713)

The Burbot relies on flooded wetlands to spawn, and the drainage of the Fens appears to have seriously affected it. A seventeenth-century broadside ballad called 'The Powte's Complaint' protests this

Burbot records, 1519–1772. There are records from the North, Midlands and part of the South of England.

drainage from the perspective of a burbot: 'For they do mean all Fens to drain, and waters overmaster, / All will be dry, and we must die, 'cause Essex calves want pasture' (in Everard 2021: 82–5).

The Burbot went extinct in Britain in the twentieth century, with a widespread decline from around 1850 (Worthington, 2010). The last reliable record is from 1969, with unverified records continuing into the 1970s (Everard 2021: 88–9; Marlborough 1970; Maitland & Campbell 1992). The most probable reasons for its extinction are pollution and habitat loss, and perhaps overexploitation at a local scale (Marlborough 1970). In other parts of its range, the Burbot requires very cold weather to spawn, but this does not seem to be as important for the Western European populations. The fish is still found elsewhere in Europe and beyond, and a reintroduction project has been proposed by Norfolk Wildlife Trust (Everard 2021). Burbot have already been reintroduced in Belgium.

LING
Molva molva

NATIVE STATUS Native

MODERN CONSERVATION STATUS	
Europe	Least Concern
UK	Not Evaluated
ROI	Not Evaluated
Trend since 1772	Probably no change

The Ling was attested by early modern recorders to occur around Scotland especially, but also Ireland, the west coast of England and Wales and off north-east England.

RECOGNITION
This species was called *ling* in Early Modern English and *Asellus varius* or *Asellus longus* in New Latin. These terms seem to be specific to the Ling, although *Asellus* as a genus included many different whitefishes. The term *ling* was apparently later used for the Burbot (p. 262), but none of the early modern records using this term refer to freshwater populations, so this appears to have been a later development.

DISTRIBUTION
The distribution of top-quality records for this species is statistically a poor fit with the known levels of recorder effort. It may have been locally distributed, locally abundant or of special local interest. This is surprising since habitat suitability modelling based on the sites where the Ling was recorded in the early modern period suggests that the species might have been well adapted to conditions all around the coasts of Britain and Ireland.

Like the other whitefishes, Ling were commonly eaten in early modern Britain and Ireland, particularly in the north and west. Ling was a commercial fishery, and fish were caught for the market (Harland and Kirkwall 2016; Hamilton-Dyer 2007; Serjeantson and Woolgar 2006; Kowaleski 2003).

Two primary sources (Leigh via Marr in Mitchell 1908: 252; Brand 1701: 30) suggest that the Ling may have declined in abundance in the Northern Isles at the height of the Little Ice Age at the end of the seventeenth century. This would fit with the archaeological record, which shows that people living in Orkney abruptly switched from eating larger whitefishes (especially Cod but also including Ling and others) to eating smaller Saithe in the early modern period (Harland and Kirkwall 2016). However, for Ling at least this decline may have been short-lived. Gifford was still able to buy 11,000 fresh Ling from fishers in Unst in 1733:

> The commodity of Shetland, which the merchants do for the most part trade with them is ling and cod, which they take with hooks and lines in small boats called yawls, about the beginning of Gravesend ocean … I bought of fisherman merchants of the island of Ounst 11655 [larger] gild ling, and two pence the gild cod … The said ling and cod being very good and merchantable, were salted on board the ship that landed me, and within seven weeks after my landing, I sent her for London with the said fish to the Earl of Pembroke.   (ed. Gifford 1879: viii)

Overall, the distribution of Ling records compares well to the species' current distribution. It is generally now common around the South West, and South Wales, Scotland and much of the island of Ireland (Wood 2018: 229; Henderson 2014: 92–3). It is more rarely found around north-west

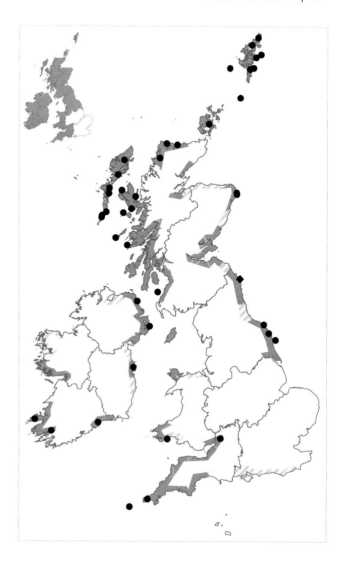

Ling records, 1519–1772. There are coastal records from every region of Britain and Ireland except the South and Midlands of England.

England and the east Midlands and south-east England. This almost exactly maps onto our records from the early modern period, with the possible exception that there are few modern records from Anglesey and the Isle of Man.

COD

Gadus morhua

NATIVE STATUS Native

MODERN CONSERVATION STATUS

World	Vulnerable
UK	Not Evaluated
ROI	Not Evaluated
Trend since 1772	No change

Cod were known to occur widely around the coasts of Britain and Ireland in the early modern period. They are recorded as being found everywhere that our contributors lived.

RECOGNITION

The most common term for this species in the early modern period was *cod* or *codfish*. It was also called *keeling* (or *killing*, *keiling*) especially in Scots. Morris attests the term *codfyn*, which might be Welsh or Cambro-English (Morris 1747), and K'Eogh (1739: 18) attests the Irish *trosk*. All these terms seem to exclusively refer to the Cod.

DISTRIBUTION

The distribution of top-quality records for this species is statistically a poor fit with the known levels of recorder effort. It may have been locally distributed, locally abundant or of special local interest. But habitat suitability modelling based on the sites where the Cod was recorded in the early modern period suggests that the species would have been well adapted to conditions all around the coasts of Britain and Ireland. In this case, there are far more records from the north of Scotland than we would expect, suggesting the recording bias is due to a local abundance of Cod in this region during the early modern period.

The Cod is one of the ten best-recorded species in the early modern sources. Cod was already one of the most commonly eaten fishes in Britain and Ireland by the end of the medieval period (Breen 2016; Harland and Kirkwall 2016; Locker 2016; Hamilton-Dyer 2007; Serjeantson and Woolgar 2006; Butlin 1991). In Britain, Cod subsequently declined in popularity through the early modern period, to the extent that it was rarely eaten by the end of the eighteenth century (Locker 2016). On the contrary, in Ireland and elsewhere in Europe the demand (particularly air-dried Cod, or 'stockfish') grew significantly in the same time period, and Cod replaced Herring as the most eaten fish from around 1530 (Holm et al. 2019, 2022).

Having said that, Cod is also one of the species most vulnerable to changes in temperature. The fish needs temperatures of 3–7°C to breed, and at temperatures significantly below this they will die. At the height of what is called the Little Ice Age, temperatures in the north Atlantic became colder than this. Cod fisheries are known to have failed off Norway, Iceland, the Faeroes and also Shetland, especially between 1685 and 1704 (Grove 2004: 607; Fagan 2000: chap. 4). The North Atlantic Cod fishery declined by more than a quarter during these years (Holm et al. 2022). There was also a shift from Cod to Saithe fishing in the Northern Isles more generally through the early modern period (Harland and Kirkwall 2016). The effect of this seems to be summarised by Hugh Leigh, minister of Bressay, Shetland, most likely as a contribution to Robert Sibbald's (1684) *Scotia Illustrata*:

> In old time the sea about this Coast was well stored with all common sort of fishes, as Mackrels, Herrings, Lings, Cods, Haddocks, Whiting, Sheaths, but especially with Podlines, i.e. young

Sheaths called by the Inhabitants Pelltacks, which in fair weather would come so near to the shore, that men yea and children, from the Rocks with Fishing-rods, could catch them in abundance. But all kinds of Fishing is greatly decayed here notwithstanding that greater pains is taken by the Fishers now than ever before.   (Leigh via Marr in Mitchell 1908: 252).

A very similar observation is made by Brand (1701: 30), who wrote about the Northern Isles on a trip two decades later, except that Brand focuses specifically on Cod and Ling as the species that have declined. This decline affected fisheries for four decades. The numbers of Cod caught in the North Atlantic only started to rise again from 1720. By the end of the early modern period, in 1780, around one million tonnes (liveweight) of Cod and Herring were caught each year in the north Atlantic (Holm et al. 2022). This figure is only 20% smaller than the amount caught at the beginning of the twentieth century, after the adoption of steam-power technology, by which point stocks of Cod had noticeably declined in the North Sea (Cardinale et al. 2015).

From the 1960s to the 1980s, Cod dramatically increased their abundance and were caught in huge numbers, this was the gadoid outburst, a period when many of the chip-shop whitefishes increased their abundance. Those decades are now long past, and the Cod has been in steep decline (Fernandes

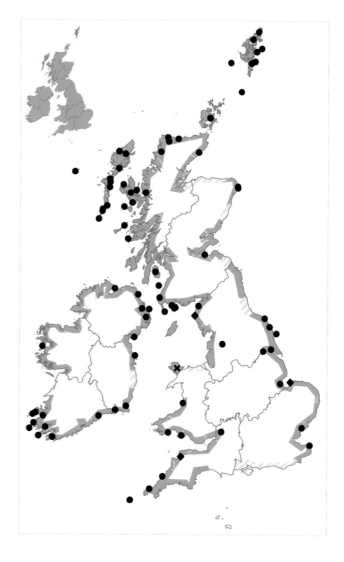

Cod records, 1519–1772. There are coastal records from every region of Britain and Ireland including the Hebrides, the Northern Isles, the Isle of Man and the Isles of Scilly.

and Cook 2013). In the 1990s and 2000s the fish also changed its distribution from being most common in the North Sea on the coasts of England and Scotland to being most common in the northern and north-eastern parts of the North Sea, due to a combination of overfishing and climate change (Engelhard et al. 2014).

The Cod recently made a brief partial recovery in the North Sea, apparently due to the decrease in fishing pressure, and the fishery there was accredited as sustainable by the Marine Stewardship Council for a short time (2017–19) (ICES 2021b; MSC 2019; Brander 2018). However, the North Sea Cod population has now once again declined back to its lowest point, and the fishery is no longer classed as sustainable.

Despite the poor abundance, Cod are still often found around Britain and Ireland today. They are rarer on the Atlantic coasts than they are elsewhere around Britain and Ireland (Wood 2018: 227; Henderson 2014: 102–3).

HADDOCK
Melanogrammus aeglefinus

NATIVE STATUS Native

MODERN CONSERVATION STATUS	
World	Vulnerable (Least Concern in Europe)
UK	Not Evaluated
ROI	Not Evaluated
Trend since 1772	No change

The Haddock was recorded by the early modern sources all around the coasts of Britain and Ireland, including in southern waters where it is no longer common.

RECOGNITION
This species was most often called the *haddock* in Early Modern English. This term appears to be specific to the Haddock and is used in every early modern record. Sometimes this term is accompanied by other terms like the Early Modern English *St Peter's fish*. In New Latin the species was most commonly called the *Asinus antiquorum*. These terms are also specific to the Haddock. It was also sometimes considered a species of *Asellus* in New Latin (e.g. *Asellus callarias*, *Asellus minor* or *Tertia asellorum*), but these names were not consistently used and thus are hard to differentiate (every fish-and-chip-shop whitefish could be called *Asellus*). The references shown on the map are all likely to be trustworthy because they all use the reliable English term *haddock*.

DISTRIBUTION
The distribution of top-quality records for this species is not statistically different from the known level of recorder effort. The gaps on the map may just reflect decreased survey effort in some regions. It may have been widespread, and was recorded in every region. It was commonly eaten in late medieval and early modern Britain and Ireland (Breen 2016; Harland et al. 2016; Hamilton-Dyer 2007; Serjeantson and Woolgar 2006; Butlin 1991).

Haddock records, 1519–1772. There are coastal records from every region of Britain and Ireland including the Hebrides, the Northern Isles and the Isle of Man.

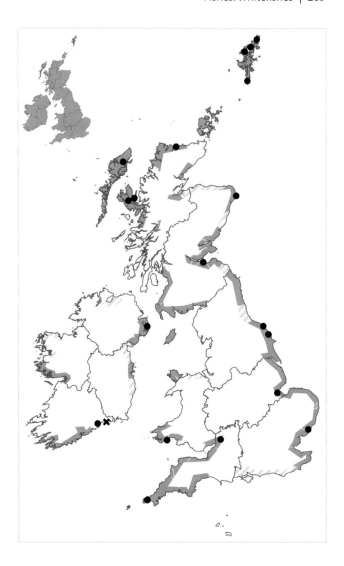

The Haddock is on the southern edge of its distribution in Britain. Today, the species is mainly encountered on the southern coasts of Britain and Ireland only in winter and rarely found in the English Channel (Henderson 2014: 104). However, this does not seem to have been the case in the past. Taylor (1730: 428), for instance, writing from Essex called it 'a common Fish in these Seas'. The species is expected to shift its distribution as a response to global warming (Jones et al. 2013), which might make us think the records were correct to suggest the fish had a more southerly distribution during the Little Ice Age. There is one example where an early modern source considers the Haddock rare. Smith, writing in 1746, recorded that Haddock could no longer be found off Co. Waterford:

> The Haddock some years ago frequented this coast, and were taken in great plenty; but, at present, there are none to be seen, scarce one being taken a year; nor can there be any tolerable reason assigned for the almost present extinction of this species of fish which formerly swarmed on the coast.
>
> (Smith 1746: 259–60)

The temperature in 1746 is likely to have been warmer than in 1710 (as described in 'Cod', p. 266). However, six years later when Pococke visited in 1752, Haddock were once again being caught (albeit 10 leagues offshore):

> There is a bank about ten leagues from Dungarvan, where they catch quantities of Hake and
> Haddock, Cod, Ling, and many other kinds of Fish. They have also a bed here of very large
> oysters.
>
> (Pococke 1891: 141)

This perhaps implies that there were temporary interruptions in the Haddock fishery in the early modern period, similar to the failure of the Herring fisheries in certain years. Yet analysis of Haddock stocks during global warming has found no reduction in range; indeed, the fish actually increased its distribution around Britain during a period of warming sea temperatures (Baudron et al. 2011; Dickey-Collas et al. 2003). It is possible that the interruptions in early modern Haddock fisheries were due to temporary depletion of local stocks by overfishing. For comparison, some local stocks in the North Sea are thought to have been depleted by longline fishing shortly after the end of the period, in the nineteenth century (Cardinale et al. 2015).

Haddock is still fished off Britain and Ireland today. It is mainly found to the west of Scotland and in the Atlantic off Cornwall, but it expanded around Ireland to the Irish and Celtic Seas in the 1980s and 1990s (Henderson 2014: 104; Dickey-Collas et al. 2003). The species is still caught (apparently sustainably) around the north of these islands, especially off the west coast of Scotland, but at far lower levels than it was in the twentieth century (ICES 2021c; Hislop 1996).

WHITING

Merlangius merlangus

NATIVE STATUS Native

MODERN CONSERVATION STATUS	
World	Least Concern
UK	Not Evaluated
ROI	Not Evaluated
Trend since 1772	No change

The Whiting was recorded in the early modern sources as occurring all around the coasts of Britain and Ireland.

RECOGNITION

This species was most commonly called the *whiting* in Early Modern English. This term does seem to be specific but needs to be carefully distinguished from *whiting-polluck* which referred to the Pollock. Similarly, Morris attests the Welsh term *chwitlyn gwyn*, which seems to be specific but needs to be distinguished from the term *chwitlyn glas* (apparently meaning the Saithe). The New Latin term *Asellus mollis major* and the rarer Early Modern English term *buckhorn* seem to be specific to the Whiting.

This map also includes generic references to the English term *whitefish*, which probably could apply to many of the species in this section of the atlas. These records are shown as diamonds.

DISTRIBUTION

The distribution of top-quality records for this species is not statistically different from the known level of recorder effort. The gaps on the map may just reflect decreased survey effort in some regions. It may have been widespread.

Whiting records, 1519–1772.
There are coastal records from every
region of Britain and Ireland including
the Hebrides and the Northern Isles.

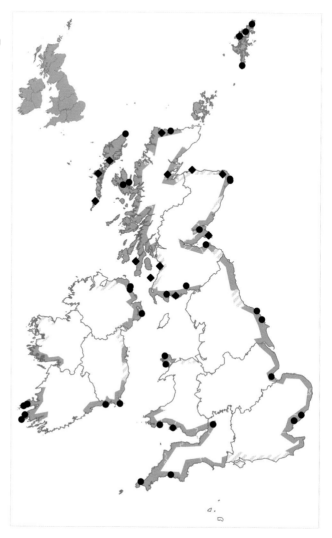

The Whiting seems to have been eaten in medieval and probably early modern Britain and Ireland (Harland et al. 2016; Hamilton-Dyer 2007; Serjeantson and Woolgar 2006). The cluster of records from Co. Kerry are all from a single text ('Mather and Bays of Kerry',[8] written by Curtis in around 1683) and do not necessarily suggest a higher abundance around Munster than elsewhere in Britain and Ireland – although the Whiting does seem to have been more commonly targeted in the west of these islands than off the east coast of England (Kowaleski 2003):

> This is certaine, which all families of note can give an account, as they have by tradition that before Terra Neuf [Newfoundland] was found the harbours bays & places in or Neere Kerry to wit: nere Dingle, Valentia, Vintree, the Skellix etc. were swarmd with French, Spanish & Biscay fishermen cheifly the 2 latter, all for cod, though herring mackerell & whiting doe there alsoe abound.
> (ed. from O'Sullivan 1971 [c.1683])

---

8   *Mather* meaning madder (apparently here *Rubia tinctorum* rather than the native *Rubia peregrina*), a plant used to make red dye.

The Whiting remains widespread in British and Irish waters, but the only population now abundant enough to support a fishery is the one in the North Sea. Here the fish has declined in abundance since the gadoid-outburst years of the 1960s and 1970s but currently seems to be recovering (ICES 2021d,e; Hislop 1996).

SAITHE

Pollachius virens

NATIVE STATUS Native

MODERN CONSERVATION STATUS	
World	Least Concern
UK	Not Evaluated
ROI	Not Evaluated
Trend since 1772	Uncertain

The Saithe was regularly recorded around early modern Britain and Ireland, but especially commonly in the north.

RECOGNITION

This species had many names. It was most often called the *coalfish* in Early Modern English and the *seath* in Middle Scots. In Ulster, the Hebrides and once on Anglesey, this species is sometimes called the *graylord* (which needs to be distinguished from the term *grayling*, *Thymallus thymallus*, although at least one early modern author confuses the two). In New Latin it was the *Asellus niger* to distinguish it from all the other species called *Asellus* (which included many of the chip-shop whitefishes). Our early modern sources also attest a few regional terms including the Norn *silluk*, the Irish *glasson*, the Welsh *chwitlyn glas* (to be distinguished from the *chwitlyn gwyn* – the Whiting). Finally, there is the term *rawlin-pollack* (possibly a mistake for *rauning-pollack*) which Ray attests from Cornwall:

> At Pensance we saw and described several sorts of fish, to wit, *Mullus major*, *Trachurus*, *Pagrus*, *Erthrinus*, hake, haddock, whistling-fish, rawlin, pollacks, holibut, conger, and tub-fish which is no other than a red gurnard. We there saw the houses in which they lay (and the manner in which they press) their fish, especially pilchards.   (ed. Ray and Derham 1846 [1658–62]: 190)

DISTRIBUTION

The distribution of top-quality records for this species is not statistically different from the known level of recorder effort. The gaps on the map may just reflect decreased survey effort in some regions. It may have been widespread. Looking at the county-level records, the species was recorded in almost every region of Britain and Ireland.

The Saithe was one of the fish eaten in medieval Britain and Ireland (Hamilton-Dyer 2007; Serjeantson and Woolgar 2006). It continued to be consumed in the early modern period. It seems to have become a more important staple of the diet in the Northern Isles, in particular following the failure or abandonment of the Cod fishery there in the early modern period (Harland and Kirkwall 2016).

The Saithe is now common in northern waters (particularly on the North Sea and Atlantic coasts) (Wood 2018: 227; Henderson 2014: 113). It is fished commercially in the North Sea (ICES 2021f;

Saithe records, 1519–1772.
There are coastal records from
every region of Britain and Ireland
except the Midlands of England.

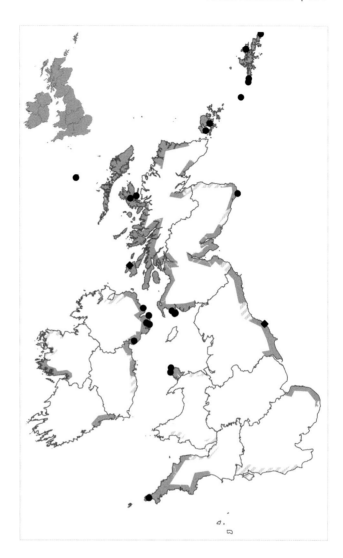

Hislop 1996). The species has declined by around two thirds from its high abundance during the gadoid outburst of the 1960s and 1970s. The distribution of the Saithe has been predicted to shift northwards between 1985 and 2050 due to the warming climate (Jones et al. 2013).

POLLOCK
Pollachius pollachius

NATIVE STATUS Native

MODERN CONSERVATION STATUS	
World	Least Concern
UK	Not Evaluated
ROI	Not Evaluated
Trend since 1772	Uncertain

The Pollock was recorded in several early modern sources as occurring frequently around Britain and Ireland.

RECOGNITION
The most certain names for this species are *pollock* in Early Modern English and *lythe* in Middle Scots. Some of our recorders treat these as different species, but the two are now understood to be the same. These names appear to be specific to the Pollock.

There are also two less obvious names. First, *podley* in Middle Scots usually seems to refer to our species but may have also applied to the Saithe. Luckily, all records of *podleys* include other more reliable names. There is also the term *whiting-pollock* in Early Modern English, which is associated with the New Latin term *Asellus virescens*. Based on both the English and New Latin terms, this appears at first sight to be a name for the Saithe. However, almost every author who uses it also separately mentions the Saithe under another term (usually *cole-fish* or *saithe* and *Asellus niger*). The detailed description given by Willughby and Ray make it clear that, surprisingly, the species intended by these names in the early modern period was actually the Pollock:

> Asellus Huitingo-Pollachius: virescens Schonfeld. A Whiting Pollack.
>
> …
>
> The lateral lines begin to curve into an arc around the first dorsal fin, and rise away from the back; around the middle section of the second fin it descends to the middle of the side, and from there it runs straight to the tail.
>
> …
>
> It differs from cod … because it lacks barbels.
>
> (translated from Willughby & Ray 1686: 167–8)

Lewis Morris attests *morlas* as the Welsh term; although this also appears to have been used for the mullet species.

DISTRIBUTION
The distribution of top-quality records for this species is not statistically different from the known level of recorder effort. The gaps on the map may just reflect decreased survey effort in some regions. It may have been widespread.

As with the other whitefishes, the Pollock is likely to have been occasionally eaten during the medieval and early modern period, although there is less archaeological evidence for this than for other species (Harland and Kirkwall 2016; Hamilton-Dyer 2007; Serjeantson and Woolgar 2006).

Pollock records, 1519–1772. There are coastal records from every region of Britain and Ireland except Connacht and the Midlands of England. There are also no records from the mainland of Highland Scotland.

Today the Pollock is widely distributed and found around every coast of Britain and Ireland (Wood 2018: 226; Henderson 2014: 114–15). It is especially commonly encountered off the western shores of Britain, which fits with where it was best recorded in the early modern period. There is no evidence this species has changed its distribution, and it is not often targeted by commercial fisheries.

MONKFISH/ANGLERFISH

Lophius piscatorius

NATIVE STATUS Native

MODERN CONSERVATION STATUS	
World	Least Concern
UK	Not Evaluated
ROI	Not Evaluated
Trend since 1772	Uncertain

The Monkfish was recorded by a handful of authors around early modern Britain and Ireland.

RECOGNITION

This species was most commonly called in New Latin *Lophius piscatorius*. In Early Modern English it had several vernacular names, commonly being called the *frog-fish* or *toad-fish*, the *devil fish* or *sea devil* or rarely *monk*. Morris attests *llyffant y mor* and *morlyffant* for Welsh (Morris 1747).

This species is also rarely called the *marmaid* in Middle Scots. Care needs to be taken with this last term, since early modern people often believed in the half-human half-fish creatures that we would still call mermaids today, and some natural history sources recorded sightings of them around Britain and Ireland (e.g. Monteith 1711: 78, and compare Erroll in Mitchell 1908: 228; Gibson 1695: 373–4; Falle 1694: 76; Miege 1691: 207). Records of *marmaids* are only identified as referring to Monkfish when they also use a more reliable term.

The records on the map shown with diamonds are uncertain. The record from Gloucestershire is a *sea tad*, assumed to be a variant spelling of *sea toad* which is used elsewhere for Monkfish. Charles Leigh provided another uncertain record:

> the Rana Piscatrix, or Sea-Toad, found frequently in the River Wire in Lancashire: It has an extream wide Mouth, and is said to be a very voracious Creature, it is not eaten as Food by the People, but I have seen them eagerly devour'd by the Sea-Gulls, and some of them almost peck'd to a perfect Skeleton; the Rows of its Teeth are not much unlike those of a Shark; wherefore I conclude its usual Food is upon small Fishes: It yields a great Quantity of Oyl but extremely fetid.
> (Leigh 1700: 186)

The description here uses terms that usually refer to the Monkfish, and the physical description is reasonable, although the presence of the Monkfish in river systems would be very unusual since they prefer to stay at depths of at least 50 metres: On the assumption that this record might refer to a specimen caught or washed up in the Wyre estuary it is indicated on the map as a diamond.

Leigh (1700) actually provided a very basic picture of the Monkfish to accompany his description. This picture was cited by John Wallis (1769: 382), who called it the *cat-fish* and described it as a marine fish with whiskers. Wallis seems to have been confused here. Leigh did elsewhere (1700: 146) refer to a *cat-fish*, but the plate was supposed to show the Monkfish. The identification of this *cat-fish* with the Monkfish is unlikely since both Leigh and Wallis described the Monkfish separately using the more reliable term *Rana piscatrix*, and because the description of the *cat-fish* also includes poisonous spines or fins, which the Monkfish does not have. The term *cat-fish* was used by other authors (e.g. Merrett 1666: 185) to describe the Stingray (*Dasyatis pastinaca*), which does have poisonous spines on its tail, and this is more likely to be the species intended by the term *cat-fish* in Leigh and Wallis – the illustration by Leigh even resembles a Stingray, explaining Wallis's confusion. Less likely identifications are the Wolffish (*Anarhichas lupus*) which was occasionally called the *cat-fish*, more commonly the *sea-cat*, never the *morcath*. The species we now think of as the ordinary catfish (*Silurus glanis* – the Wels Catfish) cannot have been the species intended because it is an obligate freshwater species and was not introduced until later, although it is worth noting that the sturgeons were occasionally referred to using the Latin terms *silurus* and *glanis* (e.g. Taylor and Dale 1730: 434). No records of *cat-fish* have been included on the map for this species.

DISTRIBUTION

The distribution of top-quality records for this species is not statistically different from the known level of recorder effort. The gaps on the map may just reflect decreased survey effort in some regions. It may have been widespread. The handful of records from Britain and single record from Ireland

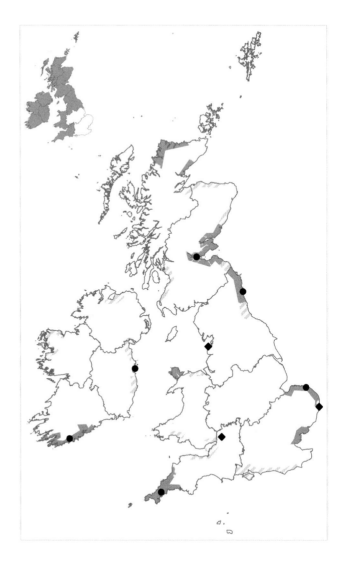

Monkfish records, 1519–1772. There are coastal records from every region of Britain and Ireland except Ulster, Connacht and the Midlands of England.

here are from those recorders most interested in marine species. This helps explain why some of the locations given (off Norfolk/Suffolk, potentially off Lancashire) are not the most likely areas to find the species today. As we have seen, the Monkfish seems to have been thought of as a special curiosity (Parsons 1750) and does not appear to have been widely eaten, although Rutty (1772: 349) does refer to it being preserved and eaten cold like a ray.

Today the Monkfish is found most commonly around Ireland and Scotland and less commonly around the west of Britain and the north east of England (Wood 2018: 247; Henderson 2014: 116–17). The species is more rarely found in the shallower seas off the Midlands and the south east of England. It is commonly targeted by commercial fisheries around Britain and Ireland, but the populations currently seem to be stable (ICES 2020a 2021g). The Monkfish is likely to shift its distribution northwards as a result of climate change (Jones et al. 2013; Solmundsson et al. 2010).

MULLETS

Chelon labrosus, Chelon ramada and *Chelon auratus*

NATIVE STATUS Native

MODERN CONSERVATION STATUS (ALL SPECIES)	
World	Least Concern
UK	Not Evaluated
ROI	Not Evaluated
Trend since 1772	Uncertain

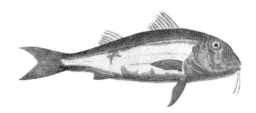

Mullets were regularly recorded off the southern coasts of early modern Britain and Ireland.

RECOGNITION

This map shows records of the English *mullet* or *grey mullet* and the Latin *mugil* and *mullus*. For the most part, these records are not distinguishable to species level, with the exception that John Ray distinguishes the *Mullus major* in his records from Penzance, St Ives and Tenby, perhaps intending *Chelon labrosus*. Having said that, the early modern recorders sometimes distinguish Red Mullet (*Mullus surmuletus*). This latter species is normally called *surmullet*, *salmoneta* or *red mullet*.

The term *mullet* was also rarely used in Early Modern English for the Atlantic Puffin. This species can usually be distinguished based on context.

DISTRIBUTION

The distribution of top-quality records for these species is statistically a poor fit with the known levels of recorder effort. They may have been locally distributed, locally abundant or of special local interest. Most of the early modern mullet records are from the Celtic Sea, Irish Sea and English Channel, with only a few from the north or east of Britain. There are no records from Scotland, where marine fishes were very well recorded. Habitat suitability modelling based on the sites where mullets were recorded in the early modern period suggests that these species may have had specific requirements. They were not recorded in areas with the coldest summer or winter temperatures.

The mullet was occasionally caught and eaten in late medieval Britain and Ireland, especially Wales (Hamilton-Dyer 2007; Serjeantson and Woolgar 2006; Kowaleski 2003). This seems to have continued into the early modern period, when one source from Essex also talks about salting the fish's roe to make bottarga:

> The Mullet. These are Inhabitants of the Sea, but the Delicacy of the Flesh, and the Difficulty of catching them enhance the Price; so that they are not Food for every one's Taste. Of the Ova or Spawn of the Females, salted and dried, is made Botarg, which quickens a depraved Appetite, excites Thirst, and a Gust to Wine.   (Taylor and Dale 1730: 431)

The early modern distribution actually fits approximately with the modern distribution of the mullets (Henderson 2014: 118–25; Maitland 2007: 266–73). *Chelon auratus* (Golden Grey Mullet) and *Chelon*

Mullet records, 1519–1772. There are coastal records from every region of Britain south of Scotland, and every region of Ireland except Connacht.

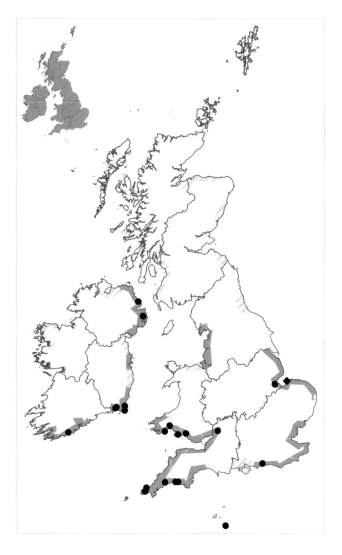

ramada (Thinlip Grey Mullet) both have a southerly distribution around Britain and Ireland and are rarer in the north. *Chelon labrosus* (Thicklip Grey Mullet) is theoretically widespread, but is only really visible in northern areas during the summer: this fish only migrates north and enters the shallow water of the estuaries there during the summer (Colclough 2004). These species are rarely targeted by commercial operations around Britain and Ireland today.

GREY GURNARD
Eutrigla gurnardus

NATIVE STATUS Native

MODERN CONSERVATION STATUS	
Europe	Least Concern
UK	Not Evaluated
ROI	Not Evaluated
Trend since 1772	Uncertain

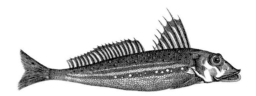

The early modern sources provided a few records of Grey Gurnard dotted around Britain and Ireland.

RECOGNITION

This species was most reliably called the *grey-gurnard* in Early Modern English, *penhaiarn gwyn* in Welsh, and *Gurnardus griseus* or *Cuculus griseus* in New Latin. These names are specific and allow identification to species level.

Our map also includes generic references to *gurnard*, *lyra* and *crooner*. These were generic terms which could refer to any kind of gurnard, although the species were more normally distinguished (e.g. *Trigla lyra*, the Piper Gurnard was called *piper*, while *Chelidonichthys lucerna*, the Tub Gurnard was called *tub*). The Welsh term *ysgyfarnog y mor* is also generic and can refer to, for example, *Chelidonichthys cuculus* (the Red Gurnard). Falle (1694: 160) attests the term *gronnard* was used for this species in Jèrriais (Jersey French), which again seems to be generic. Records using these generic terms have been indicated using diamonds, and the historical counties they are recorded in have not been shaded, unless there is also a more reliable record.

DISTRIBUTION

The distribution of top-quality records for this species is not statistically different from the known level of recorder effort. The gaps on the map may just reflect decreased survey effort in some regions. It may have been widespread.

The Grey Gurnard seems to have been eaten in the medieval and early modern periods and is sometimes found on late medieval and early modern archaeological sites from Britain and especially from Ireland (Harland et al. 2016; Hamilton-Dyer 2007; Serjeantson and Woolgar 2006).

The species remains widespread on all the coasts of Britain and Ireland (Henderson 2014: 174–5). It seems to have declined in the twentieth century, perhaps due to trawling, and has now lost its former place as one of the most commonly caught species on the western coasts (Rogers and Ellis 2000). It is not often targeted for fishing around Britain and Ireland at present and stocks seem to have been relatively stable since the early 1990s (ICES 2020b).

Grey Gurnard records, 1519–1772. There are coastal records from every region of Britain and Ireland, except that the records from Connacht, Highland Scotland and the North and Midlands of England are uncertain.

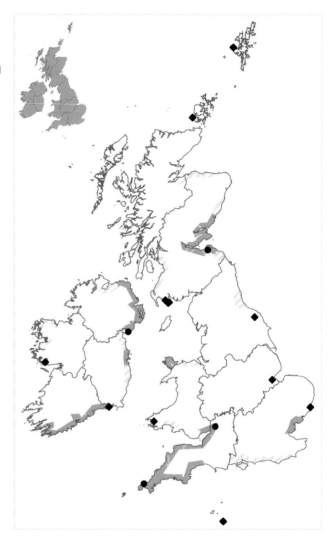

SANDEELS
Ammodytidae spp.

NATIVE STATUS Native

MODERN CONSERVATION STATUS	
World	Least Concern/Not Evaluated
UK	Not Evaluated
ROI	Not Evaluated
Trend since 1772	Uncertain

The early modern sources often recorded sandeels around the coasts of Britain and Ireland.

RECOGNITION

This map includes records of species that were referred to as *sand eels* and *launces* in Early Modern English, *llamriaid* in Welsh, *lançons* in Jèrriais, and *ammodytes* in Latin. There are two possible Middle Scots attestations – Archibald Stevenson (via Sibbald 2020: 37 (II:3)) provides *horn eel*, and more dubiously Monro (ed. Martin and Monro 2018: 356) suggests *pintill fisch* (lit: penis fish).

Ray (1713b: 38) attempts to distinguish a species called *Sandilz anglorum*, based on a description by Salvian of a species with two fins on its back, but no such fish is known today. In modern times we distinguish six species of sandeel around Britain and Ireland, but these are difficult to tell apart and the early modern authors lumped them together. For this reason, our references have been identified only as Ammodytidae species.

DISTRIBUTION

The distribution of top-quality records for these species is not statistically different from the known level of recorder effort. The gaps on the map may just reflect decreased survey effort in some regions. They may have been widespread.

There is evidence these species were eaten in medieval and early modern England (Locker 2018: 67–70; Harland et al. 2016). Our early modern texts suggest the same was true in Ireland, as in this account of Richard Pococke's travels in Co Donegal:

> On the 14th [of July 1752] I set out to go round Ennishowen and came in two miles to the strand, where I saw people at work with wooden shovels in turning up the sand as the sea left the strand and enquiring what they were about they told me they were catching sand eels; I observ'd that the moment the wave leaves the sand they run in the shovel and turn up the sand and the fish are taken; they are about 4 or 5 inches long, very small for their length, are made like a whiteing and they say are very good. (Pococke 1891: 46)

Today, there are sandeels on every coast of Britain. Most of the species are generally distributed except Corbin's Sandeel (*Hyperoplus immaculatus*) and the Smooth Sandeel (*Gymnammodytes semis-quamatus*), which are both rare on the east coast of Britain north of East Anglia (Wood 2018: 252; Henderson 2014: 204–11). The most common species is the Lesser Sandeel (*Ammodytes marinus*).

Sandeels are intensively fished and are processed into fishmeal and oil for livestock and fish-farming (ICES 2021i 2021h; Heath et al. 2012; Pullar and Low 2012: 124–5).

In the first half of the twentieth century, the Lesser Sandeel began to decline in population. This might have gone unnoticed except that the Puffin, and other seabirds which depend on the Lesser Sandeel to feed their chicks, also began to decline. At the time it was suspected that the warmer climate at the start of the twentieth century must have been partially responsible (Lockley 1953: 152, 163). One reason that sandeels are especially vulnerable to climate change is because they require a particular sandy substrate habitat; they cannot shift their distribution to find more suitable conditions as easily as other fishes to adapt to changing temperatures. The second half of the twentieth century was (initially) cooler, and sandeel populations started to recover (Grove 2004: 608; Ford 1982: 126–7). However, in the 1990s, sandeels started to decline again, this time especially in the North Sea. Initially, this decline might have been caused by overfishing, but numbers have continued to fall despite a subsequent ban on sandeel fishing across a large stretch of the east coast of England and Scotland (Heath et al. 2012). Closer analysis of this decline in the North Sea suggests that a further problem there is that although plenty of sandeels are hatched, few survive their first winter. Very young sandeels

Sandeel records, 1519–1772. There are coastal records from every region of Britain and Ireland except Connacht and the Midlands of England.

are dependent on copepod eggs for food, and the reason for their poor survival rate might be a timing mismatch between sandeel hatching and the egg production of one particular species of copepod (*Calanus helgolandicus*) (Régnier et al. 2017; MacDonald et al. 2015). Both copepod egg production and sandeel embryonic development are thought to be dictated by water temperature, so the cause for the decline of the sandeel may ultimately be climate change.

EUROPEAN BASS/SEABASS

Dicentrarchus labrax

NATIVE STATUS Native

MODERN CONSERVATION STATUS	
World	Least Concern
UK	Not Evaluated
ROI	Not Evaluated
Trend since 1772	Probably increased

The Seabass was commonly recorded around the coasts of South West England, South Wales and southern and eastern Ireland during the early modern period.

RECOGNITION

This species was generally referred to as the *bass* or *base* in Early Modern English and, less commonly, the *draenog* in Welsh and the *lupus* in Latin. None of these terms is specific: *Lupus* can refer of course to the Wolf on land and more confusingly to the Wolffish at sea. I have only included records of the *lupus* here where the name *bass* is also used. *Draenog* refers to *Erinaceus europaeus* (the Hedgehog) on land, and both the terms *draenog* and *bass* are both also sometimes applied to the Perch in freshwater habitats. It is not possible to distinguish the records based on location, because although the European Bass is normally a marine species it does frequently enter estuaries and even river systems (Maitland 2007: 292). Luckily, it has almost always been possible to reliably distinguish the two species based on context, or where a single text uses both names to describe different fishes.

In modern times, the term *bass* has been occasionally adopted for members of the *Sparidae* family (usually called the sea breams) (Maitland 2007: 292), but this does not seem to have been the case in the early modern period.

DISTRIBUTION

The distribution of top-quality records for this species is statistically a poor fit with the known levels of recorder effort. It may have been locally distributed, locally abundant or of special local interest. Habitat suitability modelling based on the coastal areas where the Bass was recorded in the early modern period suggests that the species may have had specific requirements. It was not recorded in the coldest parts of Britain and Ireland.

Most of our records are from South West England, southern and eastern Ireland and South Wales. This is approximately consistent with the current records, where Bass are most often found in the Celtic Sea and English Channel (Wood 2018: 231; Henderson 2014: 236–7). However, the species is now also regularly (although not as commonly) also found off the northern and eastern coasts of Britain. This seems to be due to a northwards extension in its range between the 1960s and 1990s, probably due to climate change along with shifts in food chains caused by fishing (Henderson 2017). In the early modern period, the records from Northumberland and Co. Down apparently represented the northern extent of the species' distribution. While the record from Co. Down suggests that the

Bass records, 1519–1772. There are coastal records from the South and South West England, Ulster, Munster and Leinster, and Wales. There is also a record of a rarity from North England.

Bass was common there, the record from Northumberland implies it may have been rare in the early modern period:

> One of those remarkable fishes, called the Basse, approaching the figure of a salmon, of a deep blackish colour on the back, and of a silvery white on the belly, was taken near Cresswell, and brought to me. It is a firm well-tasted fish. (Wallis 1769: 382)

As suggested in the record above, the Bass was occasionally eaten in medieval and early modern Britain and Ireland (Hamilton-Dyer 2007; Serjeantson and Woolgar 2006; Kowaleski 2003), perhaps especially on the Celtic Sea coast. The population continues to be targeted by both commercial and sport fishers and has significantly declined since the 2010s, probably due to overfishing and low recruitment (ICES 2022b; Beraud et al. 2018).

BLUEFIN TUNA
Thunnus thynnus

NATIVE STATUS Native

MODERN CONSERVATION STATUS	
World	Endangered
UK	Not Evaluated
ROI	Not Evaluated
Trend since 1772	Uncertain

The early modern naturalists recorded Bluefin Tuna as an occasional but well-known fish, most common off the western Atlantic shores, especially in Cornwall.

RECOGNITION

This species was most often called the *tunny* in Early Modern English and *thunnus* in Latin. Other species in the genus *Thunnus* are occasionally found in the seas around Britain and Ireland (i.e. *Thunnus alalunga*, the Albacore; *Thunnus obesus*, the Bigeye Tuna) but *Thunnus thynnus* (Bluefin Tuna) is the most common in British and Irish waters (Henderson 2014: 268), so this is the most likely identification for all of the records on the map.

Our authors also use the related terms *stoer-mackrel* in Middle Scots and *Spanish mackrel* in Early Modern English to describe this species. These terms have to be carefully distinguished from the *mackrel* (Mackerel) and the *horse-mackrel* (the shads). The term 'Spanish mackerel' today is the name for the species in the scomberomorini tribe. Sibbald actually distinguishes his North Sea *Spanish mackrel* from the true Atlantic *tunny* (ed. Sibbald 1803: 124), which might make us think that another species is intended; however, since none of the scomberomorini species actually occurs around Britain or Ireland, it is likely that the *Spanish mackrel* and *tunny* were the same early modern species, which is the way that the other early modern naturalists understood the terms.

DISTRIBUTION

There are not enough top-quality records to statistically analyse the distribution of this species.

The archaeological record suggests that Bluefin Tuna have historically been mainly caught in southern parts of Europe, and in eastern areas of the North Sea around Norway, Sweden and Denmark (Andrews et al. 2022). The species seems to have been only occasionally eaten in early modern Britain and Ireland and was mainly mentioned by our recorders as a rarity, or as a species especially found in Cornwall (almost half of the early modern presence records are county-level records attesting that tuna are caught around Cornwall). On the other hand, the Tuna was sufficiently well known to be recognised:

> August the 29th [1667], at Pensance we saw a large tunny which was taken in the pilchard nets. They call them Spanish mackrel. It was about seven feet long and of a great bigness; his stomach was full of pilchards.
> (Ray and Derham 1846: 189–90)

Bluefin Tuna records, 1519–1772.
There are coastal records from the
South West of England, Connacht,
and Lowland and Highland Scotland.

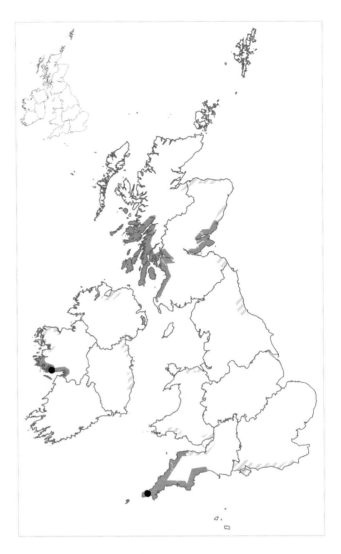

Commercial and sport fisheries for Bluefin Tuna began in the North Sea in the early decades of the twentieth century, and intensified in the 1930s and 1940s (Fromentin 2009; MacKenzie and Myers 2007). The species subsequently disappeared from the North Sea around the 1960s and 1970s and was then extremely rare for some time. Since the 2010s it has shown some signs of recovery, perhaps due to fishing restrictions (Nøttestad et al. 2020).

MACKEREL
Scomber scombrus

NATIVE STATUS Native

MODERN CONSERVATION STATUS	
Europe	Least Concern
UK	Not Evaluated
ROI	Not Evaluated
Trend since 1772	No change

The Mackerel was recorded widely around the coasts of Britain and Ireland by the early modern sources.

Mackerel records, 1519–1772. There are coastal records from every region of Britain and Ireland, including the Hebrides, the Northern Isles and the Isle of Man.

RECOGNITION

This species was ordinarily called *mackrel* in Early Modern English and *scomber* in Latin. Morris (1747) attests the Welsh term *macrell*. Identification is usually straightforward, but the English term has to be carefully distinguished from what was called the *Spanish mackerel* (the Tuna) and from the *horse mackerel* (various shads).

DISTRIBUTION

The distribution of top-quality records for this species is not statistically different from the known level of recorder effort. The gaps on the map may just reflect decreased survey effort in some regions. It may have been widespread.

> At Polton [in Galloway], in the months of July, August, and September, are sometimes great quantities of herring and mackreels taken with nets. On the coast of Whithern, Glasserton, and Mochrum, they take cronands, codlings, lyths, seathes, or glassons, mackreels by hook and bait in boats &c. On the mouth of the water of Luce, they take salmon, herring, and mackreels, in a fish-yard belonging to Sir Charles Hay of Park Hay, as I formerly said.   (ed. Symson 1823 [1684]: 93)

Mackerel is not the most common fish on medieval and early modern archaeological sites in Britain and Ireland but does seem to have been occasionally eaten (Locker 2018: 67–70; Harland et al. 2016; Mac Laughlin 2010: 263; Hamilton-Dyer 2007; Serjeantson and Woolgar 2006; Kowaleski 2003). It has become much more popular since the early modern period and today is commonly targeted by sport anglers around these islands. The species' abundance is generally stable, but the North Sea population drastically declined in the 1960s and 1970s and has still not recovered – so there is no commercial Mackerel fishery off much of the eastern coast of Britain (ICES 2021j; Iversen 2002).

SWORDFISH

Xiphias gladius

NATIVE STATUS Native?

MODERN CONSERVATION STATUS	
World	Least Concern
UK	Not Evaluated
ROI	Not Evaluated
Trend since 1772	Uncertain

The Swordfish was not usually considered to be found around Britain and Ireland. It was, however, recorded by a few authors in the early modern period, especially in the North Sea and Celtic Sea.

RECOGNITION

This species was called the *sword-fish* in Early Modern English and *xiphias* or *gladius* (or *gladius-piscis*) in New Latin. It is the only species that was known by these names around early modern Britain and Ireland.

DISTRIBUTION

There are not enough top-quality records to statistically analyse the distribution of this species.

Swordfish records, 1519–1772.
There are coastal records from South,
South West and North England,
Lowland Scotland and Wales.

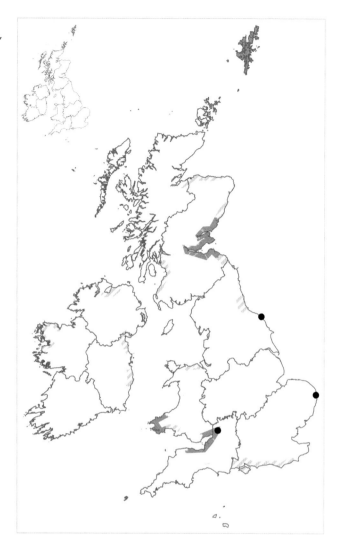

The Swordfish does not seem to have been regularly eaten in the early modern period. Remains of the species from archaeological sites are rare, but a Swordfish sword was excavated from early modern Alderman's Walk in London (Pipe 1992). However, the species does seem to have been relatively well known. In the medieval period it was one of the bestiary species (White 1954: 201), and the names used for it in the early modern texts are consistent. The species is sometimes described as one of ordinary fishes of the region (as in Berkeley, Pembrokeshire and the Firth of Forth) and sometimes as an interesting rarity, as in Norfolk and Yorkshire:

> A sword fish or Xiphias or Gladius intangled in the Herring netts at Yarmouth agreable unto the Icon in Johnstonus with a smooth sword not vnlike the Gladius of Rondeletius about a yard & half long, no teeth eyes very remarkable enclosed in an hard cartilaginous couercle about ye bignesse of a good apple.
> (ed. Browne 1902 [c.1662–68]: 36)

Today, the Swordfish seems to be on the northern edge of its distribution in Britain and Ireland (Quigley et al. 1994; Hoey et al. 1989). It nevertheless continues to be occasionally recorded here, especially on the Atlantic coast, much like the Leatherback Turtle (p. 321). It is too rare for a fishery to be established here, but it is targeted elsewhere in the world.

TURBOT
Scophthalmus maximus

NATIVE STATUS Native

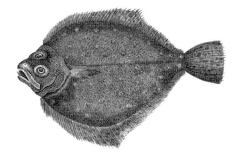

MODERN CONSERVATION STATUS	
World	Not Evaluated (Vulnerable in Europe)
UK	Not Evaluated
ROI	Not Evaluated
Trend since 1772	Certainly declined

The Turbot was widely recorded by early modern sources around Britain and Ireland. It seems to have been present off every coast including around Shetland, which is outside of the modern distribution of the species.

RECOGNITION
This species can be most easily recognised in early modern texts by the vernacular name *turbot* in Early Modern English and *torbwt* in Welsh. This name is almost always attached to *Scophthalmus maximus*, although there are exceptions. The New Latin names *Rhombus maximus* and *Rhombus maximus asper non squamosus* are also uniquely used for this fish.

However, other names in Latin and the vernacular languages were in use as well, most of which were not unique to this species. There is the Welsh *lleden chwith* (left-eyed flatfish) the Irish *liothoig muiri* (perhaps also used for the sole) and the Cornish-English *midge* and *pilchard-midge* (the former of which more normally refers to the little biting flies in the *Culicoides* genus). Finally, the very common terms *brit* and *bret* could refer either to the Turbot or, more rarely, the Brill. It is usually possible to find which of these is intended in any given text by looking at other synonyms used and checking to see if the other species is differentiated under another name.

DISTRIBUTION
The distribution of top-quality records for this species is not statistically different from the known level of recorder effort. The gaps on the map may just reflect decreased survey effort in some regions. It may have been widespread.

The early modern distribution recorded for the Turbot is very interesting because it supports the observation of Kerby et al. (2013) that the species has declined in the twentieth century, possibly due to the use of mechanised beam trawlers. At present the species is on the edge of its range in British and Irish waters. It is less common around Scotland and around north Ulster than the rest of Britain and Ireland (Henderson 2014: 282–3). This has been the case since the 1960s, but before then the Turbot seems to have been more widely distributed around Britain and Ireland. It was especially common on Turbot Bank off the coast of Aberdeenshire in north-east Scotland, where it is now rare (Kerby et al. 2013).

From this perspective, the most interesting of the early modern records are the two from Shetland, where the Turbot is no longer usually found at all. The species is included in an early modern list of

fish encountered in Dunrossness (Monteith 1711: 46), but the record from Bressay is more enter-
taining (and perhaps dubious!):

> They tell a pleasant story of an Eagle and a Turbot. About six years since an Eagle fell doun on a
> Turbot Sleeping on the surface of the Water on the East side of Brassa, and having fastned his
> Claws in her he attempted to flee up, but the Turbot awakning and being too heavy for him to
> flee up with endeavoured to draw him down beneath the Water, thus they strugled for some
> time, the Eagle labouring to go up and the Turbot to go down till a Boat that was near to them
> and beheld the sport took them both, selling the Eagle to the Hollanders then in the Countrey.
> (Brand 1701: 177)

Flatfishes were regularly eaten in medieval and early modern Britain and Ireland, although the Turbot
was not as commonly eaten as Flounder or Plaice (Locker 2018: 67–70; Harland et al. 2016; Mac
Laughlin 2010: 80; Hamilton-Dyer 2007; Serjeantson and Woolgar 2006). Today the Turbot continues
to be caught in commercial and sport fisheries in the North Sea (Henderson 2014: 282–3). The species'
population declined through the second half of the twentieth century but now seems to be stable
(ICES 2022c; Kerby et al. 2013).

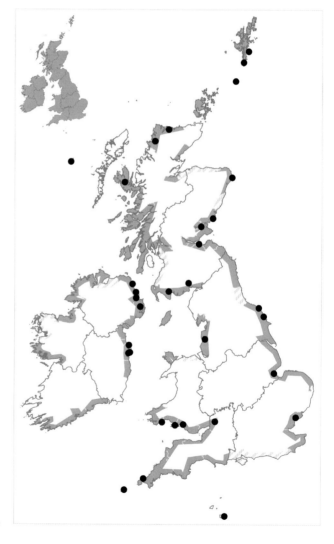

Turbot records, 1519–1772. There are
coastal records from every region
of Britain and Ireland including the
Northern Isles, the Isle of Man, the
Isles of Scilly and the Channel Islands.

BRILL

Scophthalmus rhombus

NATIVE STATUS Native

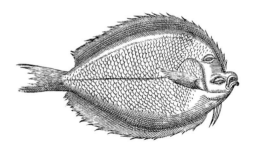

MODERN CONSERVATION STATUS

Europe	Least Concern
UK	Not Evaluated
ROI	Not Evaluated
Trend since 1772	Uncertain

The Brill was only occasionally recorded by the early modern authors most interested in marine species. It is best attested in early modern England and Wales.

RECOGNITION

This species was most reliably called by two common names: the *lug-aleaf* and the *pearl* or *pril*. These names are usually reliable identifiers, although of course, references using the term *pearl* have to be distinguished from records of the Freshwater Pearl Mussel (p. 330). In New Latin our species was called the *Rhomboides* and *Rhombus* (*non aculeatus*) *squamosus*. Where present, the New Latin name is also enough to identify it; the term theoretically distinguishes it from *Rhombus aculeatus*, also called *Rhombus maximus asper non squamosus* (meaning the Turbot).

It was also given various names terminating in variations of *-flook* (e.g. *spotted fleuk*), but these were not consistently used from one text to the other and the term *flook* by itself usually referred to the Flounder. It was most ambiguously occasionally called the *bret* or *brit*, as for example by Browne (ed. 1902: 45), but this name is also used for the Turbot. It is sometimes possible to tell which species is intended by the term *bret* where multiple names are used for a single record or where both species are included under different names. But in some instances, even when the species are differentiated, the description provided is ambiguous – as for example in the description of the *bret* in Co. Waterford, which is the only Irish record:

> We make a distinction between a halibut, a turbot and a bret on this coast … The bret, though exactly of the same shape with the turbot is distinguished from iᵗ 1st, By the smoothness of the skin; the other being rough and prickly on the back 2dly, By its being spotted, like a fluke; the turbot being without spots. 3dly, It is never so large as a turbot, nor so thick; and when dressed, eats more watry somewhat like a fluke or plaise; and therefore, not so much esteemed.
>
> (Smith 1746: 263–4n)

The description of the size and pattern here shows confusion with the Turbot and this species is only mentioned in the footnotes, making it uncertain. All uncertain records are shown as diamonds on the map.

DISTRIBUTION

There are not enough top-quality records to statistically analyse the distribution of this species.

Today, the Brill has the same distribution as the Turbot. It is widely distributed around Britain south of Scotland and around Ireland except the northern coast of Ulster, but rarer north of this range, and does not normally seem to occur around Shetland (Henderson 2014: 284–5). It is tempting to point out the lack of reliable local or county-level records of the Brill from early modern Scotland and suggest that the species had the same distribution in Britain in the past. This interpretation would

Brill records, 1519–1772. There are coastal records from every region of Britain and from Munster in Ireland, but only the records from Wales and South, South West and North England are reliable.

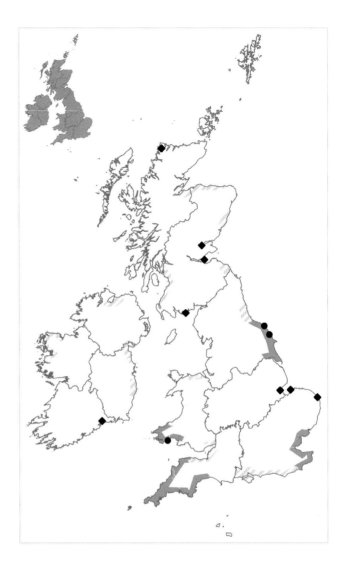

be supported by the fact that, unlike the Turbot, the Brill is known to have shifted its distribution only very slightly during the twentieth century (Kerby et al. 2013). However, it is only fair to point out as well that there are also no reliable records from Ireland, which would not fit with the modern distribution. The lack of early modern records from Scotland could just as easily have been due to the Brill being poorly recorded in the early modern period. So, it is not possible to be sure whether the species had the same distribution in the past.

The Brill does not seem to have been commonly eaten in the medieval and early modern periods but has been found occasionally during archaeological excavations of medieval English sites (Serjeantson and Woolgar 2006). At present the fish is caught by commercial fisheries mainly as bycatch. Brill populations declined in the second half of the twentieth century but are now stable (ICES 2022d; Kerby et al. 2013).

FLOUNDER

Platichthys flesus

NATIVE STATUS Native

MODERN CONSERVATION STATUS	
World	Least Concern
UK	Not Evaluated
ROI	Least Concern
Trend since 1772	No change

The Flounder was commonly recorded around Britain and Ireland in our early modern sources, including up rivers at some freshwater sites.

RECOGNITION

This species is normally called the *flounder* or *flook* (spelt variably, e.g. *fluke*, *fleuk*) or occasionally the *but* in Early Modern English (Maitland 2007: 314) and most usually *Passer fluviatilis* or occasionally *Passer amphibius* in New Latin. The Flounder seems to have been the nominal *flook* species, so these names are specific and allow the reference to be identified to species level. Other names like *Passer laevis* are also used for other similar species so need to be treated cautiously.

Morris attests the Welsh term *lleden*, but distinguishes this from the *lleden goch* (the Plaice) (Morris 1747). Similarly, K'Eogh attests the Irish term *lighoge*, which he distinguishes from the *liboge-brac* (again the Plaice), although O'Sullivan (2009: 178) uses the suspiciously similar term *liothoig* to refer to the Sole. Since all the references on this map provide at least one reliable name, they should all be records of the Flounder.

DISTRIBUTION

The distribution of top-quality records for this species is not statistically different from the known level of recorder effort. The gaps on the map may just reflect decreased survey effort in some regions. The Flounder was recorded in every region, so was certainly widespread in early modern Britain and Ireland, as it is today (Henderson 2014: 294–5; Peaty 2004).

The Flounder spends most of its life in estuaries, and sometimes moves into the freshwater sections of rivers (Maitland 2007: 313; Peaty 2004). Flounders can sometimes swim far upriver, although they always have to return to the sea to breed. There are early modern records of Flounders in the freshwater sections of the River Thames above London, the River Foyle between Dunnalong and Lifford, in Loch Lomond and, less plausibly, in the River Severn in Shropshire. The following is the description of the fish of Loch Lomond by Alexander Graham (1724):

> This loch [Lomond] abounds with salmond, trout, pike, perch, eell, fflounder, brase, and a most delicious fish called Powan or Pollacks only peculiar to itself.   (Graham in Mitchell 1906: 345–6)

Flatfishes were commonly eaten in medieval and early modern Britain and Ireland, and the Flounder is one of the two most frequently seen flatfishes in archaeological excavations (Harland et al. 2016; Hamilton-Dyer 2007; Serjeantson and Woolgar 2006). Today the Flounder continues to be caught by sport anglers but is now less commonly the target of commercial operations around Britain and Ireland. Almost a third of the Flounders caught in the North Sea are now discarded, with others often used for bait (ICES 2022e).

Flounder records, 1519–1772.
There are coastal records from every
region of Britain and Ireland, and
also inland records from Lowland
Scotland, Ulster, and the South, South
West and Midlands of England.

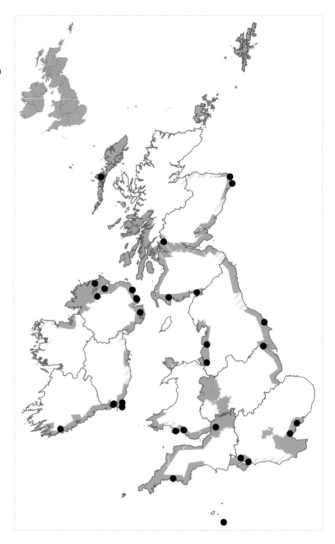

The Flounder appears to have briefly gone extinct in the Thames Estuary in the first half of the twentieth century, but the population became re-established when the river was cleared of pollution (Henderson 2017; Wheeler 1979: 163–7).

PLAICE

Pleuronectes platessa

NATIVE STATUS Native

MODERN CONSERVATION STATUS	
World	Least Concern
UK	Not Evaluated
ROI	Not Evaluated
Trend since 1772	No change

The Plaice was widely eaten in the early modern period and was recorded around the coasts of early modern Britain and Ireland.

RECOGNITION

This species was called *plaice* in Early Modern English, which was specific to our species and is luckily used in all of our records. The Welsh term *lleden goch* and the Irish term *liboge-brac* are both carefully distinguished from the more generic terms used for the Flounder (*lleden*, and *lighoge*).

The New Latin terms were *Passer*, or *Platessa* or occasionally *Passer laevis* or the *Passer* of Belon. All of these, including the multinomial New Latin names were more generic and were used for other flatfish species, as for instance *Passer asper* (Dab) and *Passer fluviatilis* (Flounder) as well as, confusingly, *Passer domesticus* (House Sparrow, which retains the genus name today).

DISTRIBUTION

The distribution of top-quality records for this species is statistically a poor fit with the known levels of recorder effort. It may have been locally distributed, locally abundant or of special local interest. There are more records than expected from Ulster and the South West of England, and less than expected from Highland Scotland. But habitat suitability modelling based on the sites where the Plaice was recorded in the early modern period suggests that the species would have been relatively well adapted to conditions all around the coasts of Britain and Ireland. This fits with the county-level records coming from every region as of Britain and Ireland as well as the modern distribution of the species (Wood 2018: 260; Henderson 2014: 296–7). One possible explanation for the bias in the recording is that it might be related to the cultural importance of the Plaice fishery in certain regions of Britain and Ireland. One such location was the northern coast of Ulster, especially Co. Antrim:

> This Bay [Carnlough] is about a mile long and a fine small white sand … hard by the Boats are three large caves in the Rocks of Freestone open to the sea, wherein poor people commonly live and one of them hath paid hearth money. This Bay yields great plenty of fish, as Salmon, Turbet, Plaice, Sole, Codd, Whiting, Mackerel, Ling, Hallybutt, a fish somewhat like a Turbet, and herrings; I have seen the people stand upon the shore, some wading a little way out, and draw in small nets upon the shore in a dark night, and the lookers on with small bags in their hands, some would throw sand in the faces and eyes of the Fishers, others with their bare feet or toes make holes in the nets for the herrings to slip out and so whip them into their Bags and away with them.
> (ed. Hill 1873 [1683]: 382)

Flatfishes were regularly consumed in medieval and early modern Britain and Ireland, with Plaice and Flounder the two most commonly eaten species (Harland et al. 2016; Hamilton-Dyer 2007; Serjeantson and Woolgar 2006; Kowaleski 2003). Since the early modern period, the Plaice has become the most important commercial flatfish in Europe. There are fisheries in the English Channel, Irish Sea, Bristol Channel and Celtic Sea and especially in the North Sea (ICES 2022e). The species

Plaice records, 1519–1772. There are coastal records from every region of Britain and Ireland, including the Isles of Scilly and Channel Islands. There is a cluster of records on the east coast of Ulster.

actually seems to have increased in abundance during the twentieth century, perhaps due to the introduction of legal minimum mesh size on trawlers after the Second World War, which allowed smaller fish to escape from trawls, and also due to the prevention of fishing in certain waters, first by minefields placed during the war, and then following the establishment of 12-mile exclusive territorial waters (Rijnsdorp and Millner 1996). The species shifted its range northwards and moved into deeper water in the twentieth century as a response to climate change (Engelhard et al. 2011). Following the increase in Plaice, there was a brief decline in the 1980s and 1990s but Britain and Ireland's populations now seem to be stable and the fish are still caught in greater numbers than at the beginning of the twentieth century.

DOVER SOLE

Solea solea

NATIVE STATUS Native

MODERN CONSERVATION STATUS	
World	Data Deficient (Least Concern in Europe)
UK	Not Evaluated
ROI	Not Evaluated
Trend since 1772	No change

The early modern sources recorded the Sole occurring around much of Britain and Ireland, apparently including around the north of Scotland where it is now rarely found.

RECOGNITION

The records presented on our map are for a species called *sole* in Early Modern English, sometimes *solefleuk* in Middle Scots and *solea* or sometimes *buglossus* in New Latin. O'Sullivan attests the word *liothoig* in Irish, which is suspiciously similar to the term *lighoge* used for the Flounder. Morris attests a Welsh term *tafod yr ych* (Morris 1747). This might be a gloss of the New Latin term, since both *buglossus* and *tafod yr ych* mean ox-tongue; both terms are also used in the early modern period for various wild plants like *Echium vulgare* (Viper's Bugloss).

The Dover Sole was distinguished from other common flatfish like the Flounder, Plaice and Dab. The term *sole* might occasionally also have been used for other similar species like Lemon Sole (*Microstomus kitt*), but this species is less often caught by anglers and was distinguished by at least one early modern recorder (Browne 1902: 45), so most of the records on our map are likely to be reliable.

DISTRIBUTION

The distribution of top-quality records for this species is statistically a poor fit with the known levels of recorder effort. It may have been locally distributed, locally abundant or of special local interest. There are more records than expected from the Celtic and Irish Sea and English Channel.

Today, the Dover Sole is very common around Britain and Ireland, but reaches the northern edge of its range here (Wood 2018: 258; Henderson 2014: 308). It is thought to be less able to tolerate the colder conditions in the northern waters of Scotland. Habitat suitability modelling based on the top-quality records of the Dover Sole from the early modern period suggests that the species may have also had specific requirements in the past. The records are associated with coastal sites which today have relatively mild and dry winters, but this is to some extent undermined by the records from Sutherland, Orkney and Shetland. Wallace's description of the fishes in Orkney explains:

> Many Ottars and Seols are to be had every where, and oft times Spout Whales and Pellacks run in great number upon the shore and are taken, as in the Year 1691, near Kairston in the Mainland, there run in a Bay no less than a hundred and fourteen at once. The Stellae-marinae and Urtica-Marina are oft thrown in great plenty. In the Sea they catch Ling, Keeling, Haddock, Whiting, Mackrel, Turbat, Scate, Congre-Eels, Sole, Fleuks, &c. and sometimes they catch Sturgeon.
>
> (Wallace 1700: 47)

This is surprising because it suggests that although the Sole appears to be on the northern edge of its range in British and Irish waters, it might have actually been commonly caught in the north of Scotland during the coldest period of the Little Ice Age. There are two plausible explanations for this:

(i) It is possible that records from the north of Scotland are mistaken, and actually refer to another species like Lemon Sole or Solenette (*Buglossidium luteum*) which are more widely distributed in the north of Scotland at present. (ii) It is also plausible that the Sole, like the Turbot, has truly moved southwards: during the twentieth century the Sole is known to have shifted its North Sea distribution southwards in response to global warming and fishing mortality (Engelhard et al. 2011). The Dover Sole was not common around the north of Scotland even at the beginning of the twentieth century, but it might possibly have undertaken southern shifts in distribution during the eighteenth or nineteenth centuries as well.

Today the Sole is an important commercial fish and is caught in the Irish Sea and on Ireland's Atlantic coast, in the Bristol Channel and the Celtic Sea, the English Channel and most importantly in the North Sea. Stocks of Sole declined at the start of the twenty-first century, but have since stabilised. There seems to have been a slight recovery in the early 2020s resulting from the reduction in fishing pressure (ICES 2022f).

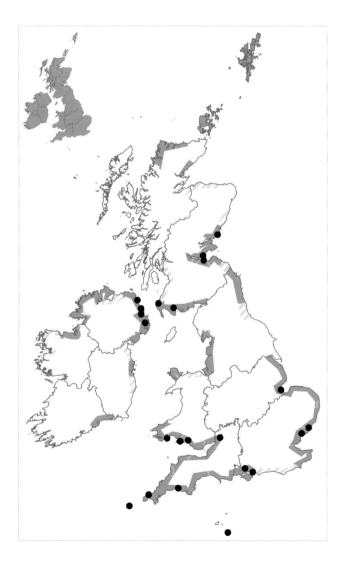

Dover Sole records, 1519–1772. There are coastal records from every region of Britain and Ireland, including the Northern Isles, the Isles of Scilly and the Channel Islands. There is a cluster of records on the east coast of Ulster.

HALIBUT

Hippoglossus hippoglossus

NATIVE STATUS Native

MODERN CONSERVATION STATUS	
World	Endangered (Vulnerable in Europe)
UK	Not Evaluated
ROI	Not Evaluated
Trend since 1772	Certainly declined

The Halibut was recorded by early modern sources around Britain and Ireland, including in the Thames Estuary where it is now absent.

RECOGNITION

The Halibut was most commonly called something like the *holybut* (spelling varies) in Early Modern English and *hippoglossus* in New Latin. These terms are specific to our species. The Halibut seems to also sometimes have been referred to as the *turbot*, which is unfortunate because that obviously invites confusion with the species now called the Turbot. Our map only includes records that refer to the species as either *holybut* or *hippoglossus* in addition to any other terms used, so the distribution is as reliable as it can be but it is possible that some of the references on the map are mistaken.

DISTRIBUTION

The distribution of top-quality records for this species is not statistically different from the known level of recorder effort. The gaps on the map may just reflect decreased survey effort in some regions. It may have been widespread.

Today, the Halibut is found most commonly in the deep northern waters around Scotland, Ulster and north Connacht. It is found less often in North England and off Munster and the western coasts of the South West of England and south-west Wales (Henderson 2014: 302). Most of the early modern records would fit into the modern distribution, as long as it is assumed that the species was present but unrecorded in some additional areas where it is now found, such as around the Northern Isles and Hebrides. However, there are some exceptions. The records from the Isle of Man, Co. Waterford and the Severn Estuary are peripheral to the Halibut's modern distribution. The record from Co. Dublin is outside of the species' modern distribution, but need not trouble us since it is of a rarity. The record from the Thames Estuary is the most interesting. The species is included in a list of fishes caught in the River Thames east of London (meaning in this case the estuary):

> The Halybut
>
> The Halybut is the largest of all Flatfish, that we are acquainted with. It exceeds the Turbot greatly and is of a longer Make; one of these Fish of a Yard long is about half a Yard broad. The upper Part of his Body is of a dusky Green, almost black; the Scales are small and there is no Roughness on the upper Part nor Prickles at the Root of the Fins; the Eyes are placed on the right Side or to the Left of the Mouth; the Fins at a greater Distance from the Head than other Flat-Fish.
>
> It is found in the German and British Ocean and likewise in the Irish Sea; it is thought to be nearly as good as a Turbot for the Fineness of its Taste.　　　(Griffiths and Binnell 1758: 218)

The description here is quite specific and suggests that this record is not a mistake. The 'German and British Ocean[s]' refers to the North Sea and English Channel respectively and, considering that this description comes from a list of fishes found in the Thames Estuary, this suggests that the Halibut formerly had a much wider distribution around southern England than it has today. It is likely to have been lost from this area during the modern period.

As the quotation above suggests, the Halibut was also occasionally eaten in the early modern period and its remains have been very occasionally excavated from pre-industrial Britain and Ireland (Harland et al. 2016; Mac Laughlin 2010: 226, Hamilton-Dyer 2007). The species was caught more regularly through the nineteenth century, and this may have been the cause of its decline. Early Swedish longline fishery data suggests that stocks of Halibut were already depleted in the North Sea by the beginning of the twentieth century (Cardinale et al. 2015). The species continued to be caught elsewhere and stocks across the whole of Europe declined through the second half of the twentieth century (Glover et al. 2006). There is no longer a viable fishery for the Halibut around Britain and Ireland, but the species is still caught by recreational anglers around the north of these islands (Henderson 2014: 302–3).

Halibut records, 1519–1772.
There are coastal records from every region of Britain and Ireland except Connacht, Highland Scotland and the Midlands of England.

COMMON TOAD

Bufo bufo

NATIVE STATUS Native to Britain, not Ireland

MODERN CONSERVATION STATUS	
World	Least Concern
UK	Not Evaluated
ROI	Not Present
Trend since 1772	Uncertain

The Common Toad was recorded in the same parts of Britain as today in the early modern period. It was recorded in Jersey but not Guernsey, parts of Orkney but not Shetland and Skye but not in the Outer Hebrides. It was not recorded in Ireland.

RECOGNITION

Both the Early Modern English term *toad* and the Latin *bufo* usually seem to have applied to the Common Toad in the early modern period. Unfortunately, however, our sources do not distinguish the Common Toad from the rarer Natterjack Toad (*Epidalea calamita*), so some of the records on the map may be records of the latter species.

DISTRIBUTION

The distribution of top-quality records for this species is not statistically different from the known level of recorder effort. The gaps on the map may just reflect decreased survey effort in some regions. The lack of records from Wales should probably be taken as an oversight rather than as suggesting that the Common Toad was absent, just like with most of the other amphibians and reptiles – although in this case there are also absence records that make it clear that the Common Toad was absent from Ireland and several smaller islands, but present across the mainland of Britain.

The Toad was of considerable interest to people in the early modern period. The species often had negative connotations as an ugly, hateful creature, and one that was associated with witchcraft and the devil (DeGraaff 1991: 100–14; Thomas 1991: chap. 2). These ideas and others were highlighted in 1579–81 when there was an outpouring of political satires against Queen Elizabeth I's suitor, Duke Francis of Alençon, which were encoded in stories about Toads in order to escape censorship (Adler 1981). At the same time, though, the Toad did also have some positive connotations. The bones of Toads were believed to have special virtues and were collected for use in amulets. So-called toadstones were thought to be useful antidotes to poison (Chambley 2000; DeGraaff 1991: 147f). Such early modern cultural beliefs about Toads are unlikely to have significantly affected the real Toad's abundance or distribution in Britain, even when they included hunting the species, but might have boosted the cultural currency of the Toad enough to encourage authors to record it where it occurred.

As well as its cultural currency, the idea that the Toad was present in Britain but not Ireland made the presence and absence of this amphibian on the various smaller islands around the British and Irish mainlands a matter of special curiosity to early modern writers. Early naturalists theorised that

Toad records, 1519–1772.
There are records from every
region of Britain except Wales.

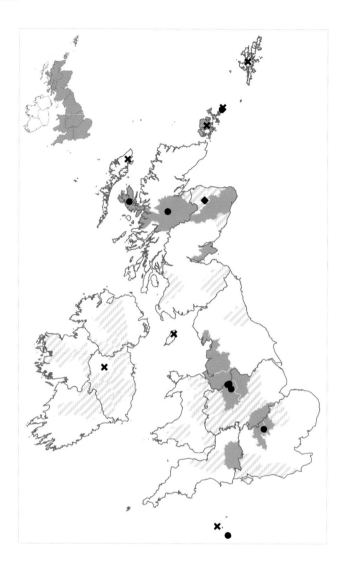

the reason for the species' absence was its poisonous nature, just like that of the Adder, which was thought to be incompatible with the soil in places like Ireland. Although obviously a false hypothesis, the interest in the species is helpful because it allows us to closely reconstruct the early modern distribution of the Toad. It was attested to be absent from Lewis in the early modern period but was present in Skye, corroborating the idea that the colony there is not a modern introduction (O'Brien et al. 2021). The Toad was also found in Jersey but not Guernsey and was also absent from both Shetland and the Isle of Man. This all approximately agrees with the current distribution of the species in Britain and Ireland, although it has now colonised the Outer Hebrides (McInerny and Minting 2016: 40; Beebee 2013: 137; Beebee and Griffiths 2000: 102–5).

There is one point of confusion for the early modern authors. The Toad was attested to be present on Orkney by James Wallace, a seventeenth-century minister of the islands:

> Frogs are seen but seldom, yet there are some Toads, tho', as it is thought, they are not
> poisonous, as indeed there are few, if any, poisonous Animals in all this Country [Orkney].
>
> (Wallace 1700: 37)

However, Jo. Ben, writing in the sixteenth century, remarked that the Toad was absent from North Ronaldsay and Damsay at least.

> No frogs, mice nor toads live here [on North Ronaldsay], and if a ship does bring mice here, they quickly die as if poisoned …
>
> No frogs, toads nor any other terrestrial pests are here [on Damsay].
>
> (Ben in Mitchell 1908: 302, 307)

Ben is probably the one mistaken here. His record is contradicted in part by a record in Blaeu's *Atlas Maior*, which attests that no venomous animal appears on North Ronaldsay *except* for the Toad (Blaeu 1662: 160). It is, in any case, a less reliable record than Wallace's because Ben appears to have been predisposed to find that the Toad was absent so as to fit with the 'poisonous nature' theory explained above. The species was recorded further on Orkney in the nineteenth century (McInerny and Minting 2016: 40).

At present the Toad is still widespread, but it is not doing well. Despite benefitting from the creation of additional garden ponds in the twentieth century (Williamson 2013: 174), it seemed to decline slightly along with the Common Frog during the 1950s and 1960s, and has certainly been declining in Britain (and elsewhere in Europe) since the 1980s (Petrovan and Schmidt 2016; Beebee et al. 2009; Carrier and Beebee 2003; Prestt et al. 1974). Since the 2010s, the Toad has also been found in a few places in Ireland. In at one of these locations (Long Lough, Co. Donegal, in Ulster) the Toad was recorded breeding, suggesting that the species is starting to become established in Ireland.

COMMON FROG

Rana temporaria

NATIVE STATUS Native to Britain, uncertain in Ireland

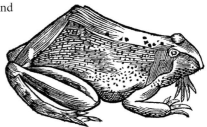

MODERN CONSERVATION STATUS	
World	Least Concern
UK	Least Concern
ROI	Least Concern
Trend since 1772	Appears to have increased

Common Frogs were recorded by a few early modern sources across Britain but were generally considered (perhaps wrongly) to be absent from Ireland.

RECOGNITION

This map shows records using the term *frog* in Early Modern English and *rana* in Latin. Most authors only distinguished one species of frog, and it seems most likely that the frog they were describing was our Common Frog. However, given the vague references, some of the populations described under this name might also have been other species that were previously present, like the Pool Frog or Tree Frog. Some authors also use the Middle Scots term *paddock*. This name is also specific in our texts and is always accompanied by the term rana or frog.

This species actually got its taxonomic name (*Rana temporaria* – the temporary frog) from the early modern belief that the species might be spontaneously generated from the soil when the rain hits it and then disappear again soon afterwards (presumably based on seeing more frogs on or after wet days) (Sleigh 2012; Browne 1646, sec. III.13). However, no reliable early modern record uses this term.

Common Frog records, 1519–1772.
There are records from every region
of Britain except Wales. There is
also a single record from Dublin.

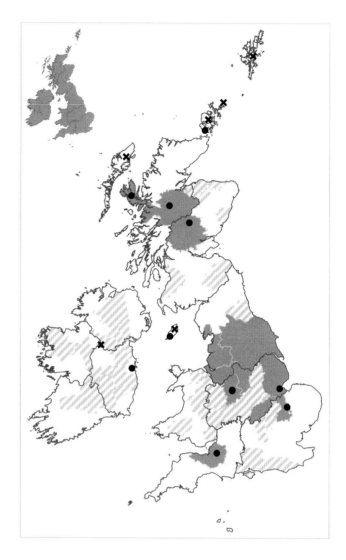

DISTRIBUTION

The distribution of top-quality records for this species is not statistically different from the known level of recorder effort. The gaps on the map may just reflect decreased survey effort in some regions. It may have been widespread. However, the absence records suggest that this species was restricted in its range. The species was recorded in almost every region of mainland Britain but was recorded as absent from Shetland and the Outer Hebrides, where it has recently been introduced (McInerny and Minting 2016: 18; Beebee 2013: 137; Beebee and Griffiths 2000: 83–5). But beyond this core range, the evidence for the distribution of the Frog in early modern Britain and Ireland is contradictory.

It is usually thought that Common Frog is not native to Orkney and was only introduced there in the twentieth century (McInerny and Minting 2016: 18; Beebee 2013: 137; Beebee and Griffiths 2000: 84). However, John Brand, a minister from mainland Scotland, recorded the Common Frog as already present in the early modern period.

> No Poddocks or Froggs are to be seen [in Shetland], tho many in Orkney.　(Brand 1701: 77)

Other early modern authors argue to the contrary that the Common Frog was absent from Orkney in the period (Sutton 2010: sec. 0.37; Lesley 1675: 38, 1888: 64). Wallace, the seventeenth-century minister actually resident in the Northern Isles quoted previously on p. 306, described the Common Frog as 'seen but seldom' compared to the Toad (Wallace 1700: 37). These records are contradictory. It is therefore possible that Brand was mistaken to attest the presence of the Frog (he may have been thinking of the Toad). There is, however, evidence to support his statement. The Common Frog is fairly well represented in the Orkney archaeological record from the Neolithic to the early medieval period (Swandro-Orkney Coastal Archaeology Trust 2020; Davidson and Henshall 1989: 130). There are also records of populations from the nineteenth century (McInerny and Minting 2016: 30). The most simple explanation is that the Frog first colonised Orkney after the end of the last glacial period, or was introduced in the Neolithic, it then went extinct in or after the eighteenth century before being reintroduced in the nineteenth century, going extinct again, and being reintroduced for a third time in the twentieth century! The disagreement by early modern recorders about whether the species was present might imply an additional introduction and extinction between these dates.

Similarly, the Common Frog is widespread today on the Isle of Man, but it is unclear how old the population there is. At the beginning of the eighteenth century, William Sacheverell attested that the Frog was absent from Curraghs, the biggest wetland on the Isle of Man (Sacheverell 1859: 18) and another source suggests that the Frog was only introduced or reintroduced shortly before 1765 (Burke 1766: 74). This may be the origin of the modern population.

Most surprisingly, five of our sources attest that the Frog was absent from Ireland in the early modern period. This for years led to the narrative that the Common Frog was not native to Ireland but introduced in the seventeenth century (McCormick 1999). This historical hypothesis was challenged by Amber Griffiths based on the genetic evidence that some Irish Frogs form a unique phylogenetic group, which might suggest that the Common Frog survived through the glacial period in a refugium in Munster (Teacher et al. 2009). This would imply that the modern population is native and has been in place for over 100,000 years. However, the genetic distinctiveness of the Irish haplotypes has been challenged by further research, and it seems that the Irish population is part of the Western clade, which recolonised Britain and Ireland after the last glacial period (Stefani et al. 2012). This theoretically suggests that there is no barrier to the Common Frog having been introduced since the early modern period, although the genetic diversity in Ireland implies that if this was the case there must have been multiple introduction events. Alternatively, the early sources could of course have been biased by the conviction that Ireland was free from all poisonous and unholy animals, which were thought to include the Frog in the same way as the Toad (Scharff 1893).

The most reliable source that does refer to the Common Frog in early modern Ireland is provided in Rutty's *Essay Towards a Natural History of the County of Dublin*:

> *Rana.* The Frog.
> It was brought into this kingdom, A.D. 1699, by Dr Guithers.   (Rutty 1772: 290)

Rutty goes on to quote a passage about the culinary use of 'green frogs' which likely refers to overseas populations of *Pelophylax* kl. *esculentus* or *Hyla arborea*, but this earlier part of the passage appears to refer to a local population. If we are to support the idea that the Common Frog was native to Ireland, we might suggest that Rutty was mistaken in believing that the population in Dublin was introduced. Dublin remains the place where Common Frogs are best recorded today, although of course this is likely to be because Dublin is better recorded than other counties due to its population density. Stringer, a hunter based in Ulster, does make a comparable reference to Frogs being a common prey of the Otter in *The Experienc'd Huntsman* (1714: 209). However, this reference lacks a location, and possibly might have been based on knowledge of Otters from outside of Ireland.

Historical drainage of wetlands in the nineteenth and twentieth centuries may have led to severe declines in Common Frogs, although the declines may have been partially mitigated by the creation of drainage ditches, which themselves became important spawning sites, as they likely did during early modern reclamations of wetlands (Reid et al. 2014; Williamson 2013: 32–4, 101–3). More recently the Common Frog is known to have declined significantly in parts of the South and Midlands of England (Prestt et al. 1974). This decline seems to have been halted by the growing popularity of wildlife ponds since the 1960s, and the species' population now seems to have stabilised (Reid et al. 2013; Williamson 2013: 174; Beebee et al. 2009; Carrier and Beebee 2003; Beebee and Griffiths 2000: 83). Long-term population decline has continued in some ponds due to ranavirus, which was introduced in the 1980s (Teacher et al. 2010). Since the 2010s chytrid (a serious fungal disease affecting amphibians) has also been confirmed in Britain but not yet in Ireland.

OTHER FROGS

Pool Frog (*Pelophylax lessonae*), Tree Frog (*Hyla arborea*) and others

NATIVE STATUS Pool Frog at least is native to Britain

MODERN CONSERVATION STATUS	
World	Least Concern
UK	Not Evaluated
ROI	Not Present

One early modern source in the atlas database refers to additional species of frog in early modern Britain. Because this is a national rather than local record, there are no maps for the other frog species.

RECOGNITION

The Pool Frog was commonly called *Rana aquatica* in New Latin and the *water frog* in Early Modern English (Raye 2017b). This term was exclusively used for the *Pelophylax* genus, of which the Pool Frog seems to have been the only species native to Britain. However, shortly after the end of the early modern period *Rana aquatica* is occasionally listed as a synonym for the Common Frog (e.g. Pennant 1776c: 9).

The Tree Frog was referred to in New Latin as *Dryopetis*, *Rana viridis* and *Ranunculus viridis*. All these terms seem to be specific to this species. The Early Modern English form of that last Latin phrase [*little*] *greene frog* also usually referred to the Tree Frog, but occasionally the term might also have been used for the *Pelophylax* genus (Raye 2017b).

Gleed-Owen (2000) has suggested that two other frogs might also have been present in Britain based on the archaeological evidence. These are Moor Frog (*Rana arvalis*) and most importantly Agile Frog (*Rana dalmatina*), which is still present in Jersey. These two species do not seem to be distinguished in any early modern historical sources but may have been recorded generically as 'frogs'. Gleed-Owen suggests that the Moor Frog may have been unable to bear the cold temperatures of Britain during the Little Ice Age, so might have gone extinct before or during our period. Given the strong association of the Moor Frog with Beaver ponds in Finland, this species is also likely to have been negatively impacted by the extinction of the Beaver (p. 31) in England during the medieval period (Vehkaoja and Nummi 2015). The extinction of an Agile Frog population prior to the early modern period is harder to explain. Perhaps this species was still present but overlooked or lumped together with the Common Frog.

DISTRIBUTION

Pool frogs are known to have been present in Britain in the historical period, based on archaeological evidence, genetic evidence, call analysis and a search of historical naturalist's accounts from the nineteenth century (Beebee et al. 2005; Snell et al. 2005; Kelly 2004; Wycherley et al. 2002; Gleed-Owen 2000 2001).

The native status of the European Tree Frog in Britain is unclear, but there were well-established populations here until the twentieth century, and they are described by historical naturalists. They appear to have been regularly used in medicines in early modern Britain (Raye 2017b; Snell 1991 2006).

The species are considered together here because the same early modern source refers to both of them. This is Christopher Merrett's list of species found in Britain:

> *Rana*, a frog … it is either a large yellow land frog, or smaller and black,
> Along with: [*Rana*] *aquatica vulgaris* or *maculata*.
> *Ranunculus viridis* or *Dryopetes*.   (translated from Merrett 1666: 169)

I take this passage to refer to three species, all of which Merrett thought were present in Britain: (i) the Common Frog, (ii) the Pool Frog and (iii) the Tree Frog.

This source must be set against another source by John Jonston which attests the absence of a species:

> In England, green frogs are not to be seen. In Germany and Italy there are very many, especially
> in the territory of Bologna.   (translated from Jonston 1657: 130)

This quotation appears to be based in part on an extract from Conrad Gessner, and they make more sense together:

> Frogs of a green colour are common everywhere in Flanders and in all of Germany, if I am not
> deceived. In England they are not, as far as I have heard, which is why they do not eat frogs
> there.   (translated from Gessner 1554b: 47)

It is not entirely clear which species either naturalist is referring to. The quotation from Jonston comes from a section entitled '*Rana aquaticus*' (usually referring to the *Pelophylax* spp.), and Gessner's idea that the English do not eat frogs because they lack green ones suggest he was thinking especially of frogs like *Pelophylax* kl. *esculentus* (Edible Frog), which is closely related to the Pool Frog. However, the term 'green frog' (here *[ranas] virides*, and *[rana] colore viridi*) more normally referred to the Tree Frog, as explained above. Jonston, in fact, uses it to mean this shortly afterwards (1657: 133) in the form of *Ranunculus viridis*, which is helpfully depicted. Today though, the term 'green frog' is used synonymously with 'water frog' to refer to species like the Pool Frog. Most probably, therefore, this is an early version of that usage, and these quotations are suggesting that the water frogs were absent from England. If so, as explained previously, the archaeological, genetic and other historical evidence cited above implies that Merrett was right, and Jonston and Gessner were wrong, since the Pool Frog did survive in Britain through the early modern period into the twentieth century.

Literature reviews have revealed other corroborating historical references for both the Pool Frog and the Tree Frog in Britain (Raye 2017b; Snell 2006; Kelly 2004). A few of these references are from the early modern period. For example, the Tree Frog was celebrated as a home-grown medicine by Timothie Bright (1580: 46). However, these records do not meet the requirements for evidence in this atlas since they relate to the frogs as products rather than wild animals, and they are not geographically specific enough in their scope. Importantly, this includes the references from three other early modern natural history sources (Lovell 1660: 52; Browne 1646: 138 (III.13); Wotton 1552: fol. 93v).

The Pool Frog has now been reintroduced to two sites in Norfolk, including Thompson Common. These frogs are spreading to nearby ponds and may be introduced to additional sites in the coming years (Snell 2016).

The Tree Frog is extinct in Britain and has not been reintroduced since it is not yet clear whether the species is native to Britain, or whether it was only introduced for use in medicine (Raye 2017b). Snell (2006) has suggested that the population previously found in the New Forest in the twentieth century might have been a relict native population. There is no evidence of this population in our records, but frogs are poorly recorded so a population might have been missed. Gilbert White of Selborne (1977: 50), writing just after the end of our period 40 miles from the New Forest, did not know of any populations in England, and thought the species non-native.

NEWTS

Smooth Newt (*Lissotriton vulgaris*), Palmate Newt (*Lissotriton helveticus*), Great Crested Newt (*Triturus cristatus*)

NATIVE STATUS Native in Britain, in Ireland *Lissotriton vulgaris* only is native

MODERN CONSERVATION STATUS	
World	Least Concern
UK	Least Concern
ROI	Least Concern
Trend since 1772	Uncertain

The early modern sources recorded the newt species living in several regions across Britain and Ireland, including on the Isle of Skye in the Inner Hebrides.

RECOGNITION

Identifying newt species from early modern records is difficult. While the Smooth Newt is the only species native to Ireland, the Smooth Newt, Palmate Newt and Great Crested Newt are all found in Britain. To complicate things further, most recorders did not distinguish the newt species from the lizards (of which the Common or Viviparous Lizard (*Zootoca vivipara*) is native in Ireland, this and the Sand Lizard (*Lacerta agilis*) are native to Britain, with two others, the Green Lizard (*Lacerta bilneata*) and the Wall Lizard (*Podarcis muralis*) native to the Channel Islands). This means that the Early Modern English terms *lizard*, *swift*, *newt* and *eft*, Middle Scots *ask*, Irish *airk-luachra*, Late Cornish *padzher pou* and Latin *lacerta* were all generic terms that applied to any of the newts and lizards listed above.

There were some terms that were more distinctive. Whenever Early Modern English *water-* or New Latin *-aquatica* is added to the name, as *water eft*, *water lizard*, *water newt*, *Lacerta aquatica*, these terms refer to what we would call the newt species. Both the terms *swift* and *eft* are sometimes generic, but the Early Modern English terms *swift eft* and *land eft* seem to be specific in early modern Britain and Ireland to the lizards, in practice meaning the Common Lizard. Beyond this, the description given sometimes allows an identification at species level (particularly for the Great Crested Newt),

Newt records, 1519–1772. There are records from the North, the South and the Midlands of England as well as from the Inner Hebrides and from Leinster in Ireland.

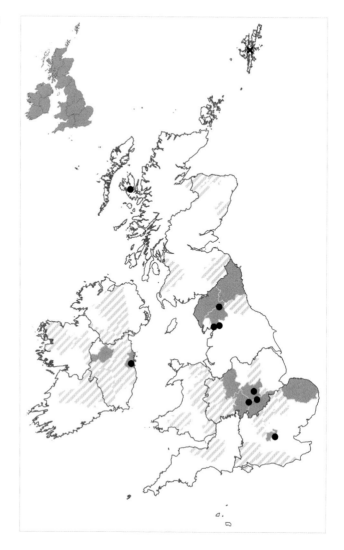

and some authors distinguish multiple species, allowing us to identify their records more easily. For example, Morton in his *Natural History of Northampton-shire* distinguishes the Great Crested Newt from both the Smooth Newt and Palmate Newt and Common Lizard.

> The greater Verrucose or warty Water-Newt … the lesser or smooth Water-Newt, the slow Eft or Newt: and the swift Eft. (Morton 1712: 440)

Here the *warty newt* is the Great Crested Newt, the *water newt* is the aquatic form of a *Lissotriton* species (Smooth or Palmate Newt), the *slow eft* is the terrestrial form of these species, and the *swift eft* is the Viviparous Lizard. Despite this, there is still more interpretation involved in identifying these records than for other species, and some of the records included here might actually refer to *Zootoca vivipara*.

DISTRIBUTION

The distribution of top-quality records for these species is statistically a poor fit with the known levels of recorder effort. They may have been locally distributed, locally abundant or of special local interest. Habitat suitability modelling based on the sites where the newts were recorded in the early modern period suggests that the species may have had specific requirements. They are only recorded on sites that today have moderate winter temperatures (neither too warm nor too cold). In this case, the distribution recorded by the early modern naturalists might be consistent with the modern distribution which includes Britain and Ireland, including the Inner Hebrides but excluding the Northern Isles, just as is shown in the early modern texts. The lack of records from Lowland Scotland, Wales and South West England actually reflects the distribution of the Great Crested Newt and Smooth Newt, which are both much more common in lowland England, although the Palmate Newt is particularly well recorded in Cornwall, West Wales and parts of Scotland (Beebee and Griffiths 2000: 54, 61, 70; Prestt et al. 1974). The lack of records from Ireland outside of Leinster is possibly due to the small number of records overall.

As with the Common Frog, the newts are likely to have suffered from historical drainage of wetlands but have benefitted in the twentieth century with the growth in popularity of the wildlife pond (Williamson 2013: 170–4; Prestt et al. 1974). There is a small amount of evidence that all three native newt species declined in the 1970s and 1980s, and the Great Crested Newt still seems to be declining on a European level (O'Brien et al. 2021; Beebee et al. 2009). This last species is especially threatened on the edge of its range in Scotland, where populations have become fragmented.

SLOW WORM
Anguis fragilis

NATIVE STATUS Native to Britain, not Ireland

MODERN CONSERVATION STATUS	
World	Least Concern
UK	Least Concern
ROI	Not Evaluated
Trend since 1772	Uncertain

The Slow Worm was very poorly recorded in the early modern texts but seems to have lived across England and Scotland, but not Ireland.

RECOGNITION

The records on the map refer to the Early Modern English *slow-worm* and *blind-worm* and, less commonly, the New Latin *caecilia* and *typhlos* or *typhlops*. These New Latin terms have become modern taxonomic genus names, and in the case of *caecilia* also the modern English word 'caecilian', but in the early modern period they were specific to the Slow Worm.

The species is also occasionally referred to as a kind of *serpent*; this latter term was not specific to the species and can only be identified based on description and context, as in Martin Martin's *Description of the Western Isles*, quoted on p. 321. This text appears to describe the Slow Worm as the *serpent* which is short, brown and the least poisonous (Martin 1703: 159–60).

DISTRIBUTION

The distribution of top-quality records for this species is not statistically different from the known level of recorder effort. The gaps on the map may just reflect decreased survey effort in some regions. It may have been widespread. If we include county-level records, the Slow Worm is reliably recorded at least once in every region of England and Scotland, possibly including the Inner and Outer Hebrides, but not in Ireland. This fits with the consensus that the species is native across Britain but not in Ireland (Beebee 2013: 138; Beebee and Griffiths 2000: 122–4). The lack of records from Wales seems likely to have been an oversight by the early modern recorders, just like with most of the other amphibians and reptiles.

The early modern recorders had very little knowledge of the Slow Worm. The species was considered to be vermin, and it was wrongly believed to have a venomous sting or bite like the Adder (Forster 1761). It was likely regularly persecuted, as it is today (Thomas 1991: 78n31). Celia Fiennes's reaction to finding Slow Worms in her bedroom when she visited Ely in 1698 was actually unusually practical and enlightened for the period, despite its negativity towards Ely in general and its amphibians and reptiles in particular:

> [Ely] … seems only a harbour to breed and nest vermine in, of which there is plenty Enough,
> so that tho' my Chamber was near 20 Stepps up I had froggs and slow worms and snailes in my

Slow Worm records, 1519–1772.
There are records from every region of
Britain except Wales, but the records
from Highland Scotland are uncertain.

Roome, but suppose it was brought up wth ye faggotts. But it Cannot but be infested with all
such things being altogether moorish ffenny ground which Lyes Low.
(ed. from Fiennes 1888 [1702]: 128)

Like the Common Frog, the Slow Worm seems to have benefitted from the growing number of
gardens and urban edgelands in the nineteenth and twentieth centuries (Williamson 2013: 170–4,
178–82), although the species seems to have declined in the 1970s and 1980s (Beebee et al. 2009;
Smith 1990: 9–10). At present the Slow Worm seems to be stable across much of Britain (Beebee and
Ratcliffe 2018), but it may be declining in the arable-dominated counties of south-eastern England
and has been susceptible to local declines in population elsewhere (Raye 2021d; Riddell 1996; Prestt
et al. 1974). The isolated Slow Worm population known in the Burren, Co. Clare, in Munster seems
to have been introduced in the modern period (Platenberg 1999).

ADDER
Vipera berus

NATIVE STATUS Native to Britain

MODERN CONSERVATION STATUS	
World	Least Concern
UK	Not Evaluated
ROI	Not Present
Trend since 1772	No change

The Adder was recorded by the early modern sources as being present across much of Britain, including especially southern England and the Hebrides. It was absent from Ireland.

RECOGNITION

Where the Early Modern English term *adder* and the Middle Scots term *eddir* are used, this is specific to our species. Other sources are less specific about what species they intend, employing terms such as *snake*, *serpent* and *viper* in English, *neidr* in Welsh, and *coluber*, *vipera*, *serpens* and *anguis* in Latin. These terms sometimes also refer to the Grass Snake or the Slow Worm. They can be identified where they are used alongside the term *adder*, or where they distinguish multiple species or give a detailed description.

DISTRIBUTION

The distribution of top-quality records for this species is statistically a poor fit with the known levels of recorder effort. It may have been locally distributed, locally abundant or of special local interest. The site-level records are in two clusters, one in Northamptonshire and Wiltshire and one across the West Highlands of Scotland. This matches approximately with known modern abundance hotspots – in southern England and across much of Scotland (Beebee 2013: 139; Beebee and Griffiths 2000: 154–6). The Adder was also present beyond these zones. There are county-level records from every other region of Britain (including Wales where no other amphibian or reptile was recorded in the early modern sources), but the species was recorded as absent in Ireland and a few other smaller islands. Habitat suitability modelling based on the sites where the Adder was recorded in the early modern period suggests that the species would have been well adapted to conditions across Britain, as well as Ireland (which it presumably never had the chance to colonise).

Like the other amphibians and reptiles, the Adder had a reputation as an unwholesome species with special powers in the early modern period. Professional viper-hunters were often hired to exterminate Adders and other reptiles in places where they were thought to be too abundant. The Royal Society in particular also put much scientific effort into testing viper-hunter cures for Adder bites like the use of salad oil (see, for example, Mortimer 1735). The absence of this much-maligned reptile from Ireland but not Britain was also of special interest to the early modern recorders, just as explained for the Toad (p. 305). Early modern authors sometimes suggested that the reason for the absence might be some quality of the soil or air which killed snakes. Whatever the reason, the absence of Adders from an area was seen as a selling point; a place with no snakes was more fit for humans and might even have some unknown virtue. This might explain the absence records dotted around the south of England close to known populations.

This also made the presence or absence of Adders on islands an item of special interest of the early modern recorders – would they share the quality of Ireland or of Britain? This is useful for our purposes because it means that early recorders paid additional attention to the species' distribution. The absence of the Adder from the Isles of Scilly, the Isle of Man and the Northern Isles as recorded in the early modern sources fits with the modern distribution (McInerny and Minting 2016: 131; Beebee 2013: 139; Beebee and Griffiths 2000: 154–6; Prestt et al. 1974). There are, however, also more surprising records.

The map shows one record of the Adder from Harris in the Outer Hebrides. This would extend the known native range of the species, since the Adder is now only known to occur in the Inner Hebrides (McInerny and Minting 2016: 131; Beebee and Griffiths 2000: 154). Yet the record is uncertain:

> There is no Venemous Creatures of any kind here, except a little Viper, which was not thought
> Venemous till of late, that a Woman dyed of a Wound she received from one of them.
>
> (Martin 1703: 37)

The Adder is the only venomous species of snake in Britain, but the surprised description makes it sound as if this 'little Viper' might not be the ordinary *viper* which Martin recorded as living elsewhere in the Inner Hebrides. It may be a misplaced record based on the early modern myth of the venomous Slow Worm (p. 315), since the Slow Worm is thought to be native to the Outer Hebrides.

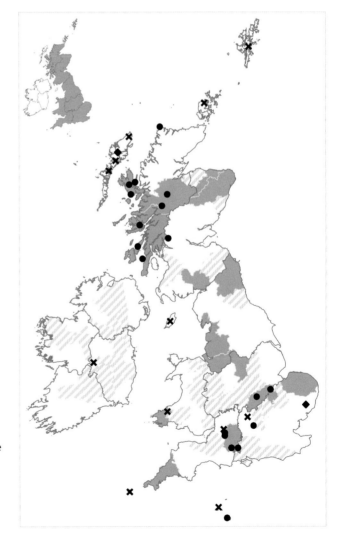

Adder records, 1519–1772. There are records from every region of Britain including Wales. There are absence records from Ireland, the Northern Isles, the Outer Hebrides, the Isle of Man, the Isles of Scilly and Guernsey in the Channel Islands.

Three early modern national sources also attest that although the Adder was absent in the Channel Islands from Guernsey, it could be found on Jersey. The first of these accounts (which the others probably drew on) was John Speed's *Theatre of the empire of Great Britaine*:

> The Aire and Climat of this Ile [of Guernsey] hath little or no difference in temper or qualitie from that of Jersey. And this deserves to be remembred of it; that in this Ile is neither Toade, Snake, Adder, or any other venemous creature, and the other [Jersey] hath great plenty.
>
> (ed. from Speed 1611: 94)

It is possible that these references record a now extinct population of Adders on Jersey, but looking at this quotation, a more conservative interpretation would be that Speed noticed the discrepancy in the distribution of the Toad and Grass Snake between the Channel Islands and extended that discrepancy to what he thought of as other similar poisonous species likely to have the same distribution.

Just like with the Toad, the Adder was sometimes portrayed more respectfully. The power of the Adder was thought to extend to healing, and snakestones (stones supposedly made by clusters of snakes) became popular in Britain following the publication of the translation of Camden's *Britannia* (Gibson 1695). Ammonites, snakestone beads and serpentinite stones were kept for medicinal use in Cornwall, Wales, Scotland and elsewhere through the eighteenth and nineteenth centuries (Pymm 2018; Morgan 1983). Adders themselves also benefitted from the deforested upland heaths and moors created and maintained by humans and livestock in the early modern period (Williamson 2013: 24–9).

At present, the Adder seems to be declining in Britain (Beebee and Ratcliffe 2018). This decline may have been affecting the species for some time (Beebee et al. 2009; Prestt et al. 1974). Larger populations currently seem to be stable or increasing but smaller populations are shrinking and disappearing (Gardner et al. 2019).

GRASS SNAKE

Natrix helvetica

NATIVE STATUS Native to Britain

MODERN CONSERVATION STATUS	
World	Least Concern
UK	Least Concern
ROI	Not Present
Trend since 1772	Uncertain

The Grass Snake was recorded by early modern sources in every region of England as well as on the Isle of Skye in Scotland.

RECOGNITION
The most specific terms for this species in the early modern period were *grass snake* and *water snake* in English and *natrix* and *hydra* in Latin. When used for a British or Irish species, these terms were unique to the Grass Snake. However, this species was more often referred to as the *snake* or *serpent* in English and *serpens*, *anguis* or *coluber* in Latin. All of these latter terms were generic and could also refer to the Adder and Slow Worm. The species intended can be identified if a description is given, or if the other species are distinguished using other names. Records of absence using generic names

can also be taken to include all the species called by that name, so that, for example, if *serpents* are absent this can be taken to include the Grass Snake.

DISTRIBUTION

The distribution of top-quality records for this species is not statistically different from the known level of recorder effort. In this case though, we know the Grass Snake was not generally distributed because there are absence records from several areas. The bias is not obvious in the statistical analysis because there are few records, and the regions where the Grass Snake was absent (like most of Ireland) are less well-recorded regions. There are early modern absence records from the Northern Isles, the Outer Hebrides, the Isle of Man, Scilly and the species' presence on Jersey but not Guernsey also matches the present distribution.

Today, the Grass Snake is sometimes thought of as a species of England and Wales. It is absent from Ireland and is often thought to be absent from Scotland too. The only early modern record from Lowland Scotland (Aberdeenshire) is an absence record. However, the Grass Snake is occasionally found in Scotland today (McInerny and Minting 2016: 146–9), and was also rarely recorded there in

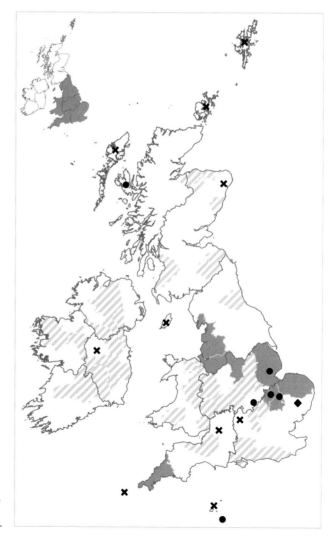

Grass Snake records, 1519–1772. There are presence records from the Inner Hebrides and every region of England. There are absence records from the Northern Isles, the Hebrides, the Isle of Man, the Isles of Scilly and from Guernsey in the Channel Islands.

the early modern period. As well as the national record by Sibbald (ed. 2020: 28 (II:3)), there is also at least one reliable local record for the Grass Snake from the Inner Hebrides:

> Serpents abound in several parts of this Isle [Skye], there are three kinds of them, the first Black and White spotted which is the most Poysonous, and if a speedy remedy be not made use of after the Wound given, the Party is in danger …
>
> The yellow Serpent with brown spots is not so poisonous, nor so long as the black and white one.
>
> The brown Serpent is of all three the least poisonous, and smallest and shortest in size.
>
> (Martin 1703: 159–60)

This is a strong record: the author clearly describes the Adder, Grass Snake and Slow Worm respectively, meaning that no confusion with other species is likely – a recorder who knew the difference between the three species believed all three were present on Skye.

Like the other amphibians and reptiles, the Grass Snake had a poor reputation in early modern Britain and Ireland (Lenders and Janssen 2014; Thomas 1991: 78n31). It is likely that the special aversion that people had for the animal is what has provided us with such a good account of its early modern distribution on the smaller islands around Britain.

Just as with the Adder, the Grass Snake benefitted in the early modern period from human-created habitats. The species is reliant on exogenic sources of heat to incubate its eggs, and so it was likely especially associated in the past with dung heaps (Lenders and Janssen 2014). The early modern sources also record that it was especially common in the Fens, which seem to have provided the Grass Snake with many amphibians to eat (Topsell 1658: 766). The species may therefore have been negatively affected by the departure of livestock and horses from urban areas and the drainage of the Fens in the nineteenth century (Almeroth-Williams 2019; Williamson 2013: 101–3). Since then, the Grass Snake appears to have declined through the second half of the twentieth century, but this decline seems to have slowed in recent years (Beebee et al. 2009; Prestt et al. 1974).

SEA TURTLES
esp. Leatherback Turtle (*Dermochelys coriacea*)

NATIVE STATUS Native

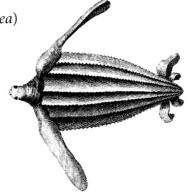

MODERN CONSERVATION STATUS (LEATHERBACK)	
World	Vulnerable (Northwest Atlantic subpopulation is Endangered)[9]
UK	Not Evaluated
ROI	Least Concern
Trend since 1772	Uncertain

Leatherback turtles were very rarely recorded as vagrants on the coasts around early modern Britain.

---

9   Despite the name, the Northwest Atlantic Ocean subpopulation is the one that visits Britain and Ireland.

Leatherback Turtle records, 1519–1772. There are records from South West England and from the Northern Isles of Scotland, but the Northern Isles records are unreliable.

RECOGNITION

The names used for this species in the early modern period were generic rather than specific. The most common names were *tortoise* and *testudo*. These terms are not specific and were used for pet land tortoises as well as sea turtles. But in the context of a wild, marine species, these names are most likely to refer to the Leatherback Turtle; this was by far the most common sea turtle around Britain and Ireland and the only one likely to be within its range and not just dead or dying (Beebee and Griffiths 2000: 176–7). The exact species can be fully confirmed by the description, or in the case of one of the Cornish records, by the illustration included here, which was provided by Borlase (1758).

DISTRIBUTION

There are not enough top-quality records to statistically analyse the distribution of this species.

There are only two secure records of Leatherback Turtles from early modern Britain and Ireland, both recorded by Borlase – one which can be securely identified from the illustration at the start of this section, and another which weighed 'six hundred and three quarters'. Presumably this means 6¾ hundredweight or about 340 kg, which would identify this as a Leatherback Turtle as well. Apart from these records there is a national record of an uncertain sea turtle from Ireland (O'Sullivan 2009: 181), and also generic sea turtle records from Orkney and Shetland.

There are six records of sea turtles from the Northern Isles, but most of these refer generally to Orkney as their location. These are not all records of Leatherback Turtles: John Brand gives one clear record from Shetland of a sea turtle which is not a Leatherback:

> Tho no Tortoises use to be found in all these Northern Seas, yet in Urie-Firth in the Parish of Northmavine, there was one found alive upon the sand in an ebb, the Shell of it was given me as a present by a Gentleman of the Countrey, it is about a Foot length, and a large half Foot in Breadth. The Inhabitants thought it is so strange, never any such having been found in these Seas formerly, which ever they came to the knowledge of, that they could not imagine what to make of it, some saying that it hath fallen out of some East-India Ship Sailing alongst by the Coasts, which looks not so probable. (Brand 1701: 174–5)

This sea turtle described by Brand is hard to identify, but Leatherbacks are much larger than this and do not have hard shells. The most probable identification is that this was a Loggerhead Turtle (*Caretta caretta*) (McInerny and Minting 2016: 224–6). On the other hand, the idea that a sea turtle could have been stranded off Orkney after being transported on a trading vessel is actually not as unlikely as it might appear. Green Turtles (*Chelonia mydas*) were regularly transported across the Atlantic for use in turtle soup. They regularly died on the way and were thrown overboard and washed up on shore, although such records were far more common from later on in the eighteenth century (Brongersma 1972: 189–94).

Another shell was acquired two decades earlier by one of the sons of Bishop Murdock Mackenzie of Orkney and passed to Robert Sibbald (Sibbald 1710: 20–1, 2020: 13[11] (II:3)). This shell is the one that later became part of the Museum of Balfour, which was called there the 'Testudo marina squamosa. The Scalie Sea-tortoise' (Sibbald 1697: 193). Sibbald borrowed this name from a description of a Hawksbill Turtle (*Eretmochelys imbricata*), which has suggested to some later naturalists that this was the species he saw (Fleming 1828: 149). Given that sea turtles were still so poorly understood, and Hawksbill Turtles are so rare around these islands, Sibbald's nomenclature seems insufficient to prove that his specimen was not simply a Loggerhead, which was far more regularly seen in Scotland the past (McInerny and Minting 2016: 224–6; Brongersma 1972: 196–7; Stephen 1953).

The Leatherback Turtle became more commonly sighted in Britain in the second half of the twentieth century. This may be partially due to increased survey effort, and perhaps improved conditions for the species, but the population also seemed to be expanding (McInerny and Minting 2016: 207 214; Prestt et al. 1974). This trend has stalled in the twenty-first century and the Northwest Atlantic population now seems to be declining (Northwest Atlantic Leatherback Working Group 2018).

CUTTLEFISH
Sepia officinalis

NATIVE STATUS Native

MODERN CONSERVATION STATUS	
World	Least Concern
UK	Not Evaluated
ROI	Not Evaluated
Trend since 1772	Uncertain

The early modern sources recorded Cuttlefishes all around the inshore regions of Britain and the south coast of Ireland.

RECOGNITION
The Cuttlefish was called *sepia* in New Latin (the colour sepia is named after the ink originally made from this species (Marren and Mabey 2010: 460)). In Early Modern English it was also called the *cuttlefish* or *cuttle* (once wonderfully the *cudle*). O'Sullivan (2009: 176) provides the Irish name *scuduil* (probably a mistake for *cuduil* (Scharff 1916), although note Scottish Gaelic *sgudal* = fish guts). These names may have been generic to some extent. *Loligo vulgaris* (the European Squid) was sometimes distinguished (as *loligo* or the *sleeve-fish*) but other species like *Rossia macrosoma* (the Stout Bobtail) and *Sepia elegans* (Elegant Cuttlefish) do not seem to have been distinguished, and the names are sometimes used generically, meaning these records are not fully reliable, although the Cuttlefish is the most common species of its kind found around Britain and Ireland. Often only the bone is recorded, which was used by goldsmiths to make moulds (as in Sibbald 2020: 26 (II:3)).

DISTRIBUTION
The distribution of top-quality records for this species is not statistically different from the known level of recorder effort. The gaps on the map may just reflect decreased survey effort in some regions. It may have been widespread.

Today, the Cuttlefish is best recorded around the southern coasts of Britain and Ireland. It is rarely recorded on the coasts of north-east England or around Scotland, but occasionally found in deeper water (Wood 2018: 155). The species is particularly rarely found in the northern parts of the North Sea, especially off the eastern coast of Scotland, but is occasionally swept into the area in years when currents from the Atlantic are exceptionally strong (Stephen 1944). This implies that the Cuttlefish may have changed its distribution because it was well recorded on the North Sea coast in the early modern period. Sibbald, for instance, records a persistent population in the Firth of Forth:

> The Sepia or Cuttle-Fish, without doubt, haunts this firth; for the bone of it is frequently cast up upon the shoars: we find not the entire animal, because, so soon as they are cast ashoar, the small crabs presently eat up all the parenchyma of them. I have found these crabs, we call Keavies, eating Slieve-fish [squids] greedily.   (ed. Sibbald 1803 [1710]: 130)

Cuttlefish records, 1519–1772.
There are coastal records from every
region of Britain, except the Midlands
of England and the mainland of
Highland Scotland. There are also
records from Munster in Ireland.

Today, the Cuttlefish is regularly caught in fisheries around the south and west of Britain and Ireland, especially in the English Channel. Landings seem to have been relatively stable over the past two decades, so the population seems to be secure but the exact fishery figures are not known (Guerra et al. 2015).

LIMPETS
esp. the Common Limpet (*Patella vulgata*)

NATIVE STATUS Native

MODERN CONSERVATION STATUS	
World	Not Evaluated
UK	Not Evaluated
ROI	Not Evaluated
Trend since 1772	Uncertain

Limpets were recorded on every coast of Britain and around the south of Ireland in the early modern texts.

RECOGNITION

Limpet species were not reliably distinguished in the early modern period. The records shown on this map include especially English *limpet* and Latin *patella*. There are also records attesting the English terms *limpin*, *flither* and *papshell* or *papfish* and the Irish term *baerneach*. All of these terms are generic and could presumably refer to, for example, the China Limpet (*Patella ulyssiponensis*) and the Black-footed Limpet (*Patella depressa*), as well as the Common Limpet (*Patella vulgata*). The Slipper Limpet (*Crepidula fornicata*) was only introduced comparatively recently and was not present in the early modern period.

The terms *Patella vulgaris*, *Patella vulgaris major* and the *Patella maculosa, fere striata, modo levior* of Lister appear to be specific to the Common Limpet. Unfortunately, these terms are rare but they allow us to be reasonably confident that the Common Limpet in particular occurred on North Sea, English Channel and Irish Sea coasts.

DISTRIBUTION

The distribution of top-quality records for these species is not statistically different from the known level of recorder effort. The gaps on the map may just reflect decreased survey effort in some regions. Limpets may have been widespread. There are records from every region of Britain, and since the lack of records from Ulster and Connacht is consistent with the level of recording the species may well have been widespread around Ireland too. The Common Limpet can still be found on every shore of Britain and Ireland today, although other limpet species are less widespread (Wood 2018: 126).

Limpets shells are common in the middens of medieval and early modern settlements in Scotland and Ireland (Harris et al. 2018; Milner et al. 2007; Murray 2007; Dodgshon 2004). They were sometimes eaten, often but not exclusively by the poorest in coastal communities, and during times of famine, but they were also commonly used as bait for catching whitefishes and the shells were incorporated into mortar (Firth 2021; Kelly 1997: 298). The shells were also collected for people with chafed nipples, particularly when this was due to breastfeeding, as Leigh explains in *The Natural History of Lancashire, Cheshire and the Peak in Derbyshire*:

> The Pap-Fish is common, so call'd from the likeness it bears to a Nipple, the Country People use them for their Nipples when sore, which by guarding them from fretting on their Cloaths, give relief.
>
> (Leigh 1700: 139)

Limpets remain widespread today. There is now no large market for them around Britain and Ireland, but they are collected locally for food, and sometimes cruelly kicked off their rocks for fun or curiosity (Firth 2021).

Limpet records, 1519–1772.
There are coastal records from
every region of Britain and from
Munster and Leinster in Ireland.

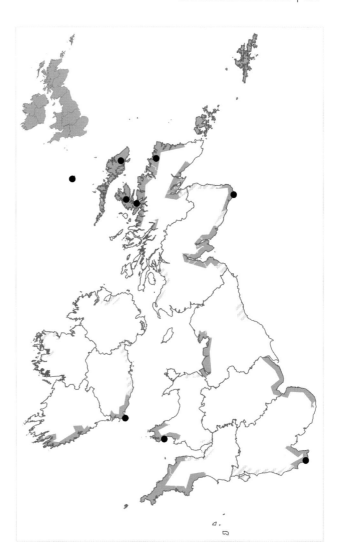

COMMON MUSSEL
Mytilus edulis

NATIVE STATUS Native

MODERN CONSERVATION STATUS	
World	Not Evaluated
UK	Not Evaluated
ROI	Not Evaluated
Trend since 1772	Uncertain

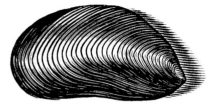

The Common Mussel was recorded widely in sources from around early modern Britain and Ireland except the Isle of Man.

RECOGNITION
This species was called the *muscle* or *mussel* in Early Modern English, *mussill* in Middle Scots and the *mytulus* or *musculus* in New Latin. These terms were also regularly used to refer to other species like the Freshwater Pearl Mussel and, presumably, *Mytilus galloprovincialis* (the hard-to-distinguish Mediterranean Mussel (Wood 2018: 148)). This means some of our records might refer to other species, although the Common Mussel was so well known and lived in such high-profile beds that other species were usually distinguished when they were discussed.

DISTRIBUTION
The distribution of top-quality records for this species is not statistically different from the known level of recorder effort. The gaps on the map may just reflect decreased survey effort in some regions. It may have been widespread, as it is today. The only absence record comes from the Isle of Man, where the second edition of Camden's Britannia adds that 'there are few or no oysters or muscles' (Gibson 1695: 1062).

Common Mussels seem to have been frequently eaten in medieval and early modern Britain and Ireland, and their shells are often found by archaeologists at human settlement sites (Noble et al. 2017; Breen 2016; Murray 2007). The species had a mixed reputation in the early modern period. Although frequently eaten, Common Mussels get their own food by filtering impurities from water. If they are exposed to raw sewage they can spread diseases like typhoid when people eat them (Marren and Mabey 2010: 441; Edwards 1997):

> Musculus ex coeruleo niger … The Muscle or Sea Musele. It is to be found in this Haven [Harwich in Essex]. The Shells are found on the Shore. It is a common Fish, and much eaten among the common People, but it is thought not to be very wholesome, being suspected of causing Sickness and Inflammations; some attributing this Quality to the Seta or Hairs, others to a small poisonous Insect found in them.　　　　　　　　(Taylor and Dale 1730: 388)

Today the Mussel is commonly found on every coast of Britain and Ireland, including around the Isle of Man. Its distribution seems to be stable (Wood 2018: 148). There continues to be a small market for mussels in these islands, but the demand is now met by the aquaculture industry, which encourages seed mussels (sprats) to grow in highly productive, clean locations and harvests them from there (Smaal 2002; Edwards 1997). The species is cultivated around Britain and Ireland but production has been decreasing since around the turn of the millennium (Avdelas et al. 2021).

Common Mussel records, 1519–1772. There are coastal records from every region of Britain and Ireland except the Midlands of England.

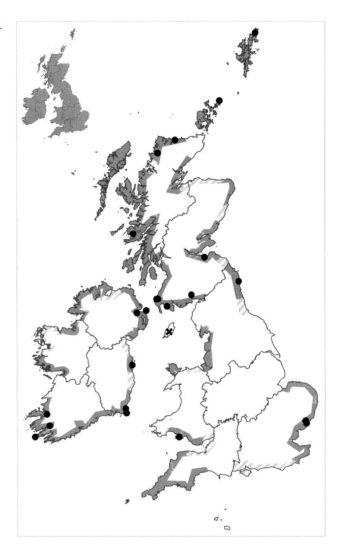

FRESHWATER PEARL MUSSEL

Margaritifera margaritifera

NATIVE STATUS Native

MODERN CONSERVATION STATUS	
World	Endangered (Critically Endangered in Europe)
UK	Not Evaluated
ROI	Not Evaluated
Trend since 1772	Uncertain

The Pearl Mussel was recorded in early modern sources from around the uplands of Ireland and northern and western Britain.

RECOGNITION

Most of the records of this species given on the map refer simply to the *pearl* in Early Modern English or *margarita* in New Latin. The record is therefore of the product obtained, not of the animal present. Care needs to be taken to separate records of rivers that produce Freshwater Pearl Mussels from settlements and people who traded and owned pearls which were produced elsewhere. Identification is also difficult since other molluscs can produce pearls too. Finally, the species needs to be distinguished from the Brill, which was also regularly called the *pearl*. Fortunately, almost all records of *pearl* describe it as a river rather than marine species, which makes possible identification of this species based on location.

The most reliable records usually explain that the *pearls* come from a *muscle*. Some authors also invent taxonomic names like *Concha margaritafera* or *Mytulus major margaratiferus*, which are good descriptions. Lhuyd provides the Welsh term *kregin diliw*, which also seems to be specific (Gibson 1695: 669–70). Where the word mussel is not used, identification needs to be based on the description of the species and its location.

To be safe, records that describe the species as an *oyster* or *horse mussel* have also been excluded except where the species is described more exactly because these are likely to refer to other molluscs. The one reference I have left in (the diamond over Pembrokeshire) refers to *horse mussels* that reliably have pearls, some with multiple pearls, which sounds like it might be the Freshwater Pearl Mussel.

DISTRIBUTION

The distribution of top-quality records for this species is not statistically different from the known level of recorder effort. The gaps on the map may just reflect decreased survey effort in some regions. It may have been widespread. However, having said that, the species today exclusively lives in fast-flowing upland streams in the north and west of Britain, and there is little evidence that it was ever present in lowland England (Marren and Mabey 2010: 445).

Despite widespread local declines, the species does not seem to have changed its distribution at regional level in Britain and Ireland since the early modern period: it is still found across upland Britain and Ireland except in South England and the Midlands. However, at local level the species seems to be declining (Skinner et al. 2003). Outside of the Highlands of Scotland, most populations are in poor condition with no recruitment of young and our maps show some populations (e.g. in

Galloway) that have since disappeared. This local decline might have started early on. For example, Daniel Defoe noted in the eighteenth century:

> I enquired much for the pearl fishery here, which Mr Cambden speaks of, as a thing well known about Ravenglass and the River Ire, which was made a kind of bubble lately. But the country people, nor even the fishermen, could give us no account of any such thing; nor indeed is there any great quantity of the shell-fish to be found here (now) in which the pearl are found, I mean the large oyster or mussel. What might be in former times, I know not.   (Defoe 1971 [1724–27]: 553)

Freshwater Pearl Mussels have continued to decline across Europe through the modern period (Marren and Mabey 2010: 443–6; Skinner et al. 2003). The increased demand for pearls in the nineteenth century led to overharvesting and rapid decline of the species. This decline continued through the twentieth century due to pollution and an increase in sediment, which means that there is often not enough oxygen at the bottom of rivers to support Freshwater Pearl Mussels, along with hydrological management and climate change. The species also relies on Brown Trout or Salmon to complete its lifecycle (Davies et al. 2004: 145; Skinner et al. 2003), which presents problems when these species in decline or are blocked from entering the upper reaches of rivers. The species is now

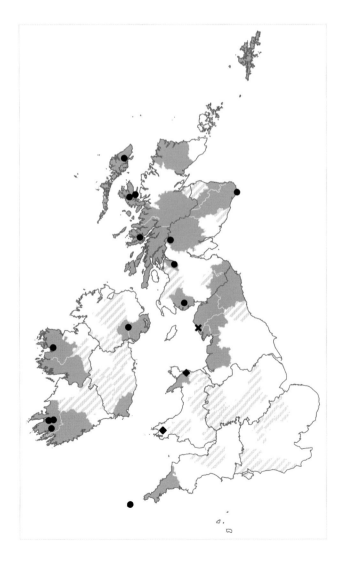

Freshwater Pearl Mussel records, 1519–1772. There are records from every region of Britain and Ireland except the South and Midlands of England.

classed as Critically Endangered in Europe. A large percentage of the world's remaining breeding population is found in Scotland, but the species continues to decline due to all the factors above, including most importantly illegal harvesting (Cosgrove et al. 2016).

OYSTER
Ostrea edulis

NATIVE STATUS Native

MODERN CONSERVATION STATUS	
World	Not Evaluated
UK	Not Evaluated
ROI	Not Evaluated
Trend since 1772	Certainly declined

Oysters were recorded by early modern sources all around the coasts of Britain and Ireland, including on the east coast of Scotland where they are now not normally found.

RECOGNITION

In Early Modern English and Middle Scots this species was called the *oyster*. K'Eogh (1739: 70) attests the Irish term *oistry*. Occasionally, New Latin names are used like *Ostreum vulgare maximum* or *Ostrea major*, but these are not consistent between texts. Despite the lack of a standard New Latin name, because the texts were written before the introduction of the Pacific Oyster (*Magallana gigas*) and the Kumamoto Oyster (*Crassostrea sikamea*) for food, records of this species can be easily recognised.

DISTRIBUTION

The distribution of top-quality records for this species is statistically a poor fit with the known levels of recorder effort. It may have been locally distributed, locally abundant or of special local interest. Habitat suitability modelling based on the sites where the Oyster was recorded in the early modern period suggests that the species would have been well adapted to conditions all around the coasts of Britain and Ireland. In this case, the Oyster is one of the ten best-recorded species in the early modern sources. The species was recorded on every coast of Britain and Ireland, especially if we include the county-level records. This suggests that the distribution pattern is likely to reflect a regional abundance. There are more records than expected from Leinster, Munster, Wales and the South of England and less than expected from Scotland, Ulster and the North and Midlands of England. The areas where it was less commonly recorded in the early modern period match approximately with the areas where the species is now absent; the Oyster may have been more vulnerable in these regions than elsewhere.

The Oyster was commonly eaten in medieval and early modern Britain and Ireland (Breen 2016; Murray 2007; Mac Laughlin 2010: 79–80; Kowaleski 2003). Oysters were so popular that specialist oyster-sellers became common figures in urban areas. Poor women often became oyster-sellers and they were often fetishised in early modern ballads and cries as hypersexual nymphs, fuelling moral panics (Taverner 2019). In the early modern period, Oysters could be obtained easily from huge Oyster beds (some purposefully seeded (Rowlands 1721)) which were routinely harvested, including in areas where they are now never normally seen, such as in the Firth of Forth (Fariñas-Franco et al. 2018):

> They take great quantities of oysters upon this shore [the Firth of Forth] also, with which they
> not only supply the city of Edinburgh but they carry abundance of them in large, open boats,
> called cobles, as far as Newcastle upon Tyne.   (Defoe 1971 [1724–27]: 573)

This use of Oysters seems to have been sustainable. However, after the end of the early modern period, the demand for Oysters rose and their abundance rapidly declined. The decline commenced in the nineteenth century, perhaps initially due to pollution along with overexploitation, and has continued since (Marren and Mabey 2010: 455; Edwards 1997). A parasite, *Bonamia ostreae*, spread through native populations. Other species of oyster were introduced to meet demand, but these have occasionally expanded beyond the fishery – as in the case of the Pacific Oyster, which can outcompete the native Oyster (Tully and Clarke 2012). Other harmful species have also been accidentally established along with them, including the American Sting-Winkle or Oyster Drill (*Urosalpinx cinerea*) which predates Oysters, and the Slipper Limpet (*Crepidula fornicata*) which outcompetes the native Oyster in parts of its range. The decline in Oysters has been Europe-wide and, even in areas where the Oyster continues to be present, populations have been declining since at least the 1970s (Tully and Clarke 2012). Oysters are now absent from much of the east coast of Britain from the Northern Isles to East Anglia (Fariñas-Franco et al. 2018; Wood 2018: 153).

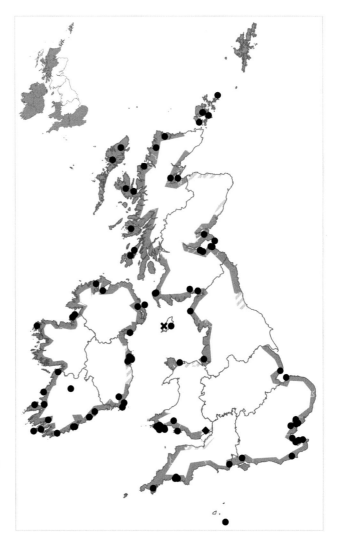

Oyster records, 1519–1772. There are records from every region of Britain and Ireland including from the Hebrides, the Northern Isles, the Channel Islands and a contested record from the Isle of Man.

SCALLOPS

Three common species: *Pecten maximus* (King), *Aequipecten opercularis* (Queen), *Chlamys varia* (Variegated)

NATIVE STATUS Native

MODERN CONSERVATION STATUS	
World	Not Evaluated
UK	Not Evaluated
ROI	Not Evaluated
Trend since 1772	Uncertain

The early modern sources provided a few records of scallop dotted around Britain and Ireland.

RECOGNITION

The map here shows references to what are called *scallops* or *escallops* in Early Modern English, *pecten* or *pectines* (plural) in Latin. O'Sullivan provides the Irish *macmuirin*. Unfortunately, these terms are not enough to identify the animal seen to species level, so this map shows all kinds of scallop together. The Latin term *pecten* forms the base of *pectunculus* (little *pecten*), the word used for the cockle species.

DISTRIBUTION

The distribution of top-quality records for these species is statistically a poor fit with the known levels of recorder effort. They may have been locally distributed, locally abundant or of special local interest. There are more records than expected from Munster and fewer records than expected from Highland and Lowland Scotland and South and South West England. However, habitat suitability modelling based on the sites where the Scallop was recorded in the early modern period suggests that the species would have been well adapted to conditions across Britain and Ireland, except possibly the areas with the coldest winters. If we include the county-level references there are records of scallops in every region except Lowland Scotland. Considering that the record from the north of Northumberland borders on Lowland Scotland, and that the modern distribution does not show any special abundance of scallops in Munster compared to, for example, Highland Scotland or South West England (Wood 2018: 151; Brand 2006), the bias in the records seems most likely to reflect increased local interest in the species due to local cultural importance of the scallop (and other seafoods) in Munster:

> About half a mile from the head of the river or bay of Kenmare, lives the rev. Mr. Orpin, rector of the parish, with a colony of protestant families, consisting of ship-carpenters, rope-makers, smiths etc. who are very necessary in supplying the fishing vessels that frequent the trade of this river; for it abounds with cod, hake, mackerel, ling, herring, and divers other kinds of fish; which are taken by a considerable number of boats, sometimes to the amount of an hundred … It abounds also with lobsters, crabs, escallops, oysters, muscles, cockles, and most kinds of shell fish, besides salmon fisheries in five or six places: but these last are in no very considerably quantity, they being much destroyed by seals and sea-dogs which are so very numerous in this river, that in summer all the rocks on the shores, are in a manner covered with them.
>
> (Smith 1756: 84)

The importance of marine fisheries for the economy of early modern Munster was previously discussed in the analysis of records of the European Pilchard (p. 255). The lack of records from Scotland might be best explained if we accept the theory that the early modern fishery for scallops described for the south coast of Munster was not taken up in mainland Scotland until the modern period, and there was therefore less interest in scallops in Scotland, despite them being present there

Scallop records, 1519–1772. There are records from every region of Britain and Ireland except Lowland Scotland.

in the early modern period (Mason in Edwards 1997). There is evidence that Scallops were eaten in England and Ireland already in the period, and this also seems to have been the case for medieval Orkney (Breen 2016; Milner et al. 2007; Sykes 2001).

After the end of the early modern period, scallops began to be caught in greater quantities (Mason 1983). By the end of the nineteenth century, a dredge fishery had started on the coast of Scotland and Irish scallops were being sold in London. There was a decline in the scallop fishery during the first half of the twentieth century, but the fishery was re-established and considerably intensified in the 1960s and 1970s. Today, scallops are found off every coast of Britain and Ireland, although less commonly on the east coast of Britain than elsewhere (Wood 2018: 151). King and Queen Scallops are now the most valuable of Britain's shellfish fisheries, and are caught in greater quantity around Britain and Ireland than anywhere else in Europe (Marren and Mabey 2010: 449; Edwards 1997). Fishing pressure has intensified since the 1990s and is now probably at an unsustainable level: the scallops are currently likely to be in decline (Duncan et al. 2016).

COCKLES

Many species of cockle dwell around Britain and Ireland. The most widespread is *Cerastoderma edule* (Common Cockle).

NATIVE STATUS Native

MODERN CONSERVATION STATUS	
World	Not Evaluated
UK	Not Evaluated
ROI	Not Evaluated
Trend since 1772	Uncertain

Cockles were recorded around all the coasts of Britain and Ireland in the early modern period.

RECOGNITION

The English term *cockle*, the Middle Scots *cockil*, the Irish term *ruacan*, and the New Latin term *pectunculus* were used in the early modern period to refer to all the cockles of Britain and Ireland, including similar-looking fossil species (these are excluded from our records). There was little effort to distinguish the different species. The most common and widespread was and continues to be the Common Cockle, so many of our records probably refer to this species.

DISTRIBUTION

The distribution of top-quality records for this species is statistically a poor fit with the known levels of recorder effort. There are more records than expected from Lowland Scotland, Leinster, Munster and South England, and fewer than expected from Ulster and the South West, Midlands and North of England. Including the county-level records, we can see that cockles were widespread. They are recorded on every coast and in every region except the Midlands; this omission is likely to be an oversight considering that cockles were recorded in Norfolk and in Yorkshire. Habitat suitability modelling based on the sites where cockles were recorded in the early modern period suggests that these species would have been well adapted to conditions all around the coasts of Britain and Ireland. As we have seen, shellfish were especially commonly eaten in Munster and South England, so the pattern of records seen here might reflect a cultural bias.

Cockles are often found at archaeological excavations of human settlements and seem to have been commonly eaten in medieval and early modern Britain and Ireland (Breen 2016; Milner et al. 2007; Murray 2007):

> The Cockle. This is an Inhabitant of these Seas [around Harwich and Dovercourt], and the dead Shell often found upon the Shore. These are plentifully sold at London, being brought chiefly from Sussex, perhaps the best comes from Selsea, from whence they are frequently called Selsea-Cockles.   (Taylor and Dale 1730: 387–8)

Cockles continue to be widespread around Britain and Ireland. The Common Cockle in particular is harvested commercially in very large numbers around these islands (Marren and Mabey 2010: 439; Edwards 1997). The numbers of cockles present has declined since the 1990s, due to overexploitation and mass mortalities (some perhaps influenced by climate change), but the fishery is now more sustainable with Total Allowable Catch numbers set each year based on abundance, and tractor- and boat-dredging banned in the UK (Mahony et al. 2020; Southall and Tully 2014).

Cockle records, 1519–1772.
There are records from every
region of Britain and Ireland
except the Midlands of England.

SHRIMPS

esp. *Crangon crangon* (Brown Shrimp)

NATIVE STATUS Native

MODERN CONSERVATION STATUS	
World	Not Evaluated
UK	Not Evaluated
ROI	Not Evaluated
Trend since 1772	Uncertain

Shrimp were recorded regularly around early modern Britain and Ireland, and seem to have been relatively widespread.

RECOGNITION

The records on this map mainly use the English term *shrimp* and the Latin term *Squilla gibba*. These terms are not necessarily specific to the Brown Shrimp, but likely would also have been used to refer to similar species – especially the hard-to-distinguish *Crangon allmani*. Therefore, these records should be considered uncertain. However, the Brown Shrimp was the most widespread, common and visible species, and also the one targeted for food, so it is likely that most of our records refer to this species.

The terms *shrimp* and *Squilla gibba* have to be especially distinguished from the species that was called *Squilla major*, which is once called the *grey shrimp* (Prawn, *Palaemon serratus*).

DISTRIBUTION

The distribution of top-quality records for this species is statistically a poor fit with the known levels of recorder effort. It may have been locally distributed, locally abundant or of special local interest. Habitat suitability modelling based on the sites where shrimp were recorded in the early modern period suggests that the species may have had specific requirements. They were not recorded on the coldest or wettest stretches of coastline. Based on the difference in the number of records between the south and north of these islands, it appears possible that the shrimp might have increased its range during the modern period. This implies that shrimps might have been restricted in their range during the early modern period during the Little Ice Age. In cold winters, Brown Shrimp hatching is delayed, and larvae are not released in time for the spring bloom in microplankton (Saborowski and Hünerlage 2022). However, the species is found in the Arctic waters of Norway and Russia, so it seems unlikely that a cold sea temperature could have been a limiting factor in British and Irish waters, even during the Little Ice Age (Ásgeirsson et al. 2007).

The Brown Shrimp has been the target of continuous commercial fisheries around Britain since the early modern period (Wheeler 1979: 81–2). Like the shellfish, there appears to have been a strong early modern fishery for shrimp in the south of Ireland:

> and in the same River [Barrow on the Waterford/Wexford border], over against the forte of Duncannon, are dredged a very good kind of oysters, alsoe Lobsters, Crabbs, Prawnes, and Shrimps, are found along that Cost to the Tower of Hooke, which lyes on the maine sea.
>
> (Hore 1859 [1684]: 453)

Shrimp records, 1519–1772. There are records from every region of Britain and Ireland except Highland Scotland.

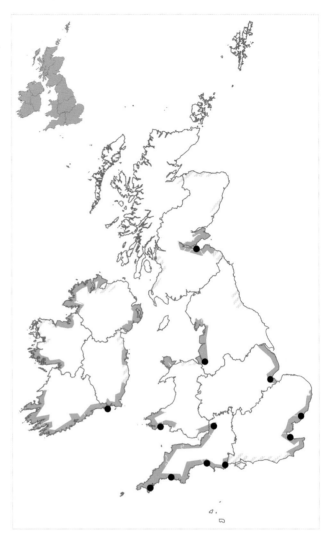

The species continues to be caught around Britain and Ireland today, but now in much lower numbers than elsewhere in the North Sea (ICES 2022g; Marren and Mabey 2010: 68). At present, no exact statistics are kept on how many shrimp are caught on a Europe-wide basis; overall stocks seem to be decreasing due to overfishing, although other populations around Britain and Ireland seem to be stable (ICES 2022g; Henderson et al. 2006). Today the Brown Shrimp is found all around Britain and Ireland, but now less commonly around Connacht and Munster than elsewhere (Wood 2018: 106).

CRABS

All crab species, perhaps esp. Edible/Brown Crab (*Cancer pagurus*),
Common Shore Crab (*Carcinus maenas*)

NATIVE STATUS Native

MODERN CONSERVATION STATUS	
World	Not Evaluated
UK	Not Evaluated
ROI	Not Evaluated
Trend since 1772	No change

Crabs were very commonly recorded by local sources around the coasts of early modern Britain and
Ireland.

Crab records, 1519–1772. There are
records from every region of Britain
and Ireland including the Hebrides,
the Northern Isles, the Isle of
Man and the Channel Islands.

RECOGNITION

The map shows references to *crab* and *punger* in Early Modern English, *partan* and *keavy* in Scots and *pagurus* in Latin. These names were not specific. They are perhaps most likely to have been most commonly applied to the Edible Crab and the Common Shore Crab, simply because these were the most visible and culturally important species. This map probably does not show any records of the Hermit Crab because this species seems to have been reliably distinguished from the other crabs by early modern recorders.

DISTRIBUTION

The distribution of top-quality records for these species is not statistically different from the known level of recorder effort. The gaps on the map may just reflect decreased survey effort in some regions. They may have been widespread. When we also include the county-level records it is clear that the crabs could most likely be found on every coast of every region of Britain and Ireland. The species seem to have been often eaten, as it is today:

> Lobsters in great number about Sheringham and Cromer [Norfolk] from whence all the country is supplyed.
>
> Astacus marinus pediculi facie found also in that place. With the advantage of ye long foreclawes about 4 inches long.
>
> Crabs large & well tasted found also in the same coast.
>
> Another kind of crab taken for cancer fluviatilis litle slender & of very quick motion found in the River running through Yarmouth & in Bliburgh River.　　(ed. Browne 1902 [*c*.1662–68]: 46)

Today, crabs are still widely distributed on every coast of Britain and Ireland; at regional level the distribution of these species has not changed since the early modern period. The Edible Crab continues to be the target of several commercial fisheries around Britain and Ireland (Marren and Mabey 2010: 78; Bannister 2009). The effect of these fisheries is tracked regionally – in some areas the fishery currently seems to be sustainable, whereas in others the stock may be declining, possibly due to overfishing (Bridges 2020; Street et al. 2020).

HERMIT CRABS

esp. *Pagurus bernhardus*

NATIVE STATUS Native

MODERN CONSERVATION STATUS	
World	Not Evaluated
UK	Not Evaluated
ROI	Not Evaluated
Trend since 1772	Uncertain

Hermit crabs were occasionally recorded separately from the other crabs by the sources most interested in marine species.

RECOGNITION

The hermit crabs were differentiated from the other crabs by the terms *Bernard the hermit*, *souldier crab* and *wrong-heir* in Early Modern English. The Latin terms vary slightly but are usually a variation of *Cancellus in turbine degens*, or sometimes just *cancellus*. All of these terms identify the hermit crabs.

Hermit crab records, 1519–1772.
There are records from the South
and South West of England,
Lowland Scotland, and Munster
and Leinster in Ireland.

The early modern sources did not distinguish the various hermit crab species occurring around early modern Britain and Ireland, but the most common and widespread today is the Common Hermit Crab, and so many or most of the records may refer to this species.

> Of the shrimp kind, great quantities are taken in Helford harbour, Mount's Bay, etc. in calm weather. Here we often find the hermit-shrimp, bernard, or cancellus, remarkable for taking possession of some empty shell, and there fixing his habitation as firmly as if it were his own native place; when it marches, it draws the shell after it; in danger retires wholly into it, and guards the mouth with one of its forcipated claws.
> (Borlase 1758: 474)

DISTRIBUTION

The distribution of top-quality records for these species are not statistically different from the known level of recorder effort. The gaps on the map may just reflect decreased survey effort in some regions. They may have been widespread. There are records from Munster and Leinster, the South and South West of England and Lowland Scotland, which are the regions where recorders are most interested in marine life, so the lack of records from Wales, the Midlands and North of England, Ulster and Connacht may well be oversights given how few records there are overall.

Today, the Common Hermit Crab is widely distributed around the whole coastline of Britain and Ireland (Wood 2018: 116). The species is not commonly eaten or commercially fished. It appears to be somewhat resilient to the effects of scallop-dredging, which negatively impact on other benthic species (Bradshaw et al. 2002). Its distribution around Britain and Ireland may not have changed since the early modern period, but it is predicted to shift its overall North Sea distribution to the south west as a result of climate change (Weinert et al. 2016).

WHITE-CLAWED CRAYFISH
Austropotamobius pallipes

NATIVE STATUS Native

MODERN CONSERVATION STATUS	
World	Least Concern
UK	Not Evaluated
ROI	Not Evaluated
Trend since 1772	Probably increased

The crayfish was recorded in early modern rivers and streams across England, but was absent from Cornwall and not recorded in Scotland, Wales, or most of Ireland.

RECOGNITION
The White-clawed Crayfish was most commonly called the *crayfish*, *craw-fish*, *cree* or *crevice* in Early Modern English and *gammarus*, or *cammarus* in Latin. All of these terms could also refer to another species: the Crawfish *Palinurus elephas*. The two species were occasionally distinguished by distinct New Latin names, the White-clawed Crayfish was *Astacus fluviatilis* and the Crawfish was *Astacus major*. The latter species was also sometimes distinguished in Early Modern English as specifically the *sea crayfish*. Where the term used is generic, the White-clawed Crayfish can still usually be identified based on context since it is a freshwater species found away from the coasts. No other freshwater crayfish species is known to have been present in the early modern period. The records on the map are therefore likely to be reliable.

DISTRIBUTION
For the obligate freshwater riverine species, records at river-level have been shown on the map by shading every county with similar topography which that river runs through.

The distribution of top-quality records for this species is statistically a poor fit with the known levels of recorder effort. It may have been locally distributed, locally abundant or of special local interest. There are far more records from the South, South West (except Cornwall) and Midlands of England than we would expect, but none at all from Highland or Lowland Scotland. Habitat suitability modelling based on the sites where the White-clawed Crayfish was recorded in the early modern period suggests that the species may have had specific requirements. It is only recorded at sites that today have relatively mild, dry winters and relatively warm summers, although, as with all the freshwater species, this might have reflected lack of opportunity rather than lack of ability to colonise other areas. Including the county-level records, this distribution pattern agrees approximately with the twentieth-century range of the White-clawed Crayfish, when the species was found across England but absent from Cornwall and Scotland, with the exception that by then the species was (and continues to be) also recorded in the east of Wales

(Holdich and Rogers 1997). However, the early modern sources also provide no records from Ireland outside of Leinster, suggesting that the species may have had a more limited distribution in the past.

The White-clawed Crayfish appears to have been caught for food in the early modern period. It was kept in ponds in Britain and Ireland and seems sometimes to have escaped from these into surrounding river systems:

> Astacus fluviatilis, sive Cammarus, The Cray-fish.
>
> It has been sometimes found in this County, chiefly in Gentlemens' ponds and lately in the river near Finglass; but said to have been brought thither from Munster, tho' the Cray-fish common in Cork is another [species], viz. the next following this, river Cray-fish being a rarity also among them, but the following is very common there, viz
>
> Cammarus, seu Asatcus major. The Sea Cray-fish.   (Rutty 1772: 372)

The two species named here are the White-clawed Crayfish and the Crawfish. This is the only reliable record from Ireland, which refers to a population introduced to Co. Dublin from Munster. This of course suggests the presence of a population in Munster, but it is not clear whether this

White-clawed Crayfish records, 1519–1772. There are records from every region of England and a single record from Dublin in Ireland.

latter population was wild or in captivity; unlike the Crawfish, the White-clawed Crayfish was not mentioned in Smith's descriptions of Cos. Cork, Kerry or Waterford (1746, 1750b, 1756), or in any of the other sources describing Munster. Like the freshwater fishes (see 'Breams', p. 208), the White-clawed Crayfish is not usually thought to have recolonised Ireland naturally after the last glacial period, but to have been introduced since then, either before or during the early modern period (Gouin et al. 2003; McCormick 1999). The genetic evidence suggests that the Irish population was established with individuals imported from western France. The lack of records from the early modern sources implies that the White-clawed Crayfish only became widespread in the wild after the end of our period. The population described in the quotation by Rutty from the 1770s was newly established.

The White-clawed Crayfish is now declining dramatically in Britain. This is due to the accidental introduction of the Signal Crayfish (*Pacifastacus leniusculus*), which was kept as a food species in the 1980s but escaped into British rivers and became established here. The Signal Crayfish carries a fungal disease, the Crayfish Plague (*Aphanomyces astaci*), which is deadly to the White-clawed Crayfish. The effect of Crayfish Plague has been to decimate the populations of White-clawed Crayfish, especially in the South and South West of England. The Signal Crayfish has now replaced the native species in many major river systems (Marren and Mabey 2010: 74; Holdich and Rogers 1997). However, at present the native species continues to be found in all the regions where it was present before the introduction of the Signal Crayfish (Chadwick 2019). The modern presence of the White-clawed Crayfish in Ulster, Connacht and Munster as well as Leinster, in Ireland, means that the species is still classed here as having 'probably increased' since 1772 despite the decimation of British populations by Crayfish Plague.

LOBSTER
Homarus gammarus

NATIVE STATUS Native

MODERN CONSERVATION STATUS	
World	Least Concern
UK	Not Evaluated
ROI	Not Evaluated
Trend since 1772	No change

Lobsters were widely recorded on almost every coast of early modern Britain and Ireland.

RECOGNITION
The map shows records of species most often referred to as *lobster* in Early Modern English, or occasionally *punger*. K'Eogh (1739: 60) provides the Irish term *glimugh*. In New Latin it was most commonly *Astacus marinus*, although one author distinguishes *Astacus marinus major* (Common Lobster) from *Astacus marinus minor* (possibly the Scampi, *Nephrops norvegicus*). All of these terms were probably most often used for the European Lobster, but the smaller squat lobsters are not distinguished in our texts so the names above may have occasionally been applied to them as well.

DISTRIBUTION
The distribution of top-quality records for this species is not statistically different from the known level of recorder effort. The gaps on the map may just reflect decreased survey effort in some regions.

It may have been widespread. If we include county-level records, the Lobster was recorded by almost every coastal recorder, on almost every coast in every region. It seems to have been regularly eaten (Kowaleski 2003). Gaps in the record around, for instance, Cardigan Bay and Argyll most likely only reflect where there was less effort applied to surveying.

> The most remarkable places for shell-fish, are the several creeks and harbours in the river of Kenmare [Co. Kerry], particularly Sneeme harbour, which affords fine oysters, lobsters and crabs. Dingle harbour is also famous for excellent oysters, as is that of Ventry for escallops; which fish also, with most other kinds of shell-fish, are to be had in great perfection in Valentia harbour. That side of the bay of Dingle is noted as having very large cray-fish, as is the northern side for abounding in lobsters. There are very fine beds of oysters also in the mouth of the Shannon, which are not valued at more than 2d. per hundred, for some thousands may be dredged up in a day's time by one boat.
>
> (Smith 1756: 370–1)

The distribution of the Lobster around Britain and Ireland does not seem to have changed between the early modern period and today; it remains widespread on all coasts (Wood 2018: 109). Lobster continues to be caught commercially by local fisheries. The effect of these fisheries is tracked regionally – just like with the crabs ('Hermit Crabs', p. 341) in some areas the fishery currently seems to be

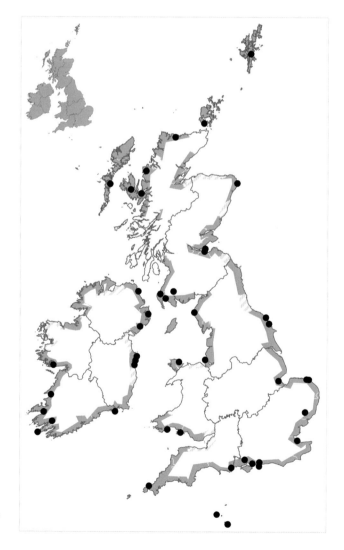

Lobster records, 1519–1772. There are records from every region of Britain and Ireland, including the Hebrides, the Northern Isles, the Isle of Man and the Channel Islands.

sustainable, whereas in others the stock may be temporarily declining, possibly due to overfishing (Bridges 2020; Street et al. 2020). The number of landings briefly dropped in the 1970s but have now recovered (Prodöhl et al. 2006). In some parts of Europe the wild Lobster fishery is being supplemented or replaced with animals hatched, grown or fattened in aquaculture settings.

GOOSE BARNACLES

esp. *Lepas anatifera*

NATIVE STATUS Thought to be native

MODERN CONSERVATION STATUS	
World	Not Evaluated
UK	Not Evaluated
ROI	Not Evaluated
Trend since 1772	Uncertain

The goose barnacles were recorded by several authors around Britain and in Munster.

RECOGNITION

The map here shows records of what are most reliably called *Concha anatifera* in New Latin, which is specific to the stalked barnacles (Lepadidae). The Common Goose Barnacle is the most common of these species and so the most likely identification for any records (Wood 2018: 102), but other species like the Small Goose Barnacle (*Lepas pectinata*) and the Buoy Barnacle (*Dosima fascicularis*) are also regularly seen around these islands today, so might also have been indicated by this term.

This species was also sometimes simply called the *barnacle* or *bernicle* in Early Modern English, but records using this term obviously have to be distinguished from records of the *Cirripedia* (barnacles like *Semibalanus balanoides*, our Acorn Barnacle; and *Pollicipes pollicipes*, our Goose-neck Barnacle) as well as the Barnacle Goose.

The theory that the Common Goose Barnacle might actually be the juvenile form of the Barnacle Goose (p. 86) was a contentious topic in the early modern period (Pastore 2021; Marren and Mabey 2010: 59–60; Kelly 1997: 300). Some authors accepted the idea while others rejected it, and the goose barnacles became key case-study species for scholars discussing the concept of spontaneous generation. This meant the goose barnacles were better recorded than most other rarely eaten marine invertebrates, but it also makes interpreting some of the records complicated. Authors who believed that goose barnacles were young Barnacle Geese might be more inclined to record both species wherever they saw one or the other. This means that we need to look for a basic description when identifying records of the species.

The records on this map are also complicated because goose barnacles tend to colonise driftwood and other floating debris but also the hulls of ships (Marren and Mabey 2010: 60). Early modern authors frequently had an idea that goose barnacles came from somewhere far away. Because of this, they often recorded the goose barnacle as if it was an exotic species brought back on the sides of ships rather than as a native species (Chynoweth et al. 2004: 109; Owen 1994: 134; Smith 1746: 346–7). This assumption needs to be treated critically. That the species was present prior to the intensification of global traffic in the early modern period is proved by the presence of remains of the Common Goose

Goose Barnacle records, 1519–1772.
There are records from every region
of Britain except the Midlands of
England, and from Munster in Ireland.

Barnacle on charcoal (probably burnt driftwood) at medieval Doonloughan, Co. Galway (Murray 2007). While many ships were likely colonised by barnacles on international voyages, others were probably colonised in British and Irish waters too.

DISTRIBUTION

The distribution of top-quality records for this species is not statistically different from the known level of recorder effort. The gaps on the map may just reflect decreased survey effort in some regions. It may have been widespread.

The Common Goose Barnacle is now found on the south and west coasts of Britain and Ireland but is rare on the east coast (Neal 2007). At a regional level, this seems to represent a more restricted distribution than the species had in the early modern period when it was also recorded on the east coast of Britain as well as in north-west England where it is not currently recorded. The following is a record of a goose barnacle from Buchan in Aberdeenshire by Alexander Gordon of Troup:

> We have nothing else cast in by the sea or any remark save firr that has lyen long in the sea, we
> find when cast in [it is] very much overgrown with a kind of shell fish which are rooted in the

stock by a trunk of flesh or resembling flesh about two inches long in so much that when cut or broke off it will bleed. The shells of this fish doe somewhat represent the wings of a fowl and in the end of it farthest from the tree it hath a membrane, which I suppose to be the Gill, but it represents the train of a fowl. These two with the trunk of flesh, which some think to be the neck, gives occasion to that conjecture of this being a kind of the Clack Geese production, but sure it is not so, for we never find this creature bigger than about the quantity of a mans nail, but we will find them much less. (Gordon (1683) in Mitchell 1907: 136–7)

If this record is accepted as a Common Goose Barnacle, it and others on the map might indicate that the species had a wider distribution in the early modern period.

COMMON SUNSTAR

Crossaster papposus

NATIVE STATUS Native

MODERN CONSERVATION STATUS	
World	Not Evaluated
UK	Not Evaluated
ROI	Not Evaluated
Trend since 1772	Uncertain

Whilst most starfishes were not distinguished in the early modern period, the Common Sunstar was identified and recorded on the south and east coasts of Ireland, the east coast of Britain and at Tenby in south-west Wales.

RECOGNITION

The Common Sunstar is the only species of starfish found around Britain and Ireland that can have more than ten arms (the Purple Sunstar, *Solaster endeca*, has nine or ten) (Wood 2018: 185). This made it distinctive to the early modern naturalists, who distinguished starfishes based on the number of arms. New Latin terms like *Stella marina undecim radiorum*, *Stella duodecim radiorum*, *Stella marina radiis tredecim* all referred specifically to the Common Sunstar. Some of the Early Modern English records are on the edge between descriptions and names, but terms like *twelve-fingers* and *fourteen-fingers* come up often and can be considered distinctive names.

These terms have to be distinguished from the terms *Stella marina quinque radiorum*, *Stella pentadactyla* (with various suffixes) and the Early Modern English *five-fingers*, all of which were applied to the more familiar starfishes like the Common Starfish (*Asterias rubens*) and Sand Star (*Astropecten irregularis*), as well as many other species.

The term *sea-rose* is used by two early modern authors to refer to the Common Sunstar but might presumably have applied to the Purple Sunstar as well. The term sun-fish is used once for this species, but could also sometimes apply to the Ocean Sunfish (*Mola mola*) as well as the Basking Shark (*Cetorhinus maximus*). The term *sea-star* is a translation of New Latin *Stella marina* and was used for all starfishes and brittlestars.

DISTRIBUTION

There are not enough top-quality records to statistically analyse the distribution of this species.

At present the Common Sunstar is widespread around these islands, but most common on the North Sea, Irish Sea and west Scotland coasts (Wood 2018: 185). This does approximately fit with the early modern pattern of the records from the southern and eastern coasts of Ireland, the east coast of Britain and from West Wales. This might suggest that the Common Sunstar has kept a similar distribution over the last few centuries.

> Tenby is famous for the largest oysters in the world, they are about seven inches over; in drudging for them we found the Echini, twelve fingers and fourteen fingers [and] of the five fingers; we saw a great number with the oysters sticking to them. (ed. Pococke 1889 [1754–57]: 191)

Common Sunstar records, 1519–1772. There are records from the North and South of England, Wales, Lowland Scotland and Munster and Leinster in Ireland.

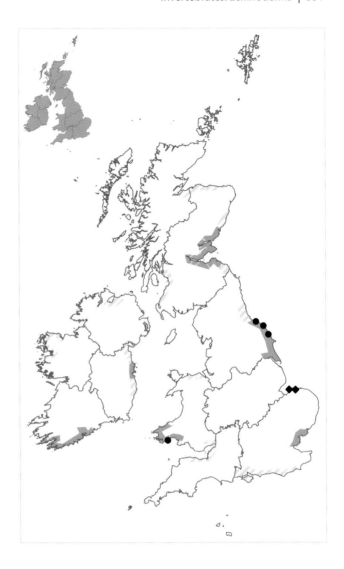

SEA URCHINS

Several species, most notably Common Sea Urchin (*Echinus esculentus*), Green Sea Urchin (*Psammechinius miliaris*), Sea Potato (*Echinocardium cordatum*)

NATIVE STATUS Native

MODERN CONSERVATION STATUS (*Echinus esculentus*)	
World	Near Threatened
UK	Not Evaluated
ROI	Not Evaluated
Trend since 1772	Uncertain

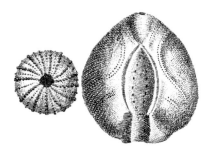

Sea urchins were recorded by authors around the coasts of early modern Britain and Ireland. There is a cluster of records on the coast of north-east England. It is usually impossible to distinguish records of different species.

RECOGNITION

These species are referred to commonly as *echinus* or *Echinus marinus* in New Latin and *sea hedgehog*, *sea urchin* and *sea button* or *round button fish* in Early Modern English. O'Sullivan (ed. 2009: 180) also attests the term *cuan mairi* for Irish and Wallace (Wallace 1700: 41) attests a local Orkney name *ivegar* (this is most likely a Norn term, perhaps related to Nynorn *julter* and Faeroese *igulker*) (Marren and Mabey 2010: 467). All these terms are used generically. They are easily distinguished from most other species, as long as they are not confused with the (land) Hedgehog (*Erinaceus europaeus*) or the Butterfish. For the most part, it is not possible to distinguish the different species of Echinoidea based on the names used by the early modern recorders alone. To reflect this, although county-level records have been shaded as normal, site-level records have been shown as diamonds except where they are more specific.

Some authors distinguish the different species. Robert Sibbald, in his *History, Ancient and Modern, of the Sheriffdoms of Fife and Kinross* (1710) distinguishes four kinds (ed. Sibbald 1803: 133):

1. 'Echinus marinus vulgaris, spinis albis, the common Sea Urchin'. This is probably the Common Sea Urchin.
2. 'Echinus marinus minor, viridis' [lit: the little green sea urchin]. This is probably the Green Sea Urchin.
3. 'Echinus marinus minor, purpureus' [lit: the little purple sea urchin]. This is probably the Northern Sea Urchin (*Strongylocentrotus droebachiensis*).
4. 'Echinus spatagus'. This is probably the Sea Potato.

DISTRIBUTION

There are not enough top-quality records to statistically analyse the distribution of this species.

At present the Common Sea Urchin is found most commonly around Ireland, Scotland and north-east England as far south as Yorkshire and around the west of Wales and South West England and the Channel Islands. It is less common around the Midlands and especially the southern England. The Green Sea Urchin is not found on the east coast of England from East Anglia to Durham (Wood 2018: 190–1). This is somewhat consistent with the early modern records, except that the records from the north coast of Wales, Norfolk and Essex come from areas where sea urchins are less commonly seen today.

Sea urchin records, 1519–1772. There are records from every region of Britain except the Midlands of England, and from Munster and Leinster in Ireland.

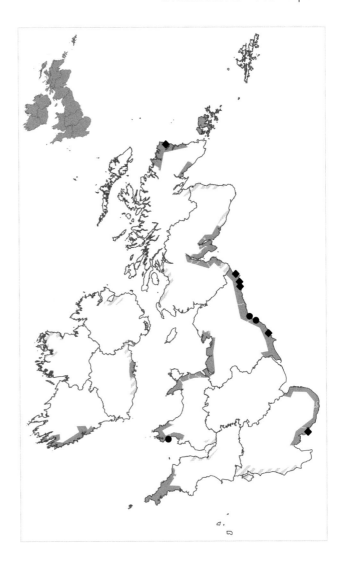

There is a cluster of early modern sea urchin records (six records from five sources) on the north-eastern coast of England. This remains an area where the Common Sea Urchin is widespread today, but is also likely to reflect a cultural interest in the species. Sea urchins may have been caught here as bycatch in trawl fisheries, but at least one author from the area specifically describes the culture of eating the Common Sea Urchin:

> The great subglobose Echinus, or Sea-Egg, is frequently taken with the fish in it by the Lobster-fishermen of Newton, near Embleton. The shells are of different colours; some red, and others purple, in lighter and darker shades … It was a great favourite at table among the antients, and was called The Ovum, or Egg. The Flesh is divided into five equal parts. it is eaten at the Turkish tables with pepper and vinegar, at the time of the full moon, when it is turgid, and esteemed a delicacy.
>
> (Wallis 1769: 293)

Although Wallis does not say here that he himself eats sea urchins, his knowledge that 'the Flesh is divided into five equal parts' might suggest a personal experience.

The Common Sea Urchin was previously the object of a fishery in Cornwall in the second half of the twentieth century. The urchins were collected by divers; their tests (shells) could then be cleaned out and sold in tourist shops or exported internationally (Marren and Mabey 2010: 468; Nichols 1981). There is now some demand for sea urchin roe (gonads) as food, but a fishery for the species does not seem to be commercially viable. The demand can be met by aquaculture, but there is no commercial production at present (Kelly et al. 2015).

The Common Sea Urchin is currently common and widespread (Wood 2018: 191). The distribution of the sea urchins is likely to change in the next century. Rising sea temperatures during the twenty-first century might mean that the Northern Sea Urchin is lost from British and Irish waters, but the Common Sea Urchin is likely to become more abundant, and the Purple Sea Urchin (*Paracentrotus lividus*) may become more widespread (Hiscock et al. 2004).

CONCLUSIONS

OVERALL ANALYSIS

Of the 151 species and groups of species covered in this atlas, almost one in five currently has a reduced distribution compared to the past, while another one in five currently has an increased distribution – meaning they have been lost or gained in one or more regions since 1772. This is a minimum estimate. It is worth emphasising that the 'Trend since 1772' tracks changes in distribution, not changes in abundance like modern trends used on red lists to assess conservation status (Gibbons et al. 1996). Species will usually decline or increase in abundance before the changes are visible at a presence/absence level. Further, even looking only at distribution changes, almost half of the species covered in this atlas lack sufficient data to determine whether they have declined, increased or remained unchanged. Many others declined or increased in the nineteenth century but have now returned to something closer to their early modern distribution, and such changes are invisible in a comparison between 1772 and the present day.

GROUPS THAT HAVE EXPANDED THEIR RANGE

Generally, the species that have increased their distribution can be divided into those that may have been previously overlooked, those that have naturally colonised and those introduced for sport. It is perhaps also worth mentioning here that while it might appear from Figure 7 that the species that have declined in Britain and Ireland have been matched by those which have increased, in reality the expansion of many of the species on Table 3 has had a negative effect on biodiversity and local businesses (perhaps especially the ungulates and the freshwater fishes).

The Common Frog, Wheatear, Ring Ouzel and Freshwater Lamprey all appear from the early modern records to have subsequently increased their distribution over the last 250 years. It is not clear why these species would have previously been absent from many areas, and it seems most probable that they were simply overlooked by the early modern naturalists.

Six species seem to have colonised new regions naturally since the early modern period. This group includes five birds – the Magpie, Oystercatcher, Black-tailed Godwit, Herring Gull and Peregrine Falcon – plus the Seabass, all of which are mobile enough to adapt to changing conditions and colonise new environments relatively quickly.

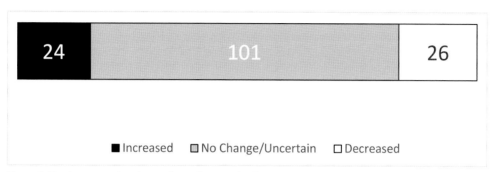

Figure 7 Graph representing the numbers of species that have increased, decreased or remained unchanged since 1772, according to the 'Trend since 1772' information included under each species. The trend for 101 of the species is either uncertain or unchanged, while 24 species have increased and 26 have decreased.

Table 3 Species that have expanded their range since 1772.

SPECIES	REASON FOR INCREASE
Common Frog, Wheatear, Ring Ouzel, Freshwater Lamprey	Previously overlooked?
Mobile species (Magpie, Oystercatcher, Godwit, Herring Gull, Peregrine, Seabass)	Natural colonisation
Game birds (Red-legged Partridge, Pheasant, Red Grouse)	Sport introduction/sport reintroduction
Freshwater fishes (Breams, Carps, Tench, Grayling, Perch) and also White-clawed Crayfish	Sport introduction and escapes and releases
Ungulates (Roe Deer, Red Deer, Fallow Deer, Wild Boar) and also Rabbit	Escapes and releases and sport reintroduction

The largest proportion of species that have expanded their range are those that have become established into new regions after being accidentally or purposefully introduced by humans for sport or sometimes food. This includes three game birds (Pheasant, Red Grouse and Red-legged Partridge) which were and continue to be restocked for shooting. It also includes ungulates (Red Deer, Fallow Deer, Roe Deer and Wild Boar) and the Rabbit, which have become established much more widely in Britain after escaping or being released for sport. All of these species have a long history in these islands, and the Fallow Deer, Rabbit and Red Deer were already relatively widespread in the early modern period, but mostly kept in captivity. Finally, this category includes at least five freshwater fishes (Bream, Carp, Tench, Grayling, Perch) and one freshwater crustacean (White-clawed Crayfish), which all probably became established in new regions after being introduced for angling or food. In the past, writers have suggested that all obligate freshwater fishes (i.e. not including migrating species like the Salmon and Brown Trout) are only native to southern England (Locker 2010; McCormick 1999; Maitland 1977 1987; Wheeler and Maitland 1973), and that all modern populations of obligate freshwater fishes in Scotland, Ireland and Wales result from modern introductions. However, based on the early modern data, several obligate freshwater species were clearly already widespread outside of the south of England in this time period. This includes especially the Grayling (surprisingly widespread in Wales; p. 235), the Perch (surprisingly widespread across the whole of Britain; p. 237) and the Pike (surprisingly widespread across the whole of Britain as well as Ireland; p. 219). So, Wales, Scotland and even Ireland may have been colonised by freshwater fishes prior to the early modern period.

GROUPS THAT HAVE REDUCED THEIR RANGE

The species that have reduced their range since the end of the early modern period can be divided into seven groups (see Table 4). The Orca seems to be the only species likely to have naturally adapted its range in reaction to environmental conditions, although it might also have been targeted by whaling vessels. On the other side, the Corncrake is the only farmland species included here and seems to have declined due to agricultural changes (in this case the increased use of mowers and tractors). The other species were more purposefully targeted. The carnivore mammals (Wolf, Lynx, Wildcat) and the raptors (Osprey, White-tailed Eagle, Golden Eagle) declined due to being directly targeted for pest control. The rodents (Red Squirrel, Ship Rat) declined due to a combination of intensive persecution and the importation of replacement species (Grey Squirrel and Brown Rat).

In contrast to the species subject to pest control, as we saw in the previous section, some quarry species such as the ungulates and most of the game birds have actually increased their distribution since the end of the early modern period. Unfortunately, though, popularity has not always been enough

Table 4 Species that have reduced their range since 1772

SPECIES	REASON FOR DECLINE
Carnivores and raptors (Wolf, Lynx, Wildcat, Osprey, White-tailed Eagle, Golden Eagle)	Pest control
Rodents (Ship Rat, Red Squirrel)	Pest control and introduction of competition species (Grey Squirrel, Brown Rat)
Freshwater fishes (Sturgeon, Burbot, Smelt, Shad, Salmon) as well as Oyster	Overfishing, river blockage, pollution, invasive species
Demersal fishes (Turbot, Halibut, Common Skate)	Overfishing and caught as bycatch in trawl fisheries
Great Auk, Bittern, Crane, Bustard, Black Grouse, Capercaillie, Greylag Goose	Overexploitation and habitat loss
Corncrake	Agricultural changes
Orca	Natural redistribution?

to protect a species from decline. The miscellaneous edible birds collected on this list (Great Auk, Bittern, Crane, Bustard, Black Grouse, Capercaillie, Greylag) declined due to constant overhunting compounded by the destruction of natural habitat. Many of these species are now recovering due to protection from hunting and habitat restoration. The freshwater fishes (Sturgeon, Burbot, Smelt, Shad, Salmon) as well as the Oyster were similarly all deeply popular and therefore overexploited. Their decline was compounded by the blockage and pollution of rivers in the twentieth century. The last category are the demersal fishes: the Turbot, Halibut and Common Skate. Like the freshwater fishes, these species have been overexploited for food, and they are also vulnerable to being caught as bycatch in trawl fisheries.

The declines have also affected different regions to different extents. The worst-hit regions in Britain and Ireland were Munster and the Midlands of England. If we include in our analysis only species that have certainly declined, then in the last 250 years most regions have lost around seven of the species we have studied. However, at least 10 of the species from this atlas seem to have been lost from Munster in that period (Wolf, Capercaillie, Bittern, Osprey, Golden Eagle, Crane, Sturgeon, Common Skate and perhaps Bustard and Black Grouse). The Midlands, meanwhile, has lost at least 12 species which were formerly present there (Ship Rat, Wildcat, Pine Marten, Black Grouse, White-tailed Eagle, Bustard, Smelt (inland), Salmon, Burbot, Oyster and at least one sturgeon and shad). It is not clear why these regions have been so negatively impacted, but the environmental changes in these areas may have happened faster than in the rest of Britain and Ireland.

PROBLEMS WITH USING THE LITTLE ICE AGE AS A BASELINE

The world is currently undergoing a biodiversity crisis, in which species are going extinct much faster than the historical average rate (Ceballos et al. 2015). One of the most commonly agreed solutions for this crisis is to conserve and restore ecosystems, which implies returning and maintaining ecosystems to a state in which they are known to have been healthy in the past (Jones et al. 2018). The state of wildlife in the early modern period is a tempting target to use as a reference baseline because it is the last period before industrialisation. But it must be remembered that the idea that nature was in a perfectly healthy state before industrialisation is a myth. As we have seen, wildlife during the early modern period was affected by human overexploitation and habitat changes, just like it is today. A pristine state of nature probably never existed. History is complicated, and nature is continually adapting; we cannot rewild our way back to an eternal Garden of Eden. This is not to say that the early modern period should not be used as a baseline: the ecosystems of Britain and Ireland *do* seem to have been, for the most part, healthier 250–500 years ago than they are today. But there were some factors influencing the state of nature in the early modern period that no longer exist, meaning that exactly recreating the ecosystems of the time period may not be possible. The most important of these factors may have been the Little Ice Age.

During the early modern period, parts of the Northern Hemisphere experienced very cold weather, a phenomenon now known as the Little Ice Age. In Britain and Ireland, the coldest period was 1550–1700 CE, which is also when the majority of sources for this atlas were written. At this time the growing season was three weeks to a month shorter than in the 1990s, and in winter there was sea ice off the coast of southern England (Lamb 2011; Fagan 2000: chap. 7). Britain and Ireland are now warmer, not just as a result of the end of the Little Ice Age, but also increasingly due to global warming brought about by climate change. It is likely that changes during the Little Ice Age had many effects on the wildlife of Britain and Ireland, but it is difficult to confidently reconstruct what these were centuries later. A good starting point is to look at which of the atlas species seem especially vulnerable to changes in climate. A decent list of these can be created based on species reactions to changes over the last century. Climate scientists working in the twentieth century traced a warmer period (the 'Amelioration') from the 1890s to the 1940s and a colder period from the 1940s to the 1970s. Species that are sensitive to shifts in climate have traditionally been sorted into groups depending on their reaction to these twentieth-century climate changes:

1. Heat-loving 'southern' species, many of which expanded their range from the south of Britain and Ireland during the warmer first half of the twentieth century. If correctly categorised, these species are likely to be expanding further in the twenty-first century due to global warming, and would presumably have reduced their northern distribution during the Little Ice Age: Black-tailed Godwit, Capercaillie, Lapwing, Curlew, White Stork, Bittern, Grey Heron, Crane, Common Carp, Pilchard.

2. Cold-loving or moisture-loving 'northern' species, many of which expanded their range from the north of Britain and Ireland during the colder years in the second half of the twentieth century. If the categorisation of these species is correct, they are likely to be reducing their range in the twenty-first century due to global warming and would presumably have increased their distribution to the south during the Little Ice Age: Lynx, Mountain Hare, Brent Goose, Puffin, Great Northern Diver, Ring Ouzel, Golden Plover, Grey Plover, Lapwing, Curlew, Snipe, Wheatear, Hooded Crow, Golden Eagle, White-tailed Eagle, Haddock, Ling, Whiting, Plaice, Eel, Conger Eel and Lesser Sandeel.

3. Species with an exact temperature 'isotherm' preference, which might move south or north in reaction to changes in climate. If the categorisation of these species was correct, they are likely to move north or to deeper depths due to global warming in the twenty-first century and would presumably have been present further south than today during the Little Ice Age: Gannet, Cod, Herring, Saithe.

In retrospect, the declines and increases of some species on this list in the twentieth century might actually have been due to chance, or to a mixture of climate and other factors. It is also now clear that the shifts in climate between 1880 and 1980 were not that major in contrast with the climate changes expected over the next 50 years due to global warming. However, this list also highlights some species that seem to have also been affected by climate changes during the Little Ice Age of the early modern period. For example, Cod (p. 266) only breed in temperatures between 3°C and 7°C, and there is known to have been a significant cooldown in the Atlantic in the seventeenth century, which culminated around 1695–1704 (Grove 2004: 607; Fagan 2000: chap. 4). Cod were commonly recorded around the whole of Britain and Ireland in the early modern period, but two of our sources seem to describe disruption to the fisheries of the Northern Isles, especially Cod and Ling, in the 1680s and the 1700s. The large whitefish fishery there was abandoned in the early modern period and replaced with a Saithe fishery (Harland and Kirkwall 2016).

Some birds were also likely affected by the end of the Little Ice Age. The Golden Plover, which bred in the early modern period in the south of Britain, has since declined into the north and the west. At least some of the decline in this species seems to be due to the reduction in craneflies (Tipulidae spp.), which are less abundant after hot summers (Pearce-Higgins et al. 2010). Likewise, the Hooded Crow has a mutually exclusive range in Britain with the Carrion Crow, except for a narrow hybridisation band through Scotland where both species co-occur. This band has moved northwards, apparently in line with the warmer climate in the twentieth and twenty-first centuries (Young 2007b; Cook 1975). Some Hooded Crows migrate south for winter, and during the Little Ice Age one of the most common terms for this species was the Royston Crow, because it was so common in Royston, Hertfordshire. Today, when the bird does migrate south, it mostly roosts on the coasts. The era of the Royston Crow is over.

Of course, the Little Ice Age might also have had indirect impacts on the wildlife of Britain and Ireland. During the famine of 'the seven ill years' in the 1690s, Robert Sibbald (1699: 18) published *Provision for the poor in times of dearth and scarcity*, in which he suggested possible sources of wild food. The section on birds began: 'There are a vast number of Fowls which frequent this Countrey which are all of them eaten by some people.' Hunting pressure from starving humans might therefore have had more of an impact on birds than the climate. Similarly, Lovegrove (2007: 81–2) has suggested that it was partly as a response to the Little Ice Age that the second Tudor Grain Act (1566, renewed 1572, 1598) was passed, which created the parish bounty system of pest control in England and Wales that did enormous damage to mammals and birds, before being repealed in 1863. These indirect impacts are likely to be impossible to quantify.

The records compiled in this atlas represent the collected effort of hundreds of people to describe what they saw as the natural riches of Britain and Ireland. The early modern recorders were writing at a time when, despite the cold, it still seemed possible that nature could provide food, wealth and medicine enough to match all human desire. Today we have perhaps tested this theory to its limits, and we are living through a time of climate change and biodiversity loss which is likely to become much more serious and permanent than the Little Ice Age. It might now be impossible for us to steal back the biodiversity of early modern Britain and Ireland in its full glory. But if we continue to record what we find and to remember what we have lost, it could still be possible to restore a vision of the treasure that our predecessors would recognise.

References

Adler, D. (1981) 'Imaginary Toads in Real Gardens', *English Literary Renaissance*, vol. 11, no. 3: 235–60. https://doi.org/10.1111/j.1475-6757.1981.tb00816.x

Aguilar, A. (2015) 'A Review of old Basque whaling and its effect on the right whales (*Eubalaena glacialis*) of the north Atlantic', in Best, P.B. and Prescott, J.H. (eds), *Right Whales, Past and Present Status*, Cambridge, International Whaling Commission: 191–9.

Albarella, U. (2010) 'Wild boar', in O'Connor, T. and Sykes, N.J. (eds), *Extinctions and Invasions*, Oxford, Oxbow Books: 59–67. https://doi.org/10.2307/j.ctv13gvg6k.14

Albarella, U. and Thomas, R. (2002) 'They dined on crane: bird consumption, wild fowling and status in medieval England', *Acta zoologica cracoviensia*, vol. 45: 23–38.

Allen, D.E. (1976) *The Naturalist in Britain*, London, Allen Lane.

Almeida, P. R, Arakawa, H., Aronsuu, K., Baker, C., Blair, S.-R., Beaulaton, L., Belo, A.F., Kitson, J., Kucheryavyy, A., Kynard, B., Lucas, M.C., Moser, M., Potaka, B., Romakkaniemi, A., Staponkus, R., Tamarapa, S., Yanai, S., Yang, G., Zhang, T. and Zhuang, P. (2021) 'Lamprey fisheries: History, trends and management', *Journal of Great Lakes Research*, vol. 47: S159–S185. https://doi.org/10.1016/j.jglr.2021.06.006

Almeroth-Williams, T. (2019) *City of Beasts: How Animals Shaped Georgian London*, Manchester, Manchester University Press. https://doi.org/10.7765/9781526126368

Alston, E.R. (1879) '2. On the Specific Identity of the British Martens', *Proceedings of the Zoological Society of London*, vol. 47, no. 1: 468–74. https://doi.org/10.1111/j.1096-3642.1879.tb02682.x

Anderson, G. (2017) *Birds of Ireland: Facts, Folklore & History*, Dublin, Collins Press.

Andrews, A.J., di Natale, A., Bernal-Casasola, D., Aniceti, V., Onar, V., Oueslati, T., Theodropoulou, T., Morales-Muñiz, A., Cilli, E. and Tinti, F. (2022) 'Exploitation history of Atlantic bluefin tuna in the eastern Atlantic and Mediterranean – insights from ancient bones', *ICES Journal of Marine Science*, vol. 79, no. 2: 247–62. https://doi.org/10.1093/icesjms/fsab261

Anon (1803) *The history and description of Colchester, vol. 2*, London, W. Keymer.

Anon (1819) *Statutes of the Realm, vol. 4 (1547–1585)*, London, Printed by Command of his Majesty King George III.

Appleby, J.C. (1990) 'A nursery of pirates: the English pirate community in Ireland in the early seventeenth century', *International Journal of Maritime History*, vol. 2, no. 1: 1–27. https://doi.org/10.1177/084387149000200103

Aprahamian, M. (2004) 'Shads – Alosa species', in Davies, C.E., Shelley, J., Harding, P.T., McLean, I.F.G., Gardiner, R., Peirson, G. and Quill, V. (eds), *Freshwater Fishes in Britain*, Colchester, Harley Books: 54–7.

Ásgeirsson, Þ., Gunnarsson, B. and Ingólfsson, A. (2007) 'The rapid colonization by *Crangon crangon* (Linnaeus 1758) (Eucarida, Caridea, Crangonidae) of Icelandic coastal waters', *Crustaceana*, vol. 80, no. 6: 747–53. https://doi.org/10.1163/156854007781360667

Atkinson, M. (2004) 'Smelt *Osmerus eperlanus*', in Davies, C.E., Shelley, J., Harding, P.T., McLean, I.F.G., Gardiner, R., Peirson, G. and Quill, V. (eds), *Freshwater Fishes in Britain*, Colchester, Harley Books: 100–2.

Aubrey, J. (1847) *The natural history of Wiltshire*, Britton, J. (ed.), Salisbury, Wiltshire Topographical Society.

Aulagnier, S., Haffner, P., Mitchell-Jones, A.J., Moutou, F. and Zima, J. (2018) *Mammals of Europe, North Africa and the Middle East*, London, Bloomsbury Publishing.

Avdelas, L., Avdic-Mravlje, E., Borges Marques, A.C., Cano, S., Capelle, J.J., Carvalho, N., Cozzolino, M., Dennis, J., Ellis, T. and Fernandez Polanco, J.M. (2021) 'The decline of mussel aquaculture in the European Union: Causes, economic impacts and opportunities', *Reviews in Aquaculture*, vol. 13, no. 1: 91–118. https://doi.org/10.1111/raq.12465

Bainbridge, I. (2007) 'Curlew (*Numenius arquata* Linnaeus)', in Forrester, R.W., Andrews, I.J., McInerny, C.J., Murray, R., McGowan, R.Y., Zonfrillo, B., Betts, M.W., Jardine, D.C., Grundy, D., Andrews, I.J. and Scott, H.I. (eds), *The Birds of Scotland*, Aberlady, Scottish Ornithologists' Club: 672–6.

Baker, P.A. and Harris, S. (2008) 'Fox Vulpes vulpes', in Harrison, S. and Yalden, D.W. (eds), *Mammals of the British Isles: Handbook*, 4th edn, Southampton, Mammal Society: 407–23.

Balharry, E.A., Jeffries, D.J. and Birks, J.D.S. (2008) 'Pine marten Martes martes', in Harris, S. and Yalden, D.W. (eds), *Mammals of the British Isles: Handbook*, 4th edn, London, Mammal Society: 447–55.

Balmer, D.E., Gillings, S., Caffrey, B.J., Swann, R.L., Downie, I.S. and Fuller, R.J. (2013) *Bird Atlas 2007–11: The Breeding and Wintering Birds of Britain and Ireland*, BTO Thetford.

Bannister, C. (2009) *On the management of brown crab fisheries*, London.

Barrett, J.H., Locker, A.M. and Roberts, C.M. (2004) 'The origins of intensive marine fishing in medieval Europe: the English evidence', *Proceedings of the Royal Society of London B: Biological Sciences*, vol. 271, no. 1556: 2417–21. https://doi.org/10.1098/rspb.2004.2885

Barrett-Hamilton, G.E.H. (1899) 'The Introduction of the Black Grouse and of Some Other Birds into Ireland', *The Irish Naturalist*, vol. 8, no. 2: 37–43.

Barrett-Hamilton, G.E.H. (1914) *A History of British mammals*, London, Gurney and Jackson, vol. 2, no. 14–21.

Barrington, R.M. (1880) 'On the introduction of the squirrel into Ireland', *Scientific Proceedings of the Royal Dublin Society*, vol. 2: 615–31.

Baudron, A.R. and Fernandes, P.G. (2015) 'Adverse consequences of stock recovery: European hake, a new "choke" species under a discard ban?', *Fish and Fisheries*, vol. 16, no. 4: 563–75. https://doi.org/10.1111/faf.12079

Baudron, A.R., Needle, C.L. and Marshall, C.T. (2011) 'Implications of a warming North Sea for the growth of haddock *Melanogrammus aeglefinus*', *Journal of Fish Biology*, vol. 78, no. 7: 1874–89. https://doi.org/10.1111/j.1095-8649.2011.02940.x

Baxter, E.V. and Rintoul, L.J. (1953) *The Birds of Scotland, vol. 1*, London, Oliver and Boyd.

Baxter, E.V. and Rintoul, L.J. (1953) *The Birds of Scotland, vol. 2*, London, Oliver and Boyd.

Beck, S., Foote, A.D., Koetter, S., Harries, O., Mandleberg, L., Stevick, P.T., Whooley, P. and Durban, J.W. (2014) 'Using opportunistic photo-identifications to detect a population decline of killer whales (*Orcinus orca*) in British and Irish waters', *Journal of the Marine Biological Association of the United Kingdom*, vol. 94, no. 6: 1327–33. https://doi.org/10.1017/S0025315413001124

Beebee, T.J.C. (2013) *Amphibians and Reptiles*, Exeter, Pelagic Publishing.

Beebee, T.J.C., Buckley, J., Evans, I., Foster, J.P., Gent, A.H., Gleed-Owen, C.P., Kelly, G., Rowe, G., Snell, C. and Wycherley, J.T. (2005) 'Neglected native or undesirable alien? Resolution of a conservation dilemma concerning the pool frog Rana lessonae', *Biodiversity & Conservation*, vol. 14, no. 7: 1607–26. https://doi.org/10.1007/s10531-004-0532-3

Beebee, T.J.C. and Griffiths, R. (2000) *Amphibians and Reptiles*, London, Collins.

Beebee, T.J.C. and Ratcliffe, S. (2018) 'Inferring status changes of three widespread British reptiles from NBN Atlas records.', *Herpetological Bulletin*, no. 143: 18–22.

Beebee, T.J.C., Wilkinson, J.W. and Buckley, J. (2009) 'Amphibian declines are not uniquely high amongst the vertebrates: trend determination and the British perspective', *Diversity*, Molecular Diversity Preservation International, vol. 1, no. 1: 67–88. https://doi.org/10.3390/d1010067

Bell, D. (2020) 'Rabbit', in Crawley, D., Coomber, F., Kubasiewicz, L., Harrower, C., Evans, P., Waggitt, J., Smith, B. and Matthews, F. (eds), *Atlas of the Mammals of Great Britain and Northern Ireland*, Exeter, Pelagic Publishing: 26–7.

Bellenden, J. (1821) *The history and chronicles of Scotland, written in Latin by Hector Boece, vol. 1*, Edinburgh, W and C Tait.

Beraud, C., van der Molen, J., Armstrong, M., Hunter, E., Fonseca, L. and Hyder, K. (2018) 'The influence of oceanographic conditions and larval behaviour on settlement success – the European sea bass *Dicentrarchus labrax* (L.)', *ICES Journal of Marine Science*, Oxford University Press, vol. 75, no. 2: 455–70. https://doi.org/10.1093/icesjms/fsx195

Birks, J. (2017) *Pine Martens*, Stansted, Whittet Books.

Birks, J. (2020) 'Polecat', in Crawley, D., Coomber, F., Kubasiewicz, L., Harrower, C., Evans, P., Waggitt, J., Smith, B. and Matthews, F. (eds), *Atlas of the Mammals of Great Britain and Northern Ireland*, Exeter, Pelagic Publishing: 76–7.

Blaeu, J. (1662) *Atlas Maior, vol. 6*, Amsterdam, John Blaeu.

Bolster, W.J. (2012) *The Mortal Sea: Fishing the Atlantic in the Age of Sail*, Cambridge MA, Harvard University Press. https://doi.org/10.4159/harvard.9780674067219

Bolton P. (2004a) 'Crucian carp *Carassius carassius*', in Davies, C.E., Shelley, J., Harding, P.T., McLean, I.F.G., Gardiner, R., Peirson, G. and Quill, V. (eds), *Freshwater Fishes in Britain*, Colchester, Harley Books: 66–8.

Bolton P. (2004b) 'Common carp', in Davies, C.E., Shelley, J., Harding, P.T., McLean, I.F.G., Gardiner, R., Peirson, G. and Quill, V. (eds), *Freshwater Fishes in Britain*, Colchester, Harley Books: 70–2.

Bom, R.A., van de Water, M., Camphuysen, K.C.J., van der Veer, H.W. and van Leeuwen, A. (2020) 'The historical ecology and demise of the iconic Angelshark *Squatina squatina* in the southern North Sea', *Marine Biology*, vol. 167, no. 7: 1–10. https://doi.org/10.1007/s00227-020-03702-0

Bond, C.J. (1988) 'Monastic Fishponds', in Aston, M. (ed.), *Medieval Fish, Fisheries and Fishponds in England*, vol. 1, Oxford, British Archaeological Reports 182: 69–112.

Booth, C., Cuthbert, M. and Reynolds, P. (1984) *The Birds of Orkney*, Stromness, The Orkney Press.

Boran, J.R., Hoelzel, A.R. and Evans, P.G.H. (2008) 'Killer whale *Orcinus orca*', in Harris, S. and Yalden, D.W. (eds), *Mammals of the British Isles: Handbook*, 4th edn, Southampton, The Mammal Society: 743–7.

Borlase, W. (1756) *Observations on the Ancient and Present State of the Islands of Scilly, and Their Importance to the Trade of Great-Britain*, Oxford, W. Jackson.

Borlase, W. (1758) *The Natural History of Cornwall*, Oxford, W. Jackson.

Botanista, T. (1757) *Rural beauties or the natural history of the four following western counties viz. Cornwall, Devonshire, Dorsetshire, and Somersetshire*, London, William Fenner.

Bourne, W.R.P. (1993) 'The story of the Great Auk *Pinguinis impennis*', *Archives of Natural History*, Edinburgh University Press, vol. 20, no. 2: 257–78. https://doi.org/10.3366/anh.1993.20.2.257

Boyd, H.A. (1974) 'Dean William Henry's Topographical Description of the Coast of County Antrim and North Down', *The Glynns*, vol. 2: 7–14.

Boyle, M. (1666) 'General Heads for a Natural History of a Countrey, Great or Small, Imparted Likewise by Mr. Boyle', *Philosophical Transactions of the Royal Society of London*, vol. 1, no. 11: 186–9. https://doi.org/10.1098/rstl.1665.0082

Boyle, R. (1692) *General Heads for the Natural History of a Country Great or Small*, London, John Taylor at the Ship in S. Paul's Church-yard, and S. Holford, at the Crown in the Pall Mall.

Bradshaw, C., Veale, L.O. and Brand, A.R. (2002) 'The role of scallop-dredge disturbance in long-term changes in Irish Sea benthic communities: a re-analysis of an historical dataset', *Journal of Sea Research*, vol. 47, no. 2: 161–84. https://doi.org/10.1016/S1385-1101(02)00096-5

Brand, A.R. (2006) 'Chapter 12 Scallop ecology: Distributions and behaviour', in Shumway, S.E. and Parsons, G.J.B.T.-D. in A. and F.S. (eds), *Scallops: Biology, Ecology and Aquaculture*, Elsevier, vol. 35: 651–744. https://doi.org/10.1016/S0167-9309(06)80039-6

Brand, J. (1701) *A brief description of Orkney, Zetland, Pightland-Firth and Caithness*, Edinburgh, George Mosman.

Brander, K.M. (2018) 'Climate change not to blame for cod population decline', *Nature Sustainability*, vol. 1, no. 6: 262–4. https://doi.org/10.1038/s41893-018-0081-5

Bray, W. (1901) *The Diary of John Evelyn, vol. 1*. London, M. Walter Dunne.

Brazier, B., Caffrey, J.M., Cross, T.F. and Chapman, D. v (2012) *A history of common carp* Cyprinus carpio (L.) in *Ireland: a review*, Oranmore, Marine Institute.

Breen, C. (2016) 'Marine fisheries and society in medieval Ireland', in Barrett, J.H. and Orton, D.C. (eds), *Cod & Herring: The Archaeology and History of Medieval Sea Fishing*, Oxford, Oxbow Books: 91–8. https://doi.org/10.2307/j.ctvh1dw0d.13

Brereton, W. (1844) *Travels in Holland the United Provinces England Scotland and Ireland 1634–1635*, Hawkins, E. (ed.), Manchester, Chetham Society.

Bridges, T. (2020) *Brown Crab Stock Assessment*, King's Lynn.

Bright, T. (1580) *A Treatise, Wherein is Declared the Sufficiencie of English Medicines, for Cure of all Diseases, Cured with Medicine*, London, Henrie Middleton for Thomas Man.

Brongersma, L.D. (1972) *European Atlantic Turtles*, Leiden, Brill.

Brown, A.W. and Brown, L.M. (2007) 'Mute Swan *Cygnus olor* (Gmelin)', in Forrester, R.W., Andrews, I.J., McInerny, C.J., Murray, R., McGowan, R.Y., Zonfrillo, B., Betts, M.W., Jardine, D.C., Grundy, D., Andrews, I.J. and Scott, H.I. (eds), *The Birds of Scotland*, Aberlady, Scottish Ornithologists' Club: 127–30.

Brown, D. (2002) 'The foulmart: what's in a name?', *Mammal Review*, vol. 32, no. 2: 145–9. https://doi.org/10.1046/j.1365-2907.2002.00104.x

Browne, T. (1646) *Pseudodoxia Epidemica*, London, Printed for Edward Dod. https://doi.org/10.1093/oseo/instance.00033939

Browne, T. (1902) *Notes and letters on the natural history of Norfolk*, Southwell, T. (ed.), London, Jarrold & Sons. https://doi.org/10.5962/bhl.title.54120

Buchan, A. (1741) *A description of St. Kilda, the most remote Western Isle of Scotland*, Edinburgh, Printed for the author, and are sold by daughter.

Buckland, F.T. (1861) *Curiosities of Natural History*, 2nd edn, London, Richard Bentley.

Buckley, K., Gorman, C.O., Martyn, M., Kavanagh, B., Copland, A. and McMahon, B.J. (2021) 'Coexistence without conflict, the recovery of Ireland's endangered wild grey partridge *Perdix perdix*', *European Journal of Wildlife Research*, vol. 67, no. 3: 58. https://doi.org/10.1007/s10344-021-01470-w

Burke, E. (ed.) (1766) 'A description of the Isle of Man, lately drawn up from the best authorities', *The Annual Register, or a View of the History, Politicks, and Literature, for the year 1765*, London, J. Dodsley, vol. 8: 70–6.

Burnett, J., Copp, C. and Harding, P. (1995) *Biological Recording in the United Kingdom-Present Practice and Future Development. Volume 1 Report*, Ruislip, Department of the Environment.

Burrough, R. (2004) 'Rudd *Scardinius erthrophthalmus*', in Davies, C.E., Shelley, J., Harding, P.T., McLean, I.F.G., Gardiner, R., Peirson, G. and Quill, V. (eds), *Freshwater Fishes in Britain*, Colchester, Harley Books: 88–90.

Butlin, R.A. (1991) 'V. Land and people, c.1600', in Moody, T.W., Martin, F.X. and Byrne, F.J. (eds), *A New History of Ireland: Volume III: Early Modern Ireland 1534–1691: Early Modern Ireland 1534–1691*, Oxford, Oxford University Press, vol. 3: 142–67.

Campbell, J.L. and Thomson, D.S. (1963) *Edward Lhuyd in the Scottish Highlands 1699–1700*, Oxford, Clarendon Press.

Campbell-Palmer, R. (2020) 'Beaver', in Crawley, D., Coomber, F., Kubasiewicz, L., Harrower, C., Evans, P., Waggitt, J., Smith, B., and Matthews, F. (eds), *Atlas of the Mammals of Great Britain and Northern Ireland*, Exeter, Pelagic Publishing: 36–7.

Carden, R.F., McDevitt, A.D., Zachos, F.E., Woodman, P.C., O'Toole, P., Rose, H., Monaghan, N.T., Campana, M.G., Bradley, D.G. and Edwards, C.J. (2012) 'Phylogeographic, ancient DNA, fossil and morphometric analyses reveal ancient and modern introductions of a large mammal: the complex case of red deer (*Cervus elaphus*) in Ireland', *Quaternary Science Reviews*, vol. 42: 74–84. https://doi.org/10.1111/faf.12074

Cardinale, M., Bartolino, V., Svedäng, H., Sundelöf, A., Poulsen, R.T. and Casini, M. (2015) 'A centurial development of the North Sea fish megafauna as reflected by the historical Swedish longlining fisheries', *Fish and Fisheries*, vol. 16, no. 3: 522–33. https://doi.org/10.1111/faf.12074

Carnell, S. (2010) *Hare*, London, Reaktion Books.

Carrier, J.-A. and Beebee, T.J.C. (2003) 'Recent, substantial, and unexplained declines of the common toad *Bufo bufo* in lowland England', *Biological Conservation*, vol. 111, no. 3: 395–9. https://doi.org/10.1016/S0006-3207(02)00308-7

Carss, D. and Murray, R. (2007) 'Great Cormorant: *Phalocrocorax carbo* (Linnaeus)', in Forrester, R.W., Andrews, I.J., McInerny, C.J., Murray, R., McGowan, R.Y., Zonfrillo, B., Betts, M.W., Jardine, D.C., Grundy, D., Andrews, I.J. and Scott, H.I. (eds), *The Birds of Scotland*, Aberlady, Scottish Ornithologists' Club: 400–3.

Carter, I. (2019) *The Red Kite's Year*, Exeter, Pelagic Publishing.

Carter, I. and Whitlow, G. (2005) *Red Kites in the Chilterns*, 2nd edn, Chinnor, English Nature & The Chilterns Conservation Board.

Carter, M. (2004a) 'Roach *Rutilus rutilus*', in Davies, C.E., Shelley, J., Harding, P.T., McLean, I.F.G., Gardiner, R., Peirson, G. and Quill, V. (eds), *Freshwater Fishes in Britain*, Colchester, Harley Books: 86–8.

Carter, M. (2004b) 'Perch *Perca fluviatilis*', in Davies, C.E., Shelley, J., Harding, P.T., McLean, I.F.G., Gardiner, R., Peirson, G. and Quill, V. (eds), *Freshwater Fishes in Britain*, Colchester, Harley Books: 131–3.

Ceballos, G., Ehrlich, P.R., Barnosky, A.D., García, A., Pringle, R.M. and Palmer, T.M. (2015) 'Accelerated modern human-induced species losses: Entering the sixth mass extinction.' *Science Advances*, vol. 1, no. 5: e1400253. https://doi.org/10.1126/sciadv.1400253

Cessford, C. (2013) 'Pine Marten and Other Animal Species in the Poem "Dinogad's Smock"', *Environmental Archaeology*, Maney Publishing.

Chadwick, D.D.A. (2019) 'Invasion of the signal crayfish, *Pacifastacus leniusculus*, in England: implications for the conservation of the white-clawed crayfish, *Austropotamobius pallipes*', PhD thesis, University College London.

Chamberlayne, E. (1683) *The present state of England Part III. and Part IV*, London, William Whitwood.

Chambley, A.D. (2000) *The Leaper Between: An Historical Study of the Toad Bone Amulet*, Hercules, Three Hands Press.

Chanin, P. (2020) 'Otter', in Crawley, D., Coomber, F., Kubasiewicz, L., Harrower, C., Evans, P., Waggitt, J., Smith, B., and Matthews, F. (eds), *Atlas of the Mammals of Great Britain and Northern Ireland*, Exeter, Pelagic Publishing: 68–9.

Childrey, J. (1662) *Britannia Baconica*, London, Printed by the author.

Chynoweth, J., Orme, N. and Walsham, A. (2004) *The Survey of Cornwall by Richard Carew*, New Series, Exeter, Devon and Cornwall Record Society.

Clark, G. (1947) 'Whales as an Economic Factor in Prehistoric Europe', *Antiquity* 2015/01/02, Cambridge University Press, vol. 21, no. 82: 84–104. https://doi.org/10.1017/S0003598X00016513

Clarke, D.V. (1998) 'The environment and economy of Skara Brae', in Lambert, R. (ed.), *Species History in Scotland*, Edinburgh, Scottish Cultural Press: 8–19.

Clarke, W.E. (1919) 'An old-time record of the breeding of the white stork in Scotland', *Scottish Naturalist*, no. 85–6: 25–6.

Coates, P. (1998) *Nature: Western Attitudes Since Ancient Times*, Cambridge, Polity Press.

Colclough, S. (2004) 'Grey mullets Mugilidae', in Davies, C.E., Shelley, J., Harding, P.T., McLean, I.F.G., Gardiner, R., Peirson, G. and Quill, V. (eds), *Freshwater Fishes in Britain*, Colchester, Harley Books: 135–7.

Coles, B. (2006) *Beavers in Britain's Past*, Oxford, Oxbow Books.

Coles, B. (2019) *Beavers in Wales*, WARP (Wetland Archaeology Research Project).

Coles, B.J. (2010) 'The European Beaver', in O'Connor, T. and Sykes, N.J. (eds), *Extinctions and Invasions*, Oxford, Windgather Press: 104–15. https://doi.org/10.2307/j.ctv13gvg6k.19

Collett, P. R (1909) 'A few notes on the whale Balaena glacialis and its capture in recent years in the North Atlantic by Norwegian whalers', *Proceedings of the Zoological Society of London*, vol. 79, no. 1: 91–103. https://doi.org/10.1111/j.1096-3642.1909.tb01858.x

Cook, A. (1975) 'Changes in the Carrion/Hooded Crow Hybrid Zone and the Possible Importance of Climate', *Bird Study*, vol. 22, no. 3: 165–8. https://doi.org/10.1080/00063657509476460

Cooper, A. (2007) *Inventing the Indigenous: Local Knowledge and Natural History in Early Modern Europe*, Cambridge, Cambridge University Press.

Corfield, H. (2021) 'Osprey *Pandion haliaetus* Gwalch pysgod', in Pritchard, R., Hughes, J., Spence, I.M., Haycock, B., Brenchley, A., Green, J., Thorpe, R. and Porter, B. (eds), *The Birds of Wales*, Liverpool, Liverpool University Press: 303–4.

Cosgrove, P., Watt, J., Hastie, L., Sime, I., Shields, D., Cosgrove, C., Brown, L., Isherwood, I. and Bao, M. (2016) 'The status of the freshwater pearl mussel *Margaritifera margaritifera* in Scotland: extent of change since 1990s, threats and management implications', *Biodiversity and Conservation*, vol. 25, no. 11: 2093–112. https://doi.org/10.1007/s10531-016-1180-0

Cove, R. (2004) 'European grayling *Thymallus thymallus*', in Davies, C.E., Shelley, J., Harding, P.T., McLean, I.F.G., Gardiner, R., Peirson, G. and Quill, V. (eds), *Freshwater Fishes in Britain*, Colchester, Harley Books: 117–18.

Cramer, J.A. (1841) *The second book of the travels of Nicander Nucius of Corcyra*, London, Camden Society.

Crawley, D., Coomber, F., Kubasiewicz, L., Harrower, C., Evans, P., Waggitt, J., Smith, B. and Matthews, F. (2020) *Atlas of the Mammals of Great Britain and Northern Ireland*, Exeter, Pelagic Publishing. https://doi.org/10.53061/XTWI9286

Croose, E. (2020) 'Pine marten', in Crawley, D., Coomber, F., Kubasiewicz, L., Harrower, C., Evans, P., Waggitt, J., Smith, B. and Matthews, F. (eds), *Atlas of the Mammals of Great Britain and Northern Ireland*, Exeter, Pelagic Publishing: 74–5.

Cross, T., Stratford, A., Hodges, J. and Haycock, B. (2021) 'Chough *Pyrrhocorax pyrrhocorax* Brân Goesgoch', in Pritchard, R., Hughes, J., Spence, I.M., Haycock, B., Brenchley, A., Green, J., Thorpe, R. and Porter, B. (eds), *The Birds of Wales*, Liverpool, Liverpool University Press: 366–9.

Crotti, M., Adams, C.E., Etheridge, E.C., Bean, C.W., Gowans, A.R.D., Knudsen, R., Lyle, A.A., Maitland, P.S., Winfield, I.J., Elmer, K.R. and Præbel, K. (2020) 'Geographic hierarchical population genetic structuring in British European whitefish (*Coregonus lavaretus*) and its implications for conservation', *Conservation Genetics*. https://doi.org/10.1007/s10592-020-01298-y

Crumley, J. (2010) *The Last Wolf*, Edinburgh, Birlinn.

Cumming, N.M. (2019) 'The relationship between meat consumption and power in late medieval and early modern England', *Sloth*, vol. 5, no. 1.

Cunningham, J. (2011) *Conquest and Land in Ireland: The Transplantation to Connacht 1649–1680*, Woodbridge, Boydell & Brewer Ltd.

Currie, C.K. (1988) 'Medieval Fishponds in Hampshire', in Aston, M. (ed.), *Medieval Fish, Fisheries and Fishponds in England*, vol. 2, Oxford, British Archaeological Reports 182: 367–89.

Currie, C.K. (1991) 'The early history of the carp and its economic significance in England', *The Agricultural History Review*: 97–107.

D'Arcy, G. (1999) *Ireland's Lost Birds*, Dublin, Four Courts Press.

Das, S. and Lowe, M. (2018) 'Nature read in black and white: Decolonial approaches to interpreting natural history collections', *Journal of Natural Science Collections*, vol. 6: 4–14.

Davidson, I. (2004) 'Brown trout and Sea trout *Salmo trutta*', in Davies, C.E., Shelley, J., Harding, P.T., McLean, I.F.G., Gardiner, R., Peirson, G. and Quill, V. (eds), *Freshwater Fishes in Britain*, Colchester, Harley Books: 110–13. https://doi.org/10.1163/9789004473515

Davidson, J.L. and Henshall, A.S. (1989) *The Chambered Cairns of Orkney: An Inventory of the Structures and their Contents*, Edinburgh University Press.

Davies, C.E., Shelley, J., Harding, P.T., McLean, I.F.G., Gardiner, R. and Peirson, G. (eds) (2004) *Freshwater Fishes in Britain the Species and their Distribution*, Colchester, Harley Books. https://doi.org/10.1163/9789004473515

de Smet, W.M.A. (1978) 'Evidence of whaling in the North Sea and English Channel during the Middle Ages', in FAO Advisory Committee on Marine Resources Research (ed.), *Mammals in the Seas Volume 3: General Papers and Large Cetaceans: Selected Papers of the Scientific Consultation on the Conservation and Management of Marine Mammals and Their Environment*, Rome, Food and Agriculture Organization of the United States: 301–9.

Deecke, V.B., Nykänen, M., Foote, A.D. and Janik, V.M. (2011) 'Vocal behaviour and feeding ecology of killer whales *Orcinus orca* around Shetland, UK', *Aquatic Biology*, vol. 13, no. 1: 79–88. https://doi.org/10.3354/ab00353

Defoe, D. (1971) *A tour through the whole island of Great Britain*, Rogers, P. (ed.), Middlesex, Penguin.

DeGraaff, R.M. (1991) *The Book of the Toad: A Natural and Magical History of Toad–Human Relations*, Park Street Press.

Dekker, W. (2003) 'On the distribution of the European eel (*Anguilla anguilla*) and its fisheries', *Canadian Journal of Fisheries and Aquatic Sciences*, NRC Research Press Ottawa, Canada, vol. 60, no. 7: 787–99. https://doi.org/10.1139/f03-066

Delahay, R., Wilson, G., Harris, S. and Macdonald, D.W. (2008) 'Badger *Meles meles*', in Harris, S. and Yalden, D.W. (eds), *Mammals of the British Isles: Handbook*, 4th edn, Southampton, Mammal Society: 425–36.

Dennis, R. (1991) *Ospreys*, Grantown-on-Spey, Colin-Baxter.

Denton, T. (2003) *A Perambulation of Cumberland 1687–1688*, Winchester, A.J.L. and Wane, M. (eds), Woodbridge, Surtees Society CCVII.

Desforges, J.-P., Hall, A., McConnell, B., Rosing-Asvid, A., Barber, J.L., Brownlow, A., de Guise, S., Eulaers, I., Jepson, P.D. and Letcher, R.J. (2018) 'Predicting global killer whale population collapse from PCB pollution', *Science*, vol. 361, no. 6409: 1373–6. https://doi.org/10.1126/science.aat1953

Dickey-Collas, M., Armstrong, M.J., Officer, R.A., Wright, P.J., Brown, J., Dunn, M.R. and Young, E.F. (2003) 'Growth and expansion of haddock (*Melanogrammus aeglefinus* L.) stocks to the west of the British Isles in the 1990s', *ICES Marine Science Symposium*, vol. 219: 271–82.

Dickie, I., Kharadi, N., Nleupauer, S., Butcher, B., Treweek, J., Judge, J., Whitbread, S., Hunt, T., Roy, D. and Harvey, M. (2021) *Mapping the Species Data Pathway: Connecting Species Data Flows in England*, London.

Dimock, J.F. (1867) *Giraldi Cambrensis Opera: Topographica Hibernia et Expugnatio Hibernica*, Longman, Greed, Reader & Dyer, vol. 5.

Dobney, K., Jaques, D., Barrett, J., Johnstone, C., Carrott, J., Hall, A., Herman, J., Nichols, C., Muldner, G. and Grimes, V. (2007) 'Exploitation of Resources and Procurement Strategies', in Dobney, K., Jaques, D., Barret, J. and Johnstone, C. (eds), *Farmers, Monks and Aristocrats: The Environmental Archaeology of Anglo-Saxon Flixborough*, Oxford, Oxbow Books, vol. Excavation: 190–212. https://doi.org/10.2307/j.ctt1cfr765.19

Dodgshon, R.A. (2004) 'Coping with risk: subsistence crises in the Scottish Highlands and Islands 1600–1800', *Rural History*, vol. 15, no. 1: 1–25. https://doi.org/10.1017/S0956793303001067

Dodsley, J. (ed.) (1775) *A History of the Island of Anglesey*, London, for J. Dodsley.

Dougall, T. (2007) 'Sky Lark *Alauda arvensis* (Linnaeus)', in Forrester, R.W., Andrews, I.J., McInerny, C.J., Murray, R., McGowan, R.Y., Zonfrillo, B., Betts, M.W., Jardine, D.C., Grundy, D., Andrews, I.J. and Scott, H.I. (eds), *The Birds of Scotland*, Aberlady, Scottish Ornithologists' Club: 976–9.

Duncan, P.F., Brand, A.R., Strand, Ø. and Foucher, E. (2016) 'The European scallop fisheries for *Pecten maximus*, *Aequipecten opercularis*, *Chlamys islandica*, and *Mimachlamys varia*', *Developments in aquaculture and fisheries science*, vol. 40: 781–858. https://doi.org/10.1016/B978-0-444-62710-0.00019-5

Dunstanville, F. (ed.) (1811) *Carew's Survey of Cornwall to which are added notes illustrative of its history and antiquities by the late Thomas Tonkin*, London, T. Bensley.

Dutton, J. (2020) 'Wild boar', in Crawley, D., Coomber, F., Kubasiewicz, L., Harrower, C., Evans, P., Waggitt, J., Smith, B. and Matthews, F. (eds), *Atlas of the Mammals of Great Britain and Northern Ireland*, Exeter, Pelagic Publishing: 80–1.

Dwelly, E. (1988) *Faclair Gàidhlig gu Beurla le Dealbhan/The Illustrated Gaelic–English Dictionary* 10th edn, Glasgow, Gairm Publications.

Easton, K. (2004) 'Burbot', in Davies, C.E., Shelley, J., Harding, P.T., McLean, I.F.G., Gardiner, R., Peirson, G. and Quill, V. (eds), *Freshwater Fishes in Britain*, Colchester, Harley Books: 119–20.

Edwards, E. (1997) 'Molluscan fisheries in Britain', in MacKenzie, C.L., Burrell, V.G., Rosenfield, A. and Hobart, W.L. (eds), *The History, Present Condition, and Future of the Molluscan Fisheries of North and Central America and Europe*, Woods Hole, NOAA Technical Reports NMFS, vol. 129: 85–100.

Emery, F.V. (1958) 'English regional studies from Aubrey to Defoe', *The Geographical Journal*, vol. 124, no. 3: 315–25. https://doi.org/10.2307/1790782

Emery, F.V. (1974) 'A new account of Snowdonia 1693, written for Edward Lhuyd', *National Library of Wales Journal*, vol. 18: 405–17.

Engelhard, G.H., Pinnegar, J.K., Kell, L.T. and Rijnsdorp, A.D. (2011) 'Nine decades of North Sea sole and plaice distribution', *ICES Journal of Marine Science*, vol. 68, no. 6: 1090–104. https://doi.org/10.1093/icesjms/fsr031

Engelhard, G.H., Righton, D.A. and Pinnegar, J.K. (2014) 'Climate change and fishing: a century of shifting distribution in North Sea cod', *Global Change Biology*, vol. 20, no. 8: 2473–83. https://doi.org/10.1111/gcb.12513

Engels, B. (2015) 'XNomial – Exact Test for Multinomial', R Studio.

Evans, P.G.H. (2008) 'Risso's dolphin *Grampus griseus*', in Harris, S. and Yalden, D.W. (eds), *Mammals of the British Isles: Handbook*, 4th edn, Southampton, Mammal Society: 740–3.

Evans, P.G.H., Lockyer, C.H., Smeenk, C.S., Addink, M. and Read, A.J. (2008) 'Harbour porpoise *Phocoena phocoena*', in Harris, S. and Yalden, D.W. (eds), *Mammals of the British Isles: Handbook*, 4th edn, Southampton, Mammal Society: 704–9.

Evans, P.G.H. and Smeenk, C.S. (2008) 'White-beaked dolphin *Lagenorhynchus albirostris*', in Harris, S. and Yalden, D.W. (eds), *Mammals of the British Isles: Handbook*, 4th edn, Southampton, Mammal Society: 724–7.

Evans, P. and Waggitt, J. (2020a) 'Killer whale or Orca', in Crawley, D., Coomber, F., Kubasiewicz, L., Harrower, C., Evans, P., Waggitt, J., Smith, B. and Matthews, F. (eds), *Atlas of the Mammals of Great Britain and Northern Ireland*, Exeter, Pelagic Publishing: 162–3.

Evans, P. and Waggitt, J. (2020b) 'Bottlenose dolphin', in Crawley, D., Coomber, F., Kubasiewicz, L., Harrower, C., Evans, P., Waggitt, J., Smith, B. and Matthews, F. (eds), *Atlas of the Mammals of Great Britain and Northern Ireland*, Exeter, Pelagic Publishing: 178–9.

Evans, P. and Waggitt, J. (2020c) 'Harbour porpoise', in Crawley, D., Coomber, F., Kubasiewicz, L., Harrower, C., Evans, P., Waggitt, J., Smith, B. and Matthews, F. (eds), *Atlas of the Mammals of Great Britain and Northern Ireland*, Exeter, Pelagic Publishing: 180–1.

Evans, R.J., O'Toole, L. and Whitfield, D.P. (2012) 'The history of eagles in Britain and Ireland: an ecological review of placename and documentary evidence from the last 1500 years', *Bird Study*, vol. 59, no. 3: 335–49. https://doi.org/10.1080/00063657.2012.683388

Everard, M. (2021) *Burbot: Conserving the Enigmatic Freshwater Catfish*, Great Easton, 5M Books.

Ewbank, J.M. (ed.) (1963) *Antiquary on Horseback*, Kendal, Titus Wilson & Son Ltd.

Fagan, B. (2000) *The Little Ice Age*, New York, Basic Books.

Fairley, J.S. (1975) *An Irish Beast Book*, Belfast, Blackstaff Press.

Fairley, J.S. (1983) 'Exports of wild mammal skins from Ireland in the eighteenth century', *The Irish Naturalists' Journal*: 75–9.

Fairweather, J. (2005) *Liber Eliensis*, Woodbridge, Boydell Press.

Falle, P. (1694) *An Account of the Isle of Jersey, the Greatest of those Islands that are now the Only Reminder of the English Dominions in France with a New and Accurate Map of the Island*, London, for John Newton at the Three Pigeons.

Fariñas-Franco, J.M., Pearce, B., Mair, J.M., Harries, D.B., MacPherson, R.C., Porter, J.S., Reimer, P.J. and Sanderson, W.G. (2018) 'Missing native oyster (*Ostrea edulis* L.) beds in a European Marine Protected Area: Should there be widespread restorative management?', *Biological Conservation*, vol. 221: 293–311. https://doi.org/10.1016/j.biocon.2018.03.010

Farran, G.P. (1944) 'The "Goaske" in Ireland', *The Irish Naturalists' Journal*, vol. 8, no. 5: 145–7.

Ferguson, A. (2004) 'The importance of identifying conservation units: brown trout and pollan biodiversity in Ireland', *Biology and Environment: Proceedings of the Royal Irish Academy*: 33–41. https://doi.org/10.1353/bae.2004.0009

Fernandes, P.G. and Cook, R.M. (2013) 'Reversal of Fish Stock Decline in the Northeast Atlantic', *Current Biology*, vol. 23, no. 15: 1432–7. https://doi.org/10.1016/j.cub.2013.06.016

Fick, S.E. and Hijmans, R.J. (2017) 'WorldClim 2: new 1-km spatial resolution climate surfaces for global land areas', *International Journal of Climatology*, vol. 37, no. 12: 4302–15. https://doi.org/10.1002/joc.5086

Fiennes, C. (1888) *Through England on a Side Saddle*, Griffiths, E.W. (ed.), London, Field and Tuer.

Finnegan, L.A. and Néill, L.Ó. (2010) 'Mitochondrial DNA diversity of the Irish otter, *Lutra lutra*, population', *Conservation Genetics*, vol. 11, no. 4: 1573–7. https://doi.org/10.1007/s10592-009-9955-4

Firth, L.B. (2021) 'What have limpets ever done for us?: On the past and present provisioning and cultural services of limpets', *International Review of Environmental History*, ANU Press Acton, ACT, vol. 7, no. 2: 5–45. https://doi.org/10.22459/IREH.07.02.2021.01

Fisher, J. (1966) *The Shell Bird Book*, London, Ebury Press and Michael Joseph.

Fitter, R.S.R. (1959) *London's Natural History*, London, Collins.

Fleming, J. (1828) *History of British Animals*, Edinburgh, Bell & Bradfute.

Fogarty, P. (2017) *Whittled Away: Ireland's Vanishing Nature*, Cork, The Collins Press.

Foote, A.D., Hooper, R., Alexander, A., Baird, R.W., Baker, C.S., Ballance, L., Barlow, J., Brownlow, A., Collins, T., Constantine, R., Dalla Rosa, L., Davison, N.J., Durban, J.W., Esteban, R., Excoffier, L., Martin, S.L.F., Forney, K.A., Gerrodette, T., Gilbert, M.T.P., Guinet, C., Hanson, M.B., Li, S., Martin, M.D., Robertson, K.M., Samarra, F.I.P., de Stephanis, R., Tavares, S.B., Tixier, P., Totterdell, J.A., Wade, P., Wolf, J.B.W., Fan, G., Zhang, Y. and Morin, P.A. (2021) 'Runs of homozygosity in killer whale genomes provide a global record of demographic histories', *Molecular Ecology*, vol. 30, no. 23: 6162–77. https://doi.org/10.1111/mec.16137

Foote, A.D., Newton, J., Piertney, S.B., Willerslev, E. and Gilbert, M.T.P. (2009) 'Ecological, morphological and genetic divergence of sympatric North Atlantic killer whale populations', *Molecular Ecology*, vol. 18, no. 24: 5207–17. https://doi.org/10.1111/j.1365-294X.2009.04407.x

Foote, A.D., Similä, T., Víkingsson, G.A. and Stevick, P.T. (2010) 'Movement, site fidelity and connectivity in a top marine predator, the killer whale', *Evolutionary Ecology*, vol. 24, no. 4: 803–14. https://doi.org/10.1007/s10682-009-9337-x

Ford, M.J. (1982) *The Changing Climate: Responses of the Natural Flora and Fauna*, London, George Allen & Unwin.

Forster, R. (1761) 'LXXIV. Observations on noxious animals in England; by the Rev. Richard Forster, M.A. Rector of Shefford in Bucks', *Philosophical Transactions of the Royal Society of London*, vol. 52: 475–6. https://doi.org/10.1098/rstl.1761.0076

Foster Evans, D. (2006) '"Cyngor y Bioden": Ecoleg a Llenyddiaeth Gymraeg', in Rowlands, J. (ed.), *Llenyddiaeth mewn theori*, Cardiff, University of Wales Press: 41–80.

Fox, A. (2010) 'Printed questionnaires, research networks, and the discovery of the British Isles 1650–1800', *Historical Journal*, vol. 53, no. 3: 593–621. https://doi.org/10.1017/S0018246X1000021X

Fromentin, J. (2009) 'Lessons from the past: investigating historical data from bluefin tuna fisheries', *Fish and Fisheries*, vol. 10, no. 2: 197–216. https://doi.org/10.1111/j.1467-2979.2008.00311.x

Fuller, E., McInery, C. and Zonfrillo, B. (2007) 'Great Auk: Pinguinus impennis (Linnaeus)', in Forrester, R.W., Andrews, I.J., McInerny, C.J., Murray, R., McGowan, R.Y., Zonfrillo, Bernard, Betts, M.W., Jardine, D.C., Grundy, D., Andrews, I.J. and Scott, H.I. (eds), *The Birds of Scotland*, Aberlady, Scottish Ornithologists' Club: 857–9.

Fychan, C. (2006) *Galwad y blaidd*, Aberystwyth, Cymdeithas Lyfrau Ceredigion Gyf.

Gainsford, T. (1618) *The Glory of England*, London, Edward Griffon for Thomas Norton.

Gardiner, M. (1997) 'The exploitation of sea-mammals in medieval England: bones and their social context', *Archaeological Journal*, vol. 154, no. 1: 173–95. https://doi.org/10.1080/00665983.1997.11078787

Gardner, E., Julian, A., Monk, C. and Baker, J. (2019) 'Make the adder count: population trends from a citizen science survey of UK adders', *Herpetological Journal*, vol. 29: 57–70. https://doi.org/10.33256/hj29.1.5770

GEBCO Compilation Group (2022) *GEBCO_2022 Grid*, British Oceanographic Data Centre. https://doi.org/10.5285/e0f0bb80-ab44-2739-e053-6c86abc0289c

Gessner, C. (1554a) *Historiae Animalium liber IV*, Frankfurt.

Gessner, C. (1554b) *Historiae Animalium liber II*, Zunch, Chrisoph. Frochoverum.

Gessner, C. (1602) *Historia Animalium*, Frankfurt.

Gibbons, D.W., Avery, M.I. and Brown, A.F. (1996) 'Population trends of breeding birds in the United Kingdom since 1800', *British Birds*, vol. 89, no. 7: 291–305.

Gibson, E. (1695) *Camden's Britannia newly translated into English*, London, F. Collins for A. Swalle.

Gifford, T. (1879) *Historical Description of the Zetland Islands in the Year 1733*, Nichols, J. (ed.), Edinburgh, Thomas George Stevenson.

Gilbert, G. (2007) 'Great Bittern *Botaurus stellaris* (Linnaeus)', in Forrester, R.W., Andrews, I.J., McInerny, C.J., Murray, R., McGowan, R.Y., Zonfrillo, B., Betts, M.W., Jardine, D.C., Grundy, D., Andrews, I.J. and Scott, H.I. (eds), *The Birds of Scotland*, Aberlady, Scottish Ornithologists' Club: 409–11.

Gilbert, G., Stanbury, A., & Lewis, L. (2021) 'Birds of Conservation Concern in Ireland 4: 2020–2026', *Irish Birds*, vol. 43: 1–22.

Gillespie, R. and Moran, G. (1991) *Longford: Essays in County History*, Dublin, Lilliput Press.

Gleed-Owen, C.P. (2001) *Further Archaeozoological Work for the Pool Frog Species Recovery Programme*, Peterborough.

Gleed-Owen, C.P. (2000) 'Subfossil records of *Rana* cf. *lessonae*, *Rana arvalis* and *Rana* cf. *dalmatina* from Middle Saxon (*c*.600–950 AD) deposits in eastern England: evidence for native status', *Amphibia Reptilia*, vol. 21, no. 1: 57–66. https://doi.org/10.1163/156853800507273

Glover, K.A., Svåsand, T., Olesen, I. and Rye, M. (2006) 'Atlantic halibut – *Hippoglossus hippoglossus*', in *Genimpact final scientific report*, Bergen, GENIMPACT: 17–22.

Gooders, J. (1983) *Birds that Came Back*, Worcester, Ebenezer Baylis and Son Ltd.

Gordon, J.C.D. and Evans, P.G.H. (2008) 'Sperm whale *Physeter macrocephalus*', in Harris, S. and Yalden, D.W. (eds), *Mammals of the British Isles: Handbook*, 4th edn, Southampton, Mammal Society: 678–83.

Gordon, R. (1813) *A Genealogical History of the Earldom of Sutherland from its Origin to the Year 1630*, Edinburgh, George Ramsay and Co.

Gouin, N., Grandjean, F., Pain, S., Souty-Grosset, C. and Reynolds, J. (2003) 'Origin and colonization history of the white-clawed crayfish, *Austropotamobius pallipes*, in Ireland', *Heredity*, vol. 91, no. 1: 70–7. https://doi.org/10.1038/sj.hdy.6800282

Gow, D. (2023) *The Hunt for the Iron Wolf*, forthcoming.

Gow, D. and Edgcumbe, C. (2016) 'A history of the white stork in Britain', *British Wildlife*, vol. 27, no. 4: 230–8.

Grafton, A. (2018) 'Philological and artisanal knowledge making in Renaissance natural history: a study in cultures of knowledge', *History of Humanities*, vol. 3, no. 1: 39–55. https://doi.org/10.1086/696301

Green, J. and Sandham, R. (2021) 'White Stork *Ciconia ciconia* Ciconia gwyn', in Pritchard, R., Hughes, J., Spence, I.M., Haycock, B., Brenchley, A., Green, J., Thorpe, R. and Porter, B. (eds), *The Birds of Wales*, Liverpool, Liverpool University Press: 282–3.

Green, J., Sandham, R. and Pritchard, R. (2021) 'Corncrake *Crex crex* Rhegen yr Ŷd', in Pritchard, R., Hughes, J., Spence, I.M., Haycock, B., Brenchley, A., Green, J., Thorpe, R. and Porter, B. (eds), *The Birds of Wales*, Liverpool, Liverpool University Press: 129–30.

Green, M. (2021) 'Curlew *Numenius arquata* Gylfinir', in Pritchard, R., Hughes, J., Spence, I.M., Haycock, B., Brenchley, A., Green, J., Thorpe, R. and Porter, B. (eds), *The Birds of Wales*, Liverpool, Liverpool University Press: 165–8.

Green, R. (2007) 'Corn crake *Crex crex* Linnaeus', in Forrester, R.W., Andrews, I.J., McInerny, C.J., Murray, R., McGowan, R.Y., Zonfrillo, B., Betts, M.W., Jardine, D.C., Grundy, D., Andrews, I.J. and Scott, H.I. (eds), *The Birds of Scotland*, Aberlady, Scottish Ornithologists' Club: 525–8.

Greenlee, J.W. (2020) 'Seeing all the Anguilles: Eels in the cultural landscape of medieval and early modern England', Cornell University.

Grew, N. (1681) *Musaeum regalis societatis*, London, W. Rawlins.

Griffin, E. (2007) *Blood Sport: Hunting in Britain Since 1066*, London, Yale University Press.

Griffiths, D. (1997) 'The status of the Irish freshwater fish fauna: a review', *Journal of Applied Ichthyology*, vol. 13, no. 1: 9–13. https://doi.org/10.1111/j.1439-0426.1997.tb00091.x

Griffiths, R. and Binnell, R. (1758) *A description of the River Thames*, London, T. Longman.

Grove, J.M. (2004) *Little Ice Ages Volume II*, 2nd edn, London, Taylor & Francis Group.

Grundy, D. (2007) 'Brent Goose *Branta bernicla* (Linnaeus)', in Forrester, R.W., Andrews, I.J., McInerny, C.J., Murray, R., McGowan, R.Y., Zonfrillo, B., Betts, M.W., Jardine, D.C., Grundy, D., Andrews, I.J. and Scott, H.I. (eds), *The Birds of Scotland*, Aberlady, Scottish Ornithologists' Club: 169–74.

Grundy, D. and McGowan, B. (2007) 'Great bustard: *Otis tarda* Linnaeus', in Forrester, R.W., Andrews, I.J., McInerny, C.J., Murray, R., McGowan, R.Y., Zonfrillo, B., Betts, M.W., Jardine, D.C., Grundy, D., Andrews, I.J. and Scott, H.I. (eds), *The Birds of Scotland*, Aberlady, Scottish Ornithologists' Club: 544–5.

Guerra, Á., Robin, J.-P., Sykes, A., Koutsoubas, D., Jereb, P., Lefkaditou, E., Koueta, N. and Allcock, A.L. (2015) '*Sepia officinalis* Linnaeus 1758', in Jereb, P., Allcock, A.L., Lefkaditou, E., Piatkowski, U., Hastie, L.C. and Pierce, G.J. (eds), *Cephalopod Biology and Fisheries in Europe: II. Species Accounts*, Copenhagen, ICES: 53–72.

Gurnell, J. (2020) 'Black rat', in Crawley, D., Coomber, F., Kubasiewicz, L., Harrower, C., Evans, P., Waggitt, J., Smith, B. and Matthews, F. (eds), *Atlas of the Mammals of Great Britain and Northern Ireland*, Exeter, Pelagic Publishing: 60–1.

Gurney, D. (1832) 'XXII. Extracts from the Household and Privy Purse Accounts of the Lestranges of Hunstanton, from AD 1519 to AD 1578', *Archaeologia*, vol. 25: 411–569. https://doi.org/10.1017/S0261340900023365

Gurney, J.H. (1921) *Early Annals of Ornithology*, London, H.F. & G. Witherby. https://doi.org/10.5962/bhl.title.55728

Haley-Halinski, K. (2021) 'Birds and Humans in the Old Norse World, *c.* 600–1500 AD', University of Cambridge.

Hall, A.J. (2008) 'Walrus *Odobenus rosmarus*', in Harris, S. and Yalden, D. (eds), *Mammals of the British Isles: Handbook*, 4th edn, Southampton, Mammal Society: 550.

Hall, J.J. (1981) 'The cock of the wood', *Irish Birds*, vol. 2: 38–47.

Hamilton-Dyer, S. (2007) 'Exploitation of birds and fish in historic Ireland', in Murphy, E.M. and Whitehouse, N.J. (eds), *Environmental Archaeology of Ireland*, Oxford, Oxbow Books: 102–18.

Hammond, P.S., Duck, C.D., Hall, A.J. and Pomeroy, P. (2008) 'Grey seal *Halichoerus gryphus*', in Harris, S. and Yalden, D.W. (eds), *Mammals of the British Isles: Handbook*, 4th edn, Southampton, Mammal Society: 538–47.

Handford, R. (2004) 'Common sturgeon *Acipenser sturio*', in Davies, C.E., Shelley, J., Harding, P.T., McLean, I.F.G., Gardiner, R., Peirson, G. and Quill, V. (eds), *Freshwater Fishes in Britain*, Colchester, Harley Books: 49–51.

Hardey, J. (2007) 'Peregrine Falcon *Falco peregrinus* (Tunstall)', in Forrester, R.W., Andrews, I.J., McInerny, C.J., Murray, R., McGowan, R.Y., Zonfrillo, B., Betts, M.W., Jardine, D.C., Grundy, D., Andrews, I.J. and Scott, H.I. (eds), *The Birds of Scotland*, Aberlady, Scottish Ornithologists' Club: 511–14.

Harding, P. and Carter, M. (2004) 'Pike *Esox lucius*', in Davies, C.E., Shelley, J., Harding, P.T., McLean, I.F.G., Gardiner, R., Peirson, G. and Quill, V. (eds), *Freshwater Fishes in Britain*, Colchester, Harley Books: 98–100.

Harland, J., Jones, A.K.G., Orton, D. and Barrett, J. (2016) 'Fishing and fish trade in medieval York: the zooarchaeological evidence', in Barrett, J.H. and Orton, D.C. (eds), *Cod & Herring: The Archaeology and History of Medieval Sea Fishing*, Oxford, Oxbow. https://doi.org/10.2307/j.ctvh1dw0d.19

Harland, J. and Kirkwall, O. (2016) 'From The Fish Middens to the Herring: Medieval and Post-Medieval Fishing in the Northern Isles of Scotland', Gabriel, S. and Reitz, E. (eds), *Fishing through time: Archaeoichthyology,*

Biodiversity, Ecology and Human Impact on Aquatic Environments: Proceedings of the 18th ICAZ Fish Remains Working Group Meeting, Lisbon.

Harris, C.M., Ambrose Jr, W.G., Bigelow, G.F., Locke V.W.L. and Silverberg, S.M.B. (2018) 'Analysis of the size, shape, and modeled age of common limpets (*Patella vulgata*) from late Norse Middens at Sandwick, Unst, Shetland Islands, UK: Evidence for anthropogenic and climatic impacts', *The Journal of Island and Coastal Archaeology*, vol. 13, no. 3: 341–70. https://doi.org/10.1080/15564894.2017.1368743

Harris, M.P. and Wanless, S. (2011) *The Puffin*, London, Bloomsbury Publishing.

Harris, M.P. and Wanless, S. (2007) 'Atlantic Puffin *Fratercula arctica* (Linnaeus)', in Forrester, R.W., Andrews, I.J., McInerny, C.J., Murray, R., McGowan, R.Y., Zonfrillo, B., Betts, M.W., Jardine, D.C., Grundy, D., Andrews, I.J. and Scott, H.I. (eds), *The Birds of Scotland*, Aberlady, Scottish Ornithologists' Club: 867–71.

Harris, S. and Yalden, D.W. (eds) (2008) *Mammals of the British Isles: Handbook*, 4th edn, Southampton, The Mammal Society.

Harris, W. and Smith, C. (1744) *The Antient and Present State of the County of Down*, Dublin, William Williamson.

Harrison, A.J., Connor, L., Morrissey, E. and Kelly, F. (2012) 'Current status of pollan *Coregonus autumnalis pollan* in Lough Ree, Ireland', *Biology and Environment: Proceedings of the Royal Irish Academy*, vol. 112, no. 2: 225–33. https://doi.org/10.3318/BIOE.2012.09

Harrod, C., Griffiths, D., McCarthy, T.K. and Rosell, R. (2001) 'The Irish pollan, *Coregonus autumnalis*: options for its conservation', *Journal of Fish Biology*, vol. 59: 339–55. https://doi.org/10.1006/jfbi.2001.1755

Hassall, C. and Thompson, D.J. (2010) 'Accounting for recorder effort in the detection of range shifts from historical data', *Methods in Ecology and Evolution*, vol. 1, no. 4: 343–50. https://doi.org/10.1111/j.2041-210X.2010.00039.x

Hawkins, I. and Hughes, J. (2021) 'Lapwing *Vanellus vanellus* Cornchwiglen', in Pritchard, R., Hughes, J., Spence, I.M., Haycock, B., Brenchley, A., Green, J., Thorpe, R. and Porter, B. (eds), *The Birds of Wales*, Liverpool, Liverpool University Press: 149–52.

Hay, D. (1938) 'A XVI century Tynemouth marvel', *The Proceedings of the Society of Antiquaries of Newcastle Upon Tyne*, vol. 8, no. 4: 111–13.

Heal, F. (2008) 'Food gifts, the household and the politics of exchange in early modern England', *Past and Present*, vol. 199, no. 1: 41–70. https://doi.org/10.1093/pastj/gtn002

Heath, M.R., Rasmussen, J., Bailey, M.C., Dunn, J., Fraser, J., Gallego, A., Hay, S.J., Inglis, M. and Robinson, S. (2012) 'Larval mortality rates and population dynamics of Lesser Sandeel (*Ammodytes marinus*) in the northwestern North Sea', *Journal of Marine Systems*, vol. 93: 47–57. https://doi.org/10.1016/j.jmarsys.2011.08.010

Henderson, P.A. (2014) *Identification Guide to the Inshore Fish of the British Isles*, Somes, J.R. (ed.), Pennington, Pisces Conservation.

Henderson, P.A. (2017) 'Long-term temporal and spatial changes in the richness and relative abundance of the inshore fish community of the British North Sea Coast', *Journal of Sea Research*, vol. 127: 212–26. https://doi.org/10.1016/j.seares.2017.06.011

Henderson, P.A., Seaby, R.M. and Somes, J.R. (2006) 'A 25-year study of climatic and density-dependent population regulation of common shrimp *Crangon crangon* (Crustacea: Caridea) in the Bristol Channel', *Journal of the Marine Biological Association of the United Kingdom*, vol. 86, no. 2: 287–98. https://doi.org/10.1017/S0025315406013142

Henry, W. (1739a) *Hints towards a Natural and Typographical History of the Counties Sligoe, Donegal, Fermanagh and Lough Erne*, Belfast, Public Records Office of Northern Ireland MIC198/1.

Henry, W. (1739b) *[Antrim & North Down]*, Belfast, Public Records Office of Northern Ireland T2521/3/5.

Henry, W. (1740) *A natural history of the parish of Killasher*, London.

Hetherington, D.A., Lord, T.C. and Jacobi, R.M. (2006) 'New evidence for the occurrence of Eurasian lynx (*Lynx lynx*) in medieval Britain', *Journal of Quaternary Science*, vol. 21, no. 1: 3–8. https://doi.org/10.1002/jqs.960

Hetherington, D.A., Miller, D.R., Macleod, C.D. and Gorman, M.L. (2008) 'A potential habitat network for the Eurasian lynx *Lynx lynx* in Scotland', *Mammal Review*, vol. 38, no. 4: 285–303. https://doi.org/10.1111/j.1365-2907.2008.00117.x

Hetherington, D. and Geslin, L. (2018) *The Lynx and Us*, Kingussie, Scotland: The Big Picture.

Hewison, A.J.M. and Staines, B.W. (2008) 'Roe *Capreolus capreolus*', in Harris, S. and Yalden, D.W. (eds), *Mammals of the British Isles: Handbook*, 4th edn, Southampton, Mammal Society: 605–17.

Hickey, K. (2000) 'A geographical perspective on the decline and extermination of the Irish wolf *Canis lupus* – an initial assessment', *Irish Geography*, vol. 33, no. 2: 185–98. https://doi.org/10.1080/00750770009478590

Hiddink, J.G., Shepperson, J., Bater, R., Goonesekera, D. and Dulvy, N.K. (2019) 'Near disappearance of the Angelshark *Squatina squatina* over half a century of observations', *Conservation Science and Practice*, vol. 1, no. 9: e97. https://doi.org/10.1111/csp2.97

Hill, G. (1873) *An Historical Account of the Macdonnels of Antrim*, Belfast, Archer & Sons.

Hill, M.O. (2012) 'Local frequency as a key to interpreting species occurrence data when recording effort is not known', *Methods in Ecology and Evolution*, vol. 3, no. 1: 195–205. https://doi.org/10.1111/j.2041-210X.2011.00146.x

Hindes, A.M. (2004) 'Tench *Tinca tinca*', in Davies, C.E., Shelley, J., Harding, P.T., McLean, I.F.G., Gardiner, R., Peirson, G. and Quill, V. (eds), *Freshwater Fishes in Britain*, Colchester, Harley Books: 90–1.

Hiscock, K., Southward, A., Tittley, I.A.N. and Hawkins, S. (2004) 'Effects of changing temperature on benthic marine life in Britain and Ireland', *Aquatic Conservation: Marine and Freshwater Ecosystems*, vol. 14, no. 4: 333–62. https://doi.org/10.1002/aqc.628

Hislop, J.R.G. (1996) 'Changes in North Sea gadoid stocks', *ICES Journal of Marine Science*, Oxford University Press, vol. 53, no. 6: 1146–56. https://doi.org/10.1006/jmsc.1996.0140

Hoey, J.J., Conser, R.J. and Bertolino, A.R. (1989) 'The western North Atlantic swordfish', *Audubon Wildlife Report 1989–1990*. https://doi.org/10.1016/B978-0-12-041003-3.50025-1

Hoffmann, R.C. (1994) 'Remains and verbal evidence of carp (*Cyprinus carpio*) in medieval Europe', *Annalen-Koninklijk Museum voor Midden-Afrika-Zoologische Wetenschappen*, vol. 274: 139–50.

Hoffmann, R.C. (1996) 'Economic Development and Aquatic Ecosystems in Medieval Europe', *The American Historical Review*, vol. 101, no. 3: 631–69. https://doi.org/10.2307/2169418

Hogg, A. (2007) 'White Stork *Cicionia ciconia* (Linnaeus)', in Forrester, R.W., Andrews, I.J., McInerny, C.J., Murray, R., McGowan, R.Y., Zonfrillo, B., Betts, M.W., Jardine, D.C., Grundy, D., Andrews, I.J. and Scott, H.I. (eds), *The Birds of Scotland*, Aberlady, Scottish Ornithologists' Club: 435–6.

Holden, P. and Gregory, R. (2021) *RSPB Handbook of British Birds*, London, Bloomsbury Publishing.

Holdich, D.M. and Rogers, W.D. (1997) 'The white-clawed crayfish, *Austropotamobius pallipes*, in Great Britain and Ireland with particular reference to its conservation in Great Britain', *Bulletin Français de la Pêche et de la Pisciculture*, no. 347: 597–616. https://doi.org/10.1051/kmae/1997050

Holinshed Project (2008a) *Holinshed's Chronicles of England, Scotland and Ireland 1577 ed.* [Online]. Available at http://english.nsms.ox.ac.uk/holinshed/toc.php?edition=1577 (Accessed 11 August 2022).

Holinshed Project (2008b) *Holinshed's Chronicles of England, Scotland and Ireland 1587 ed.* [Online]. Available at http://english.nsms.ox.ac.uk/holinshed/toc.php?edition=1587 (Accessed 11 August 2022).

Holloway, S. (1996) *The Historical Atlas of Breeding Birds in Britain and Ireland 1875–1900*, Gibbons, D.W. (ed.), London, T & A D Poyser.

Holm, P., Ludlow, F., Scherer, C., Travis, C., Allaire, B., Brito, C., Hayes, P.W., al Matthews, J., Rankin, K.J. and Breen, R.J. (2019) 'The North Atlantic fish revolution (ca. AD 1500)', *Quaternary Research*: 1–15. https://doi.org/10.1017/qua.2018.153

Holm, P., Nicholls, J., Hayes, P.W., Ivinson, J. and Allaire, B. (2022) 'Accelerated extractions of North Atlantic cod and herring 1520–1790', *Fish and Fisheries*, vol. 23, no. 1: 54–72. https://doi.org/10.1111/faf.12598

Hore, H.F. (1859) 'A Chorographic Account of the Southern Part of the County of Wexford, Written Anno 1684: By Robert Leigh, Esq., of Rosegarland, in That County. (Concluded)', *The Journal of the Kilkenny and South-East of Ireland Archaeological Society*, vol. 2, no. 2: 451–67.

Hore, H.F. (1862a) 'An Account of the Barony of Forth, in the County of Wexford, Written at the Close of the Seventeenth Century', *The Journal of the Kilkenny and South-East of Ireland Archaeological Society*, vol. 4, no. 1: 53–84.

Hore, H.F. (1862b) 'Particulars Relative to Wexford and the Barony of Forth: By Colonel Solomon Richards 1682', *The Journal of the Kilkenny and South-East of Ireland Archaeological Society*, vol. 4, no. 1: 84–92.

Howard-McCombe, J., Ward, D., Kitchener, A.C., Lawson, D., Senn, H.V. and Beaumont, M. (2021) 'On the use of genome-wide data to model and date the time of anthropogenic hybridisation: An example from the Scottish wildcat', *Molecular Ecology*, vol. 30, no. 15: 3688–702. https://doi.org/10.1111/mec.16000

Hunter, M. (2007) 'Robert Boyle and the early Royal Society: a reciprocal exchange in the making of Baconian science', *British Journal for the History of Science*, vol. 40, no. 1: 1–23. https://doi.org/10.1017/S0007087406009083

Hutchins, J. (1774) *The History and Antiquities of the County of Dorset, vol. 1*, London, W. Bowyer and J. Nichols.

HWDT (2018) *Hebridean Marine Mammal Atlas. Part 1: Silurian 15 years of marine mammal monitoring in the Hebrides*, Scotland, A Hebridean Whale and Dolphin Trust Report (HWDT).

ICES (2020a) 'White anglerfish (*Lophius piscatorius*) in Subarea 7 and divisions 8.a–b and 8.d (Celtic Seas, Bay of Biscay)'. https://doi.org/10.17895/ices.advice.5925

ICES (2020b) 'Grey gurnard (*Eutrigla gurnardus*) in Subarea 4 and divisions 7.d and 3.a (North Sea, eastern English Channel, Skagerrak and Kattegat)'. https://doi.org/10.17895/ices.advice.5822

ICES (2021a) 'Herring (*Clupea harengus*) in Subarea 4 and divisions 3.a and 7.d, autumn spawners (North Sea, Skagerrak and Kattegat, eastern English Channel)'. https://doi.org/10.17895/ices.advice.7770

ICES (2021b) 'Cod (*Gadus morhua*) in Subarea 4, Division 7.d, and Subdivision 20 (North Sea, eastern English Channel, Skagerrak)'. https://doi.org/10.17895/ices.advice.7746

ICES (2021c) 'Haddock (*Melanogrammus aeglefinus*) in Subarea 4, Division 6.a, and Subdivision 20 (North Sea, West of Scotland, Skagerrak)'. https://doi.org/10.17895/ices.advice.7759

ICES (2021d) 'Whiting (*Merlangius merlangus*) in Subarea 4 and Division 7.d (North Sea and eastern English Channel)'. https://doi.org/10.17895/ices.advice.7885

ICES (2021e) 'Whiting (*Merlangius merlangus*) in Division 7.a (Irish Sea)'. https://doi.org/10.17895/ices.advice.7887

ICES (2021f) 'Saithe (*Pollachius virens*) in subareas 4 and 6, and in Division 3.a (North Sea, Rockall and West of Scotland, Skagerrak and Kattegat)'. https://doi.org/10.17895/ices.advice.7827

ICES (2021g) 'Anglerfish (*Lophius budegassa*, *Lophius piscatorius*) in subareas 4 and 6, and in Division 3.a (North Sea, Rockall and West of Scotland, Skagerrak and Kattegat)'. https://doi.org/10.17895/ices.advice.7723

ICES (2021h) 'Sandeel (*Ammodytes* spp.) in divisions 4.a–b, Sandeel Area 4 (northern and central North Sea)'. https://doi.org/10.17895/ices.advice.7675

ICES (2021i) 'Sandeel (*Ammodytes* spp.) in divisions 4.b–c, Sandeel Area 1r (central and southern North Sea, Dogger Bank)'. https://doi.org/10.17895/ices.advice.7672

ICES (2021j) 'Mackerel (*Scomber scombrus*) in subareas 1–8 and 14, and in Division 9.a (the Northeast Atlantic and adjacent waters)'. https://doi.org/10.17895/ices.advice.7789

ICES (2022a) 'Hake (*Merluccius merluccius*) in subareas 4, 6, and 7, and in divisions 3.a, 8.a–b, and 8.d, Northern stock (Greater North Sea, Celtic Seas, and the northern Bay of Biscay)'. https://doi.org/10.17895/ices.pub.18667901.v1

ICES (2022b) 'Sea bass (*Dicentrarchus labrax*) in divisions 4.b–c, 7.a, and 7.d–h (central and southern North Sea, Irish Sea, English Channel, Bristol Channel, and Celtic Sea)'. https://doi.org/10.17895/ices.advice.19447796.v1

ICES (2022c) 'Turbot (*Scophthalmus maximus*) in Subarea 4 (North Sea)'. https://doi.org/10.17895/ices.advice.19453871.v1

ICES (2022d) 'Brill (*Scophthalmus rhombus*) in Subarea 4 and divisions 3.a and 7.d–e (North Sea, Skagerrak and Kattegat, English Channel)'. https://doi.org/10.17895/ices.advice.19447790.v1

ICES (2022e) 'Plaice (*Pleuronectes platessa*) in Subarea 4 (North Sea) and Subdivision 20 (Skagerrak)'. https://doi.org/10.17895/ices.advice.19453586.v1

ICES (2022f) 'Sole (*Solea solea*) in Subarea 4 (North Sea)'. https://doi.org/10.17895/ices.advice.19453814.v1

ICES (2022g) 'Working Group on *Crangon* Fisheries and Life History (WGCRAN; outputs from 2021 meeting)'. https://doi.org/10.17895/ices.pub.10056

IUCN (2012) *IUCN Red List Categories and Criteria*, Gland, IUCN. https://portals.iucn.org/library/node/10315

Igoe, F., Quigley, D.T.G., Marnell, F., Meskell, E., O'Connor, W. and Byrne, C. (2004) 'The sea lamprey *Petromyzon marinus* (L.), river lamprey *Lampetra fluviatilis* (L.) and brook lamprey *Lampetra planeri* (Bloch) in Ireland: general biology, ecology, distribution and status with recommendations for conservation', *Biology and Environment: Proceedings of the Royal Irish Academy*, vol. 104, no. 3: 43–56. https://doi.org/10.1353/bae.2004.0029

Insley, H. (2007) 'Bar-tailed godwit *Limosa lapponica* Linnaeus', in Forrester, R.W., Andrews, I.J., McInerny, C.J., Murray, R., McGowan, R.Y., Zonfrillo, B., Betts, M.W., Jardine, D.C., Grundy, D., Andrews, I.J. and Scott, H.I. (eds), *The Digital Birds of Scotland*, Aberlady, Scottish Ornithologists' Club: 663–5.

Irving-Stonebraker, S. (2019) 'From Eden to savagery and civilization: British colonialism and humanity in the development of natural history, ca. 1600–1840', *History of the Human Sciences*, vol. 32, no. 4: 63–79. https://doi.org/10.1177/0952695119848623

Iversen, S.A. (2002) 'Changes in the perception of the migration pattern of Northeast Atlantic mackerel during the last 100 years'. https://doi.org/10.17895/ices.pub.8877

Jacobsen, K.-O., Marx, M. and ØIen, N. (2004) 'Two-way trans-Atlantic migration of a North Atlantic right whale (*Eubalaena glacialis*)', *Marine Mammal Science*, vol. 20, no. 1: 161–6. https://doi.org/10.1111/j.1748-7692.2004.tb01147.x

Jeffries, D.L., Copp, G.H., Maes, G.E., Lawson Handley, L., Sayer, C.D. and Hänfling, B. (2017) 'Genetic evidence challenges the native status of a threatened freshwater fish (*Carassius carassius*) in England', *Ecology and Evolution*, vol. 7, no. 9: 2871–82. https://doi.org/10.1002/ece3.2831

Jenkins, J.T. (1921) *A History of the Whale Fisheries: from the Basque Fisheries of the Tenth Century to the Hunting of the Finner Whale at the Present Date*, HF & G. Witherby. https://doi.org/10.5962/bhl.title.33378

Jepson, P.D., Deaville, R., Barber, J.L., Aguilar, À., Borrell, A., Murphy, S., Barry, J., Brownlow, A., Barnett, J., Berrow, S., Cunningham, A.A., Davison, N.J., ten Doeschate, M., Esteban, R., Ferreira, M., Foote, A.D., Genov, T., Giménez, J., Loveridge, J., Llavona, Á., Martin, V., Maxwell, D.L., Papachlimitzou, A., Penrose, R., Perkins, M.W., Smith, B., de Stephanis, R., Tregenza, N., Verborgh, P., Fernandez, A. and Law, R.J. (2016) 'PCB pollution continues to impact populations of orcas and other dolphins in European waters', *Scientific Reports*, vol. 6, no. 1: 18573. https://doi.org/10.1038/srep18573

JNCC (2021) *Biodiversity data for decision making*, [Online]. Available at https://jncc.gov.uk/our-work/ukbi-e1-biodiversity-data/.

Johnson, T. (1641) *Mercurii Botanici pars altera*, London, T&R Cotes.

Jones, H.P., Jones, P.C., Barbier, E.B., Blackburn, R.C., Rey Benayas, J.M., Holl, K.D., McCrackin, M., Meli, P., Montoya, D. and Mateos, D.M. (2018) 'Restoration and repair of Earth's damaged ecosystems', *Proceedings of the Royal Society B: Biological Sciences*, vol. 285, no. 1873: 20172577. https://doi.org/10.1098/rspb.2017.2577

Jones, M.C., Dye, S.R., Fernandes, J.A., Frölicher, T.L., Pinnegar, J.K., Warren, R. and Cheung, W.W.L. (2013) 'Predicting the Impact of Climate Change on Threatened Species in UK Waters', *PLoS ONE*, vol. 8, no. 1: e54216. https://doi.org/10.1371/journal.pone.0054216

Jonston, J. (1657) *Historiae Naturalis de Quadrupedibus libri*, Amsterdam, J.J. Schipperi. https://doi.org/10.5962/bhl.title.62511

Kämpfer, S. and Fartmann, T. (2019) 'Breeding populations of a declining farmland bird are dependent on a burrowing, herbivorous ecosystem engineer', *Ecological Engineering*, vol. 140: 105592. https://doi.org/10.1016/j.ecoleng.2019.105592

Kelly, F. (1997) 'Early Irish Farming', *Early Irish Law Series*, vol. 4: 74–106.

Kelly, G. (2004) *Literature/archive search for information relating to pool frogs Rana lessonae in East Anglia*, Peterborough, English Nature.

Kelly, M., Carboni, S., Cook, E. and Hughes, A. (2015) 'Sea urchin aquaculture in Scotland', in Brown, N.P. and Eddy, S.D. (eds), *Echinoderm Aquaculture*: 211–24. https://doi.org/10.1002/9781119005810.ch9

Kennedy, M. and Fitzmaurice, P. (1968) 'The biology of the bream *Abramis Brama* (L) in Irish waters', *Proceedings of the Royal Irish Academy. Section B: Biological, Geological, and Chemical Science*: 95–157.

Kennedy, M. and Fitzmaurice, P. (1970) 'The biology of the tench *Tinca tinca* (L.) in Irish waters', *Proceedings of the Royal Irish Academy. Section B: Biological, Geological, and Chemical Science*: 31–82.

Kennedy, M. and Fitzmaurice, P. (1974) 'Biology of the Rudd *Scardinius erythrophthalmus* (L) in Irish Waters', *Proceedings of the Royal Irish Academy. Section B: Biological, Geological, and Chemical Science*, vol. 74: 245–303.

K'Eogh, J. (1739) *Zoologia Medicinalis Hibernica*, Dublin, S. Powell.

Kerby, T.K., Cheung, W.W.L., van Oosterhout, C. and Engelhard, G.H. (2013) 'Entering uncharted waters: Long-term dynamics of two data limited fish species, turbot and brill, in the North Sea', *Journal of Sea Research*, vol. 84: 87–95. https://doi.org/10.1016/j.seares.2013.07.005

King, A.P. (2021) 'Peregrine *Falco peregrinus* Hebog tramor', in Pritchard, R., Hughes, J., Spence, I.M., Haycock, B., Brenchley, A., Green, J., Thorpe, R. and Porter, B. (eds), *The Birds of Wales*, Liverpool, Liverpool University Press: 354–7.

King, C.S. (1892) *Henry's Upper Lough Erne in 1739*, Dublin, William McGee.

King, J., Marnell, F., Kingston, N., Rosell, R., Boylan, P., Caffrey, J., FitzPatrick, Ú., Gargan, P., Kelly, F., O'Grady, M., Poole, R., Roche, W. and Cassidy, D. (2011) *Ireland Red List No. 5: Amphibians, Reptiles & Freshwater Fish*, Dublin.

Kitchener, A.C. (1998) 'Extinctions, introductions and colonisations of Scottish mammals and birds since the last Ice Age', in Lambert, R.A. (ed.), *Species history in Scotland*, Edinburgh, Scottish Cultural Press: 63–92.

Kitchener, A.C. and Daniels, M.J. (2008) 'Wildcat *Felis silvestris*', in Harris, S. and Yalden, D.W. (eds), *Mammals of the British Isles: Handbook*, 4th edn, Southampton, Mammal Society: 397–406.

Kowaleski, M. (2003) 'The commercialization of the sea fisheries in medieval England and Wales', *International Journal of Maritime History*, vol. 15, no. 2: 177–231. https://doi.org/10.1177/084387140301500212

Kraus, S.D. and Evans, P.G.H. (2008) 'Northern right whale *Eubalaena glacialis*', in Harris, S. and Yalden, D.W. (eds), *Mammals of the British Isles: Handbook*, 4th edn, Southampton, Mammal Society: 658–63.

Kumar, D. (2017) 'The evolution of colonial science in India: natural history and the East India Company', in Mackenzie, J.M. (ed.), *Imperialism and the Natural World*, Manchester, Manchester University Press: 51–66. https://doi.org/10.7765/9781526123671.00007

Lamb, H.H. (2011) *Climate: Present, Past and Future, Volume 2: Climatic History and the Future*, Abingdon, Routledge.

Lambert, R. (1998) 'From exploitation to extinction, to environmental icon: our images of the great auk', in *Species History in Scotland: Introductions and Extinctions since the Ice Age*, Edinburgh, Scottish Cultural Press: 20–37.

Langbein, J., Chapman, N.G. and Putman, R.J. (2008) 'Fallow deer *Dama dama*', in Harris, S. and Yalden, D.W. (eds), *Mammals of the British Isles: Handbook*, 4th edn, Southampton, The Mammal Society: 595–604.

Langbein, J. and Smith-Jones, C. (2020) 'Fallow deer', in Crawley, D., Coomber, F., Kubasiewicz, L., Harrower, C., Evans, P., Waggitt, J., Smith, B., and Matthews, F. (eds), *Atlas of the Mammals of Great Britain and Northern Ireland*, Exeter, Pelagic Publishing: 86–7.

Langley, P.J.W. and Yalden, D.W. (1977) 'The decline of the rarer carnivores in Great Britain during the nineteenth century', *Mammal Review*, vol. 7, no. 3–4: 95–116. https://doi.org/10.1111/j.1365-2907.1977.tb00363.x

Langton, T.E.S., Jones, M.W. and McGill, I. (2022) 'Analysis of the impact of badger culling on bovine tuberculosis in cattle in the high-risk area of England 2009–2020', *Veterinary Record*, vol. 190, no. 6: e1384. https://doi.org/10.1002/vetr.1384

Lauder, A. (2007) 'Eurasian Teal *Anas crecca* Linnaeus', in Forrester, R.W., Andrews, I.J., McInerny, C.J., Murray, R., McGowan, R.Y., Zonfrillo, B., Betts, M.W., Jardine, D.C., Grundy, D., Andrews, I.J. and Scott, H.I. (eds), *The Birds of Scotland*, Aberlady, Scottish Ornithologists' Club: 195–9.

Lawton, C., Hanniffy, R., Molloy, V., Guilfoyle, C., Stinson, M. and Reilly, E. (2020) *All-Ireland Squirrel and Pine Marten Survey 2019*, *Irish Wildlife Manuals*, Ireland, vol. 121.

Leigh, C. (1700) *The natural history of Lancashire, Cheshire and the Peak, in Derbyshire*, Oxford, Printed for the author.

Leland, J. (1906) *The Itinerary in Wales of John Leland in or about the Years 1536–1539: Extracted from His Mss*, G. Bell and Sons.

Lenders, H.J. and Janssen, I.A.W. (2014) 'The grass snake and the basilisk: from pre-christian protective house god to the antichrist', *Environment and History*, White Horse Press, vol. 20, no. 3: 319–46. https://doi.org/10.3197/096734014X14031694156367

Lenders, H.J.R., Chamuleau, T.P.M., Hendriks, A.J., Lauwerier, R.C.G.M., Leuven, R.S.E.W., Verberk, W.C.E.P., Jackson, J.B.C., Lotze, H.K., Barrett, J.M., Locker, A.M., Roberts, C.M., Dijk, G.M. van, Marteijn, E.C.L., Schulte-Wülwer-Leidig, A., Groot, S.J. de, Raat, A.J.P., Hoffmann, R.C., Hoffmann, R.C., Walter, R.C., Merrits, D.J., Halard, X., Hall, C.J., Jordaan, A., Frisk, M.G., Gibbard, P., Gupta, S., Collier, J.S., Palmer-Felgate, A., Potter, G., Rogers, L.A., Parish, D.L., Behnke, R.J., Gephard, S.R., McCormick, S.D., Reeves, G.H., Picollo, J.J., Norrgård, J.R., Greenberg, L.A., Schmitz, M., Bergman, E., Curwen, E.C., Downward, S., Skinner, K., Hodgen, M.T., Stearns, S.C., Verberk, W.C.E.P., Siepel, H., Esselink, H., Wańkowski, J.W.J., Thorpe, J.E., Baglinière, J.-L., Champigneulle, A., Nielsen, E.E., Hansen, M.M., Loeschcke, V., Schneider, J., Beechie, T., Buhle, E., Ruckelshaus, M., Fullerton, A., Holsinger, L., Eliason, E.J., Taylor, E.B., Schaffer, W.M., Elson, P.F., Gresh, T., Lichatowich, J., Schoonmaker, P., Kohler, A.A., Rugenski, A., Taki, D., Cederholm, C.J., Kunze, M.D., Murota, T., Sibatani, A., Doughty, C.E., Woude, A.M. van der, Groenewoudt, B., Smit, B. and Deelden, C.L. (2016) 'Historical rise of waterpower initiated the collapse of salmon stocks', *Scientific Reports*, vol. 6: 29269. https://doi.org/10.1038/srep29269

Lesley, J. (1675) *De Origine Moribus & Rebus Gestis Scotorum*, 2nd edn, Rome, In aedibus populi Romani.

Lesley, J. (1888) *The Historie of Scotland*, Dalrymple, J. and Cody, E.G. (eds), Edinburgh, William Blackwood and Sons.

Lever, C. (2009) *The Naturalized Animals of Britain and Ireland*, London, New Holland.

Lewis, B.J. (2022) 'An Englyn on the Wolf from the Hendregadredd Manuscript', *Studia Celtica*, vol. 56, no. 1: 123–6. https://doi-org.abc.cardiff.ac.uk/10.16922/SC.56.6

Lewns, P. (2020) 'Badger', in Crawley, D., Coomber, F., Kubasiewicz, L., Harrower, C., Evans, P., Waggitt, J., Smith, B. and Matthews, F. (eds), *Atlas of the Mammals of Great Britain and Northern Ireland*, Exeter, Pelagic Publishing: 66–7.

Lhuyd, E. (1712) 'IV. A letter from the late Mr. Edward Lhwyd, keeper of the Ashmolean Museum in Oxford, to Dr. Tancred Robinson, F.R.S. containing several observations in natural history, made in his travels thro' Wales', *Philosophical Transactions of the Royal Society of London*, vol. 27, no. 334: 462–5. https://doi.org/10.1098/rstl.1710.0047

Lhuyd, E. (1909) *Parochialia*, Morris, R.H. (ed.), vol. 1, London, Archaeologia Cambrensis supplements.

Lhuyd, E. (1910) *Parochialia*, Morris, R.H. (ed.), vol. 2, Cardiff, Archaeologia Cambrensis supplements.

Lhuyd, E. (1911) *Parochialia*, Morris, R.H. (ed.), vol. 3, Cardiff, Archaeologia Cambrensis supplements.

Lindley, P. (2021) 'Gannet *Morus bassanus* Hugan', in Pritchard, R., Hughes, J., Spence, I.M., Haycock, B., Brenchley, A., Green, J., Thorpe, R. and Porter, B. (eds), *The Birds of Wales*, Liverpool, Liverpool University Press: 283–4.

Linnaeus, C. (1758) *Systema Naturae* 10th edition, Stockholm, Laurence Salvus.

Lister, M. (1752) *Historiae sive Synopsis Methodicae Conchyliorum*, Oxford, Clarendon Press.

Liukkonen-Anttila, T., Uimaniemi, L., Orell, M. and Lumme, J. (2002) 'Mitochondrial DNA variation and the phylogeography of the grey partridge (*Perdix perdix*) in Europe: from Pleistocene history to present day populations', *Journal of Evolutionary Biology*, vol. 15, no. 6: 971–82. https://doi.org/10.1046/j.1420-9101.2002.00460.x

Llwyd, H. (1572) *Commentarioli Britannicae descriptionis fragmentum*, Cologne, Ioannem Birckmannum.

Llwyd, H. and Powel, D. (1784) *The historie of Cambria now called Wales*, London, Richard Iohnes.

Llwyd, H. and Twyne, T. (1573) *The Breuiary of Britayne*, London, Richard Iohnes.

Locker, A. (2016) 'The decline in the consumption of stored cod and herring in post-medieval and early industrialised England: a change in food culture', in Barrett, J.H. and Orton, D.C. (eds), *Cod & Herring: The Archaeology and History of Medieval Sea Fishing*, Oxford, Oxbow Books. https://doi.org/10.2307/j.ctvh1dw0d.14

Locker, A. (2018) *Freshwater Fish in England: a Social and Cultural History of Coarse Fish from Prehistory to the Present Day*, Oxford, Oxbow Books. https://doi.org/10.2307/j.ctv13nb80x

Locker, A.M. (2010) 'Freshwater fish', in O'Connor, T. and Sykes, N.J. (eds), *Extinctions and Invasions: a Social History of British Fauna*, Oxford, Windgather Press: 166–74. https://doi.org/10.2307/j.ctv13gvg6k.25

Lockley, R.M. (1953) *Puffins*, London, J.M. Dent & Sons Ltd.

Logan, J. (1971) 'Tadhg O Roddy and Two Surveys of Co. Leitrim', *Breifne*, vol. 4, no. 14: 318–34.

Love, J.A. (1984) *The Return of the Sea-Eagle*, Cambridge, Cambridge University Press.

Love, J.A. (2007) 'White-tailed Eagle *Haliaeetus albicilla* Linnaeus', in Forrester, R.W., Andrews, I.J., McInerny, C.J., Murray, R., McGowan, R.Y., Zonfrillo, B., Betts, M.W., Jardine, D.C., Grundy, D., Andrews, I.J. and Scott, H.I. (eds), *The Birds of Scotland*, Aberlady, Scottish Ornithologists' Club: 451–5.

Lovegrove, R. (1990) *The Kite's Tale: The Story of the Red Kite in Wales*, Sandy, Royal Society for the Protection of Birds.

Lovegrove, R. (2007) *Silent Fields: The Long Decline of a Nation's Wildlife*, Oxford, Oxford University Press.

Lovell, R. (1660) *Panzooryktologia*, Oxford, Henry Hall for Joseph Godwin.

Ludwig, A., Makowiecki, D. and Benecke, N. (2009) 'Further evidence of trans-Atlantic colonization of Western Europe by American Atlantic sturgeons', *Archaeofauna*, vol. 18: 185–92.

Lunham, T.A. (1909) 'Bishop Dive Downes' visitation of his diocese 1699–1702 (concluded)', *Cork Historical and Archaeological Society*, vol. 15, no. 84: 163–80.

Lyons, J. (2004) 'Common bream *Abramis brama*', in Davies, C.E., Shelley, J., Harding, P.T., McLean, I.F.G., Gardiner, R., Peirson, G. and Quill, V. (eds), *Freshwater Fishes in Britain*, Colchester, Harley Books: 59–61.

MacBrody, A. (1669) *Propugnaculum Catholicae Veritatis*, Prague, Charles University.

Maccarinelli, A. (2021) 'Was pike on the menu? Exploring the role of freshwater fish in medieval England', *Archaeological and Anthropological Sciences*, vol. 13, no. 8: 133. https://doi.org/10.1007/s12520-021-01373-6

MacDonald, A., Heath, M., Edwards, M., Furness, R., Pinnegar, J.K., Wanless, S., Speirs, D. and Greenstreet, S.P.R. (2015) 'Climate driven trophic cascades affecting seabirds around the British Isles', *Oceanography Marine Biology Annual Review*, vol. 53: 55–80. https://doi.org/10.1201/b18733-3

Macdonald, B. (2019) *Rebirding: Rewilding Britain and Its Birds*, Exeter, Pelagic Publishing Ltd.

MacKenzie, B.R. and Myers, R.A. (2007) 'The development of the northern European fishery for north Atlantic bluefin tuna *Thunnus thynnus* during 1900–1950', *Fisheries Research*, vol. 87, no. 2–3: 229–39. https://doi.org/10.1016/j.fishres.2007.01.013

Mac Laughlin, J. (2010) *Troubled Waters: a Social and Cultural History of Ireland's Sea Fisheries*, Dublin, Four Courts Press.

MacLysaght, E. (1939) *Irish Life in the Seventeenth Century*, London, Longmans, Green, & Co.

Magalotti, L. (1821) *The Travels of Cosmo the Third, Grand Duke of Tuscany*, London, J. Mawman.

Magennis, E. (2002) '"A Land of Milk and Honey": The Physico-Historical Society, Improvement and the Surveys of Mid-Eighteenth-Century Ireland', *Proceedings of the Royal Irish Academy. Section C: Archaeology, Celtic Studies, History, Linguistics, Literature*, vol. 102C, no. 6: 199–217. https://doi.org/10.1353/ria.2002.0001

Mahony, K.E., Lynch, S.A., Egerton, S., Cabral, S., de Montaudouin, X., Fitch, A., Magalhães, L., Rocroy, M. and Culloty, S.C. (2020) 'Mobilisation of data to stakeholder communities. Bridging the research-practice gap using a commercial shellfish species model', *PLoS ONE*, vol. 15, no. 9: e0238446. https://doi.org/10.1371/journal.pone.0238446

Maitland, P.S. (1977) 'Freshwater fish in Scotland in the 18th 19th and 20th centuries', *Biological Conservation*, vol. 12, no. 4: 265–78. https://doi.org/10.1016/0006-3207(77)90046-5

Maitland, P.S. (1987) 'Fish introductions and translocations-their impact on the British isles', *Angling and wildlife in fresh waters*, NERC/ITE, vol. ITE Sympos: 57–65.

Maitland, P.S. (2004) 'Ireland's Most Threatened and Rare Freshwater Fish: An International Perspective on Fish Conservation', *Biology and Environment: Proceedings of the Royal Irish Academy*, vol. 104B, no. 3: 5–16. https://doi.org/10.3318/BIOE.2004.104.3.5

Maitland, P.S. (2007) *Scotland's Freshwater Fish: Ecology, Conservation & Folklore*, Oxford, Trafford Publishing.

Maitland, P.S. and Campbell, R.N. (1992) *Freshwater Fishes of the British Isles*, London, Harper Collins.

Maitland, T. (1821) *The History and Chronicles of Scotland: Written in Latin by Hector Boece … and translated y John Ballenden*, Edinburgh, W and C Tait.

Mäkeläinen, P., Esteban, R., Foote, A.D., Kuningas, S., Nielsen, J., Samarra, F.I.P., Similä, T., van Geel, N.C.F. and Víkingsson, G.A. (2014) 'A comparison of pigmentation features among North Atlantic killer whale (*Orcinus orca*) populations', *Journal of the Marine Biological Association of the United Kingdom* 2014/04/29, Cambridge University Press, vol. 94, no. 6: 1335–41. https://doi.org/10.1017/S0025315414000277

Manning, A.D., Coles, B.J., Lunn, A.G., Halley, D.J., Ashmole, P. and Fallon, S.J. (2014) 'New evidence of late survival of beaver in Britain', *The Holocene*, vol. 24, no. 12: 1849–55. https://doi.org/10.1177/0959683614551220

Manning, A.D., Gordon, I.J. and Ripple, W.J. (2009) 'Restoring landscapes of fear with wolves in the Scottish Highlands', *Biological Conservation*, vol. 142, no. 10: 2314–21. https://doi.org/10.1016/j.biocon.2009.05.007

Marlborough, D. (1970) 'The status of the burbot *Lota lota* (L.) (Gadidae) in Britain', *Journal of Fish Biology*, vol. 2, no. 3, pp. 217–22. https://doi.org/https://doi.org/10.1111/j.1095-8649.1970.tb03277.x

Marquiss, M. (2007) 'Grey Heron: Ardea cinerea (Linnaeus)', in Forrester, R.W., Andrews, I.J., McInerny, C.J., Murray, R., McGowan, R.Y., Zonfrillo, B., Betts, M.W., Jardine, D.C., Grundy, D., Andrews, I.J. and Scott, H.I. (eds), *The Birds of Scotland*, Aberlady, Scottish Ornithologists' Club: 427–31.

Marren, P. and Mabey, R. (2010) *Bugs Britannica*, London, Chatto & Windus.

Martin, M. (1698) *A Late Voyage to St. Kilda*, London, D. Brown and T. Goodwin.

Martin, M. (1703) *A Description of the Western Isles of Scotland*, London, Andrew Bell.

Martin, M. and Monro, D. (2018) *A Description of the Western Isles: Circa 1695*, Edinburgh, Birlinn.

Mason, J. (1983) *Scallop and Queen Fisheries in the British Isles*, Farnham, Fishing News Books.

Matheson, C. (1941) 'The Rabbit and the Hare in Wales', *Antiquity* 2015/01/02, Cambridge University Press, vol. 15, no. 60: 371–81. https://doi.org/10.1017/S0003598X0001588X

Mathews, F. and Coomber, F. (2020) 'Vagrant species and those without established populations in the UK', in Crawley, D., Coomber, F., Kubasiewicz, L., Harrower, C., Evans, P., Waggitt, J., Smith, B. and Matthews, F. (eds), *Atlas of the Mammals of Great Britain and Northern Ireland*, Exeter, Pelagic Publishing: 185–90.

Mathews, F., Kubasiewicz, L.M., Gurnell, J., Harrower, C.A., McDonald, R.A. and Shore, R.F. (2018) *A Review of the Population and Conservation Status of British Mammals*, Natural Resources Wales and Scottish Natural Heritage, Brighton.

McBrien, J. (2016) 'Accumulating extinction', in Moore, J.W. (ed.), *Anthropocene or Capitalocene*, Oakland, Kairos Books. PM Press: 116–37.

McCormick, F. (1999) 'Early evidence for wild animals in Ireland', Benecke, N. (ed.), *The Holocene history of the European vertebrate fauna: modern aspects of research*, Rahden, Verlag Marie Leidorf GmbH: 355–71.

McCormick, F. (2007) 'Mammal bone studies from prehistoric Irish sites', in *Environmental archaeology in Ireland*, Oxford, Oxbow Books: 77–101.

McGowan, B. (2007) 'Great Northern Diver *Gavia immer* (Brünnich)', in Forrester, R.W., Andrews, I.J., McInerny, C.J., Murray, R., McGowan, R.Y., Zonfrillo, B., Betts, M.W., Jardine, D.C., Grundy, D., Andrews, I.J. and Scott, H.I. (eds), *The Birds of Scotland*, Aberlady, Scottish Ornithologists' Club: 334–7.

McHugh, B., Law, R.J., Allchin, C.R., Rogan, E., Murphy, S., Foley, M.B., Glynn, D. and McGovern, E. (2007) 'Bioaccumulation and enantiomeric profiling of organochlorine pesticides and persistent organic pollutants in the killer whale (*Orcinus orca*) from British and Irish waters', *Marine Pollution Bulletin*, vol. 54, no. 11: 1724–31. https://doi.org/10.1016/j.marpolbul.2007.07.004

McInerny, C. and Minting, P. (2016) *The Amphibians and Reptiles of Scotland*, Glasgow, Glasgow Natural History Society.

McKay, C.R. (2007) 'Red-billed Chough *Pyrrhocorax pyrrhocorax* (Linnaeus)', in Forrester, R.W., Andrews, I.J., McInerny, C.J., Murray, R., McGowan, R.Y., Zonfrillo, B., Betts, M.W., Jardine, D.C., Grundy, D., Andrews, I.J. and Scott, H.I. (eds), *The Birds of Scotland*, Aberlady, Scottish Ornithologists' Club: 1345–8.

Mearns, R. (2007) 'Common Raven *Corvus corax* (Linnaeus)', in Forrester, R.W., Andrews, I.J., McInerny, C.J., Murray, R., McGowan, R.Y., Zonfrillo, B., Betts, M.W., Jardine, D.C., Grundy, D., Andrews, I.J. and Scott, H.I. (eds), *The Birds of Scotland*, Aberlady, Scottish Ornithologists' Club: 1364–6.

Menzel, U. (2021) 'Exact Multinomial Test: Goodness of Fit Test for Discrete Multivariate Data', R Studio.

Merrett, C. (1666) *Pinax rerum naturalium Britannicarum*, London, T. Warren.

Miege, G. (1691) *The New State of England under their Majesties K. William and Q. Mary*, London, Jonathan Robinson.

Milner, N., Barrett, J. and Welsh, J. (2007) 'Marine resource intensification in Viking Age Europe: the molluscan evidence from Quoygrew, Orkney', *Journal of Archaeological Science*, vol. 34, no. 9: 1461–72. https://doi.org/10.1016/j.jas.2006.11.004

Mitchell, A. (1906) *Geographical Collections Relating to Scotland made by Walter MacFarlane, vol. 1*, Edinburgh, Publications of the Scottish History Society vol. LI, Edinburgh University Press.

Mitchell, A. (1907) *Geographical Collections Relating to Scotland made by Walter MacFarlane, vol. 2*, Edinburgh, Publications of the Scottish History Society vol. LII, Edinburgh University Press.

Mitchell, A. (1908) *Geographical Collections Relating to Scotland made by Walter MacFarlane, vol. 3*, Edinburgh, Publications of the Scottish History Society vol. LIII, Edinburgh University Press.

Mitchell, C. (2007) 'Eurasian Wigeon *Anas penelope* Linnaeus', in Forrester, R.W., Andrews, I.J., McInerny, C.J., Murray, R., McGowan, R.Y., Zonfrillo, B., Betts, M.W., Jardine, D.C., Grundy, D., Andrews, I.J. and Scott, H.I. (eds), *The Birds of Scotland*, Aberlady, Scottish Ornithologists' Club: 185–8.

Moffat, R., Spriggs, J. and O'Connor, S. (2008) 'The Use of Baleen for Arms, Armour and Heraldic Crests in Medieval Britain', *The Antiquaries Journal*, Cambridge University Press, vol. 88: 207–15. https://doi.org/10.1017/S0003581500001402

Molyneux, W. (ed.) (1685) 'The natural history of Ireland', Dublin, Trinity College Dublin MS 883.

Monro, D. (1774) *Description of the Western Isles of Scotland called Hybrides*, Edinburgh, William Auld.

Monteith, R. (1711) *Description of the islands of Orkney and Zetland*, Sibbald, R. (ed.) 2nd edn, Edinburgh, Thomas G. Stevenson.

Montgomery, W. (1896) 'Description of Ardes Barony, in the County of Down 1683', in Young, R.M. and Pinkerton, W. (eds), *Historical Notices of Old Belfast and its Vicinity*, Belfast, Marcus Ward & Co: 138–43.

Montgomery, W.I., Provan, J., McCabe, A.M. and Yalden, D.W. (2014) 'Origin of British and Irish mammals: disparate post-glacial colonisation and species introductions', *Quaternary Science Reviews*, vol. 98: 144–65. https://doi.org/10.1016/j.quascirev.2014.05.026

Moore, A.B.M. and Hiddink, J.G. (2022) 'Identifying critical habitat with archives: 275-year-old naturalist's notes provide high-resolution spatial evidence of long-term core habitat for a critically endangered shark', *Biological Conservation*, vol. 272: 109621. https://doi.org/10.1016/j.biocon.2022.109621

Morgan, P. (1983) 'A Welsh snakestone, its tradition and folklore', *Folklore*, vol. 94, no. 2: 184–91. https://doi.org/10.1080/0015587X.1983.9716276

Moriarty, C. and Fitzmaurice, P. (2000) 'Origin and diversity of freshwater fishes in Ireland', *SIL Proceedings 1922–2010*, vol. 27, no. 1: 128–30. https://doi.org/10.1080/03680770.1998.11901210

Morris, L. (1747) *Lewis Morris' De Historia Piscium*, Aberystwyth, National Library of Wales MS 24052E [Online]. Available at http://hdl.handle.net/10107/4627942.

Mortimer, C. (1735) 'VI. A narration of the experiments made June 1 1734 before several members of the Royal Society, and others, on a man, who suffer'd himself to be bit by a viper, or common adder …', *Philosophical Transactions of the Royal Society of London*, vol. 39, no. 443: 313–20. https://doi.org/10.1098/rstl.1735.0069

Morton, J. (1712) *The Natural History of Northampton-shire*, London, R. Knaplock.

Moryson, F. (1617) *An Itinerary Written by Fynes Moryson*, London, Printed by John Beale.

Moss, R. (2007) 'Western Capercaillie Tetrao urogallus (Linnaeus)', in Forrester, R.W., Andrews, I.J., McInerny, C.J., Murray, R., McGowan, R.Y., Zonfrillo, B., Betts, M.W., Jardine, D.C., Grundy, D., Andrews, I.J. and Scott, H.I. (eds), *The Birds of Scotland*, Aberlady, Scottish Ornithologists' Club: 305–10.

MSC (2019) *North Sea cod to lose sustainability certification* [Online]. Available at https://www.msc.org/media-centre/press-releases/press-release/north-sea-cod-to-lose-sustainability-certification (Accessed 20 June 2022).

Mulville, J. (2010) 'Red deer on Scottish islands', in O'Connor, T. and Sykes, N.J. (eds), *Extinctions and Invasions: a Social History of British Fauna*, Oxford, Windgather Press: 43–50. https://doi.org/10.2307/j.ctv13gvg6k.12

Murphy, S., Evans, P.G.H. and Collet, A. (2008) 'Common dolphin Delphinus delphis', in Harris, S. and Yalden, D.W. (eds), *Mammals of the British Isles: Handbook*, 4th edn, Southampton, Mammal Society: 719–24.

Murray, E. (2007) 'Molluscs and middens: The archaeology of "Ireland's early savage race"', in *Environmental Archaeology in Ireland*, Oxford, Oxbow Books: 119–35.

Murray, R. and Taylor, K. (2007) 'Common coot Fulica atra Linnaeus', in Forrester, R.W., Andrews, I.J., McInerny, C.J., Murray, R., McGowan, R.Y., Zonfrillo, B., Betts, M.W., Jardine, D.C., Grundy, D., Andrews, I.J. and Scott, H.I. (eds), *The Birds of Scotland*, Aberlady, Scottish Ornithologists' Club: 533–7.

National Biodiversity Data Centre (2022) *Biodiversity Maps: Mapping Ireland's Wildlife* [Online]. Available at https://maps.biodiversityireland.ie (Accessed 1 April 2022).

Neal, K.J. (2007) *Lepas (Anatifa) anatifera Common goose barnacle* [Online]. Available at https://www.marlin.ac.uk/species/detail/2058 (Accessed 21 July 2022).

Neville, F. (1713) 'XXIX. Some observations upon Lough-Neagh in Ireland. In a letter from Francis Nevill Esq; to the Lord Bishop of Clogher', *Philosophical Transactions of the Royal Society of London*, vol. 28, no. 337: 260–4. https://doi.org/10.1098/rstl.1713.0029

Newson, S.E., Marchant, J.H., Ekins, G.R. and Sellers, R.M. (2007) 'The status of inland-breeding great cormorants in England', *British Birds*, vol. 100, no. 5: 289.

Nichols, D. (1981) 'The Cornish sea-urchin fishery', *Cornish Studies*, vol. 9: 5–18.

Noble, G., Turner, J., Hamilton, D., Hastie, L., Knecht, R., Upex, B., Stirling, L., Sveinbjarnarson, O. and Milek, K. (2017) 'Early medieval shellfish gathering at the Sands of Forvie, Aberdeenshire: feast or famine?', *Proceedings of the Society of Antiquaries of Scotland*, vol. 146: 121–52. https://doi.org/10.9750/PSAS.146.1214

Northwest Atlantic Leatherback Working Group (2018) *Northwest Atlantic leatherback turtle (Dermochelys coriacea) status assessment*, Wallace, B. and Eckert, K. (eds), Godfrey [Online]. Available at WIDECAST Technical Report No. 16.

Nøttestad, L., Boge, E. and Ferter, K. (2020) 'The comeback of Atlantic bluefin tuna (*Thunnus thynnus*) to Norwegian waters', *Fisheries Research*, vol. 231: 105689. https://doi.org/10.1016/j.fishres.2020.105689

Ó Muraíle, N. (2002) 'Downing's Description of County Sligo, *c.*1684', in Timoney, M.A. (ed.), *A Celebration of Sligo: First Essays for Sligo field club*, Sligo, Sligo Field Club: 231–42.

O'Brien, D., Hall, J.E., Miró, A., O'Brien, K., Falaschi, M. and Jehle, R. (2021) 'Reversing a downward trend in threatened peripheral amphibian (*Triturus cristatus*) populations through interventions combining species, habitat and genetic information', *Journal for Nature Conservation*, vol. 64: 126077. https://doi.org/10.1016/j.jnc.2021.126077

O'Brien, D., Hall, J.E., O'Brien, K., Smith, D., Angus, S., Joglekar, R.V. and Jehle, R. (2021) 'How did the toad get over the sea to Skye? Tracing the colonisation of Scottish inshore islands by common toads (*Bufo bufo*)', *Herpetological Journal*, vol. 31: 204–13. https://doi.org/10.33256/31.4.204213

O'Brien, M. (2007a) 'Eurasian Oystercatcher *Haematopus ostralegus* Linnaeus', in Forrester, R.W., Andrews, I.J., McInerny, C.J., Murray, R., McGowan, R.Y., Zonfrillo, B., Betts, M.W., Jardine, D.C., Grundy, D., Andrews, I.J. and Scott, H.I. (eds), *The Birds of Scotland*, Aberlady, Scottish Ornithologists' Club: 546–9.

O'Brien, M. (2007b) 'Northern Lapwing *Vanellus vanellus* Linnaeus', in Forrester, R.W., Andrews, I.J., McInerny, C.J., Murray, R., McGowan, R.Y., Zonfrillo, B., Betts, M.W., Jardine, D.C., Grundy, D., Andrews, I.J. and Scott, H.I. (eds), *The Birds of Scotland*, Aberlady, Scottish Ornithologists' Club: 589–93.

O'Brien, M. (2007c) 'Common snipe *Gallinago gallinago*', in Forrester, R.W., Andrews, I.J., McInerny, C.J., Murray, R., McGowan, R.Y., Zonfrillo, B., Betts, M.W., Jardine, D.C., Grundy, D., Andrews, I.J. and Scott, H.I. (eds), *The Birds of Scotland*, Aberlady, Scottish Ornithologists' Club: 643–7.

O'Connor, T.P. (2013) *Animals as Neighbours*, East Lansing, Michigan State University Press.

O'Connor, T.P. (2017) 'Animals in urban life in medieval to early modern England', in Albarella, U., Rizzetto, M., Russ, H., Vickers, K., and Viner-Daniels, S. (eds), *The Oxford Handbook of Zooarchaeology*, Oxford, Oxford University Press: 214–29. https://doi.org/10.1093/oxfordhb/9780199686476.013.13

O'Connor, T.P. (1988) *Bones from the General Accident Site, Tanner Row*, The Archaeology of York, York, vol. 15.

O'Dalaigh, B. (1998) *The Strangers Gaze: Travels in County Clare 1534–1950*, CLASP Press.

O'Flaherty, R. (1846) *A Chorographical Description of West Or H-Iar Connaught: Written AD 1684*, Hardiman, J. (ed.), Dublin, Irish Archaeological Society.

Ogilvie, B.W. (2008) *The Science of Describing: Natural History in Renaissance Europe*, Chicago, University of Chicago Press.

O'Meara, D.B., McDevitt, A.D., O'Neill, D., Harrington, A.P., Turner, P., Carr, W., Desmond, M., Lawton, C., Marnell, F. and Rubalcava, S. (2018) 'Retracing the history and planning the future of the red squirrel (*Sciurus vulgaris*) in Ireland using non-invasive genetics', *Mammal Research*, vol. 63, no. 2: 173–84. https://doi.org/10.1007/s13364-018-0353-5

O'Meara, J. (1982) *Gerald of Wales: The History and Topography of Ireland*, London, Penguin Books.

O'Neill, M. and Sharplin, J. (2020) 'Mountain hare', in Crawley, D., Coomber, F., Kubasiewicz, L., Harrower, C., Evans, P., Waggitt, J., Smith, B. and Matthews, F. (eds), *Atlas of the Mammals of Great Britain and Northern Ireland*, Exeter, Pelagic Publishing: 30–1.

O'Regan, H.J. (2018) 'The presence of the brown bear *Ursus arctos* in Holocene Britain: a review of the evidence', *Mammal Review*, vol. 48: 229–44. https://doi.org/10.1111/mam.12127

O'Sullivan, D.C. (2009) *The Natural History of Ireland by Philip O'Sullivan Beare*, Cork, Cork University Press.

O'Sullivan, W. (1971) 'William Molyneux's Geographical Collections for Kerry', *Journal of the Kerry Archaeological and Historical Society*, vol. 4: 28–47.

Owen, G. of H. (1994) *The Description of Pembrokeshire*, Miles, D. (ed.), Llandysul, Gomer Press.

Owen, M. (2009) 'The Animals in the Law of Hywel', *The Carmarthenshire Antiquary*, vol. 41: 5–28.

Paris, J.R., King, R.A. and Stevens, J.R. (2015) 'Human mining activity across the ages determines the genetic structure of modern brown trout (*Salmo trutta* L.) populations', *Evolutionary Applications*, vol. 8, no. 6: 573–85. https://doi.org/10.1111/eva.12266

Parish, C. (2007) 'Grey Partridge *Perdix perdix* (Linnaeus)', in Forrester, R. and Andrews, I. (eds), *The Birds of Scotland*, Aberlady, Scottish Ornithologists' Club: 313–15.

Parslow, J. (1973) *Breeding birds of Britain and Ireland*, London, T & A D Poyser.

Parsons, J. (1750) 'VII. Some account of the *Rana piscatrix*', *Philosophical Transactions of the Royal Society of London*, vol. 46, no. 492: 126–31. https://doi.org/10.1098/rstl.1749.0024

Pastore, C.L. (2021) 'The science of shallow waters: connecting and classifying the early modern atlantic', *Isis*, vol. 112, no. 1: 122–9. https://doi.org/10.1086/713566

Patterson, I. (2007) 'Common Shelduck *Tadorna tadorna* (Linnaeus)', in Forrester, R.W., Andrews, I.J., McInerny, C.J., Murray, R., McGowan, R.Y., Zonfrillo, B., Betts, M.W., Jardine, D.C., Grundy, D., Andrews, I.J. and Scott, H.I. (eds), *The Birds of Scotland*, Aberlady, Scottish Ornithologists' Club: 177–81.

Payne, R. (1841) *A Brief Description of Ireland*, Smith, A. (ed.), Dublin, Irish Archaeological Society.

Pearce-Higgins, J., Dennis, P., Whittingham, M.J. and Yalden, D.W. (2010) 'Impacts of climate on prey abundance account for fluctuations in a population of a northern wader at the southern edge of its range', *Global Change Biology*, vol. 16, no. 1: 12–23. https://doi.org/10.1111/j.1365-2486.2009.01883.x

Peaty, S. (2004) 'Flounder Platichthys flesus', in Davies, C.E., Shelley, J., Harding, P.T., McLean, I.F.G., Gardiner, R., Peirson, G. and Quill, V. (eds), *Freshwater Fishes in Britain*, Colchester, Harley Books: 139–41.

Pedreschi, D., Kelly-Quinn, M., Caffrey, J., O'Grady, M. and Mariani, S. (2014) 'Genetic structure of pike (*Esox lucius*) reveals a complex and previously unrecognized colonization history of Ireland', *Journal of Biogeography*, vol. 41, no. 3: 548–60. https://doi.org/10.1111/jbi.12220

Pedreschi, D. and Mariani, S. (2015) 'Towards a balanced view of pike in Ireland: a reply to Ensing', *Ecology*, vol. 16: 3069–83. https://doi.org/10.1111/jbi.12472

Peirson, G. (2004) 'Silver bream *Abramis bjoerkna*', in Davies, C.E., Shelley, J., Harding, P.T., McLean, I.F.G., Gardiner, R., Peirson, G. and Quill, V. (eds), *Freshwater Fishes in Britain*, Colchester, Harley Books: 57–9.

Pennant, T. (1768) *British Zoology, vol. 1: Quadrupeds & Birds*, 4th edn, London, For Benjamin White, vol. 1. https://doi.org/10.5962/bhl.title.62499

Pennant, T. (1776a) *British Zoology, vol. 2 Water Fowl*, London, Printed for Benj. White. https://doi.org/10.5962/bhl.title.62481

Pennant, T. (1776b) *A Tour in Scotland 1769*, London, Benj. White, vol. 1.

Pennant, T. (1776c) *British Zoology, vol. 3: Reptiles & Fish*, London, Warrington. https://doi.org/10.5962/bhl.title.62481

Petrovan, S.O. and Schmidt, B.R. (2016) 'Volunteer conservation action data reveals large-scale and long-term negative population trends of a widespread amphibian, the common toad (*Bufo bufo*)', *PLoS one*, vol. 11, no. 10: e0161943. https://doi.org/10.1371/journal.pone.0161943

Piers, H. (1786) *A Chorographical Description of the County of West-Meath*, Vallancey, C. (ed.), Dublin, Collectanea de rebus hibernicis vol. i, Pat Wogan.

Pipe, A. (1992) 'A note on exotic animals from medieval and post-medieval London', *Anthropozoologica*, vol. 16: 189–91.

Platenberg, R.J. (1999) 'Population Ecology and Conservation of the Slow-Worm *Anguis fragilis* in Kent', University of Kent.

Plot, R. (1686) *The Natural History of Stafford-shire*, Oxford, Printed in the Theatre.

Pluskowski, A. (2010) 'The Wolf', in O'Connor, T. and Sykes, N.J. (eds), *Extinctions and Invasions*, Oxford, Windgather Press: 68–74. https://doi.org/10.2307/j.ctv13gvg6k.15

Pluskowski, A. (2018) 'The Medieval Wild', in Gerard, C.M. and Gutiérrez, A. (eds), *The Oxford Handbook of Later Medieval Archaeology in Britain*, Oxford, Oxford University Press: 141–53. https://doi.org/10.1093/oxfordhb/9780198744719.013.6

Pococke, R. (1887) *Tours in Scotland 1747 1750 1760*, Kemp, D.W. (ed.), Edinburgh, Printed at the University Press by T. and A. Constable for the Scottish History Society, vol. 1.

Pococke, R. (1889) *The travels through England of Dr. Richard Pococke*, Cambridge, Camden New Series 44, Cambridge University Press. https://doi.org/10.1017/S2042170200004320

Pococke, R. (1891) *Pococke's Tour in Ireland in 1752*, Stokes, G.T. (ed.), Dublin, Hodges, Figgis, and Co.

Poole, K. (2010) 'Bird introductions', in O'Connor, T. and Sykes, N.J. (eds), *Extinctions and Invasions: a Social History of British Fauna*, Oxford, Windgather Press: 155–65. https://doi.org/10.2307/j.ctv13gvg6k.24

Poole, K. (2015) 'Foxes and Badgers in Anglo-Saxon Life and Landscape', *Archaeological Journal*, Routledge, vol. 172, no. 2: 389–422. https://doi.org/10.1080/00665983.2015.1027871

Potts, G.R. (2012) *Partridges. Countryside Barometer*, London, New Naturalist Library 121. Collins.

Potts, G.R. and Aebischer, N.J. (1995) 'Population dynamics of the Grey Partridge *Perdix perdix* 1793–1993: monitoring, modelling and management', *Ibis*, vol. 137: S29–S37. https://doi.org/10.1111/j.1474-919X.1995.tb08454.x

Prendergast, J.P. (1865) *The Cromwellian Settlement of Ireland*, London, Longman, Green, Longman, Roberts and Green.

Prendergast, J.R., Wood, S.N., Lawton, J.H. and Eversham, B.C. (1993) 'Correcting for Variation in Recording Effort in Analyses of Diversity Hotspots', *Biodiversity Letters*, vol. 1, no. 2: 39–53. https://doi.org/10.2307/2999649

Preston, Pearman, D. and Dines, T.D. (2002) *New Atlas of the British & Irish Flora*, Oxford, Oxford University Press.

Prestt, I., Cooke, A.S. and Corbett, K.F. (1974) 'British amphibians and reptiles', in Hawksworth, D.L. (ed.), *The Changing Flora and Fauna of Britain*, London, Published for the Systematics Association by Academic Press: 229–54.

Prise, J. (1663) *A Description of Wales*, Lloyd, H. (ed.), Oxford, William Hall.

Pritchard, R. (2021a) 'Brent Goose *Branta bernicla* Gŵydd Ddu', in Pritchard, R., Hughes, J., Spence, I.M., Haycock, B., Brenchley, A., Green, J., Thorpe, R. and Porter, B. (eds), *The Birds of Wales*, Liverpool, Liverpool University Press: 57–9.

Pritchard, R. (2021b) 'Barnacle goose *Branta leucopsis* Gŵydd Wyran', in Pritchard, R., Hughes, J., Spence, I.M., Haycock, B., Brenchley, A., Green, J., Thorpe, R. and Porter, B. (eds), *The Birds of Wales*, Liverpool, Liverpool University Press: 61–2.

Pritchard, R. (2021c) 'Greylag goose *Anser anser* Gŵydd Lwyd', in Pritchard, R., Hughes, J., Spence, I.M., Haycock, B., Brenchley, A., Green, J., Thorpe, R. and Porter, B. (eds), *The Birds of Wales*, Liverpool, Liverpool University Press: 62–3.

Pritchard, R. (2021d) 'Mute Swan *Cygnus olor* Alarch Dof', in Pritchard, R., Hughes, J., Spence, I.M., Haycock, B., Brenchley, A., Green, J., Thorpe, R. and Porter, B. (eds), *The Birds of Wales*, Liverpool, Liverpool University Press: 69–71.

Pritchard, R. (2021e) 'Wigeon *Mareca penelope* Chwiwell', in Pritchard, R., Hughes, J., Spence, I.M., Haycock, B., Brenchley, A., Green, J., Thorpe, R. and Porter, B. (eds), *The Birds of Wales*, Liverpool, Liverpool University Press: 82–3.

Pritchard, R. (2021f) 'Mallard *Anas platyrhynchos* Hywaden Wyllt', in Pritchard, R., Hughes, J., Spence, I.M., Haycock, B., Brenchley, A., Green, J., Thorpe, R. and Porter, B. (eds), *The Birds of Wales*, Liverpool, Liverpool University Press: 84–5.

Pritchard, R. (2021g) 'Eider *Somateria mollissima* Hwyaden Fwythblu', in Pritchard, R., Hughes, J., Spence, I.M., Haycock, B., Brenchley, A., Green, J., Thorpe, R. and Porter, B. (eds), *The Birds of Wales*, Liverpool, Liverpool University Press: 98–9.

Pritchard, R. (2021h) 'Red Grouse *Lagopus lagopus* Grugiar goch', in Pritchard, R., Hughes, J., Spence, I.M., Haycock, B., Brenchley, A., Green, J., Thorpe, R. and Porter, B. (eds), *The Birds of Wales*, Liverpool, Liverpool University Press: 51–3.

Pritchard, R. (2021i) 'Pheasant *Phasianus colchicus* Ffesant', in Pritchard, R., Hughes, J., Spence, I.M., Haycock, B., Brenchley, A., Green, J., Thorpe, R. and Porter, B. (eds), *The Birds of Wales*, Liverpool, Liverpool University Press: 56–7.

Pritchard, R. (2021j) 'Great Northern Diver *Gavia immer* Trochydd Mawr', in Pritchard, R., Hughes, J., Spence, I.M., Haycock, B., Brenchley, A., Green, J., Thorpe, R. and Porter, B. (eds), *The Birds of Wales*, Liverpool, Liverpool University Press: 268–9.

Pritchard, R. (2021k) 'Bittern *Botaurus stellaris* Aderyn y Bwn', in Pritchard, R., Hughes, J., Spence, I.M., Haycock, B., Brenchley, A., Green, J., Thorpe, R. and Porter, B. (eds), *The Birds of Wales*, Liverpool, Liverpool University Press: 291–2.

Pritchard, R. (2021l) 'Grey heron *Ardea cinerea* Crëyr Glas', in *The Birds of Wales*, Liverpool, Liverpool University Press: 297–9.

Pritchard, R. (2021m) 'Golden Eagle *Aquila chrysaetos* Eryr Euraid', in Pritchard, R., Hughes, J., Spence, I.M., Haycock, B., Brenchley, A., Green, J., Thorpe, R. and Porter, B. (eds), *The Birds of Wales*, Liverpool, Liverpool University Press: 307–8.

Pritchard, R. (2021n) 'Grey Plover *Pluvialis squatarola* Cwtiad Llwyd', in Pritchard, R., Hughes, J., Spence, I.M., Haycock, B., Brenchley, A., Green, J., Thorpe, R. and Porter, B. (eds), *The Birds of Wales*, Liverpool, Liverpool University Press: 156.

Pritchard, R. (2021o) 'Bar-tailed godwit *Limosa lapponica* Rhostog Gynffonfraith', in Pritchard, R., Hughes, J., Spence, I.M., Haycock, B., Brenchley, A., Green, J., Thorpe, R. and Porter, B. (eds), *The Birds of Wales*, Liverpool, Liverpool University Press: 168–9.

Pritchard, R. (2021p) 'Woodcock *Scolopax rusticola* Cyffylog', in Pritchard, R., Hughes, J., Spence, I.M., Haycock, B., Brenchley, A., Green, J., Thorpe, R. and Porter, B. (eds), *The Birds of Wales*, Liverpool, Liverpool University Press: 189–91.

Pritchard, R. (2021q) 'Magpie *Pica pica* Pioden', in Pritchard, R., Hughes, J., Spence, I.M., Haycock, B., Brenchley, A., Green, J., Thorpe, R. and Porter, B. (eds), *The Birds of Wales*, Liverpool, Liverpool University Press: 365–6.

Pritchard, R. (2021r) 'Hooded Crow *Corvus cornix* Brân Llwyd', in Pritchard, R., Hughes, J., Spence, I.M., Haycock, B., Brenchley, A., Green, J., Thorpe, R. and Porter, B. (eds), *The Birds of Wales*, Liverpool, Liverpool University Press: 373–4.

Pritchard, R. (2021s) 'Crossbill *Loxia curvirostra* Gylfingroes', in Pritchard, R., Hughes, J., Spence, I.M., Haycock, B., Brenchley, A., Green, J., Thorpe, R. and Porter, B. (eds), *The Birds of Wales*, Liverpool, Liverpool University Press: 515–17.

Prodöhl, P.A., Jørstad, K.E., Triantafyllidis, A., Katsares, V. and Triantaphyllidis, C. (2006) 'European lobster – *Homarus gammarus*', in Svåsand, T., Crosetti, D., Garcia-Vázquez, E. and Verspoor, E. (eds), *Genetic Impact of Aquaculture Activities on Native Populations*, Genimpact final scientific report (EU contract n. RICA-CT-2005-022802): 91–8.

Pullar, P. and Low, M. (2012) *Fauna Scotica: Animals and People in Scotland*, Edinburgh, Birlinn.

Pymm, R. (2018) 'Snakestone Bead Folklore', *Folklore*, vol. 129, no. 4: 397–419. https://doi.org/10.1080/0015587X.2018.1515171

Quigley, D.T.G., Flannery, K. and Berrow, S. (1994) 'Swordfish *Xiphias gladius* L.', *The Irish Naturalists' Journal*, vol. 24, no. 12: 518–19.

Rackham, H. (1947) *Pliny Natural History: Books VIII–XI*, Harvard, Loeb Classical Library.

Rackham, O. (1986) *The History of the Countryside*, London, Weidenfeld & Nicolson.

Ramage, H. (1987) *Portraits of an Island: Eighteenth-Century Anglesey*, Anglesey, Anglesey Antiquarian Society and Field Club.

Rambaran-Olm, M. (2019) *Misnaming the Medieval: Rejecting 'Anglo-Saxon' Studies* [Online]. Available at https://www.historyworkshop.org.uk/misnaming-the-medieval-rejecting-anglo-saxon-studies/ (Accessed 10 May 2022).

Ramsay, L. (2012) 'Was Arthur once a Raven? The legend of "Arthur as a Chough"', *Old Cornwall*, Federation of Old Cornwall Societies, vol. 14, no. 7: 19–27.

Rao, S.J. (2017) 'Effect of reducing red deer Cervus elaphus density on browsing impact and growth of Scots pine *Pinus sylvestris* seedlings in semi-natural woodland in the Cairngorms, UK', *Conservation Evidence*, Open Book Publishers, vol. 14: 22–6.

Ratcliffe, D. (1997) *The Raven: A Natural History in Britain and Ireland*, London, T. & A.D. Poyser.

Ratcliffe, D. and Watson, D. (1993) *The Peregrine Falcon*, 2nd edn, London, T. & A.D. Poyser.

Ray, J. (1678) *The Ornithology of Francis Willugbhy of Middleton*, London, Royal Society.

Ray, J. (1693) *Synopsis Methodica Animalium Quadreupedum et Serpenti Generis*, London, Robert Southwell.

Ray, J. (1713a) *Synopsis Methodica Avium & Piscium*, London, Gulielmum Innys.

Ray, J. (1713b) *Synopsis Methodica, Volume 2: Piscium*, London, W. Innys.

Ray, J. and Derham, W. (1846) *Memorials of John Ray*, Lankester, E. (ed.), London, The Ray Society.

Raye, L. (2014) 'The early extinction date of the beaver (*Castor fiber*) in Britain', *Historical Biology*, vol. 27, no. 8: 1029–41. https://doi.org/10.1080/08912963.2014.927871

Raye, L. (2015) 'Evidence for the use of whale-baleen products in medieval Powys, Wales', *Medieval Animal Data Network (MAD) peer reviewed blog* [Online]. Available at http://mad.hypotheses.org/328 (Accessed 29 August 2015).

Raye, L. (2017a) 'The Eurasian lynx (*Lynx lynx*) in early modern Scotland', *Archives of Natural History*, vol. 44, no. 2: 321–33. https://doi.org/10.3366/anh.2017.0452

Raye, L. (2017b) 'Frogs in pre-industrial Britain', *The Herpetological Journal*, vol. 27: 368–78.

Raye, L. (2018) 'The Ugly, Greedy Crane of Medieval Wales', *Peritia*, vol. 29: 143–58. https://doi.org/10.1484/J.PERIT.5.118489

Raye, L. (2019) 'Erroll [née Drummond; married name Hay], Anne, countess of Erroll (1656–1719), Jacobite and naturalist', Oxford University Press. https://doi.org/10.1093/odnb/9780198614128.013.112756

Raye, L. (2021a) 'An 18th century reference to a Eurasian lynx (*Lynx lynx*) in Scotland', *Mammal Communications*, vol. 7: 47–52.

Raye, L. (2021b) 'Early Modern Attitudes to the Ravens and Red Kites of London', *The London Journal*, vol. 46: 1–16. https://doi.org/10.1080/03058034.2020.1857549

Raye, L. (2021c) 'Historical links between breeding Northern Wheatears *Oenanthe oenanthe* and European Rabbits *Oryctolagus cuniculus* warrens in southeast England', *Bird Study*: 1–5. https://doi.org/10.1080/00063657.2021.2003751

Raye, L. (2021d) *Six seasons of slow worms at Cardiff University*, Cardiff.

Raye, L. (2023) 'The atlas of early modern wildlife', *Mendeley Data*, V1. https://doi.org/10.17632/xkz4hzchkr.1

Reeves, R.R. (2020) 'Overview of catch history, historic abundance and distribution of right whales in the western North Atlantic and in Cintra Bay, West Africa', *Journal of Cetacean Research and Management*: 187–92. https://doi.org/10.47536/jcrm.vi.289

Reeves, R.R., Smith, T.D. and Josephson, E.A. (2007) 'Near-annihilation of a species: right whaling in the North Atlantic', in Kraus, S.D., and Rolland, R.M. (eds), *The Urban Whale: North Atlantic Right Whales at the Crossroads* Cambridge, Harvard University Press: 39–74.

Régnier, T., Gibb, F.M. and Wright, P.J. (2017) 'Importance of trophic mismatch in a winter-hatching species: evidence from lesser sandeel', *Marine Ecology Progress Series*, vol. 567: 185–97. https://doi.org/10.3354/meps12061

Reid, J.M., Bignal, E.M., Bignal, S., McCracken, D.I., Bogdanova, M.I. and Monaghan, P. (2008) 'Investigating patterns and processes of demographic variation: environmental correlates of pre-breeding survival in red-billed choughs *Pyrrhocorax pyrrhocorax*', *Journal of Animal Ecology*: 777–88. https://doi.org/10.1111/j.1365-2656.2008.01400.x

Reid, N., Dingerkus, K., Stone, R.E., Pietravalle, S., Kelly, R., Buckley, J., Beebee, T.J.C. and Wilkinson, J.W. (2013) *National Frog Survey of Ireland 2010/11*, Dublin, Irish Wildlife Manuals, No. 58, National Parks & Wildlife Service.

Reid, N., Dingerkus, S.K., Stone, R.E., Buckley, J., Beebee, T.J.C., Marnell, F. and Wilkinson, J.W. (2014) 'Assessing Historical and Current Threats to Common Frog (*Rana temporaria*) Populations in Ireland', *Journal of Herpetology*, vol. 48, no. 1: 13–19. https://doi.org/10.1670/12-053

Reyce, R. (1902) *Suffolk in the XVIIth Century: The Breviary of Suffolk by Robert Reyce 1618*, Hervey, F. (ed.), London, John Murray.

Richardson, R. (1713) 'XVIII. Several observations in natural history, made at North-Bierley in Yorkshire, by Dr. Richard Richardson (M.D.) communicated in a letter to Dr. Hans Sloane, R.S. Secr', *Philosophical Transactions of the Royal Society of London*, vol. 28, no. 337: 167–71. https://doi.org/10.1098/rstl.1713.0018

Riddell, A. (1996) 'Monitoring slow-worms and common lizards, with special reference to refugia material, refugia occupancy and individual identification', in Foster, J. and Gent, T. (eds), *Reptile survey methods: proceedings of a seminar held on 7 November 1995 at the Zoological Society of London's meeting rooms, Regent's Park, London*, London, English Nature Science no 27: 46–60.

Rielly, K. (2010) 'The black rat', in O'Connor, T. and Sykes, N.J. (eds), *Extinctions and Invasions: a Social History of British Fauna*, Oxford, Oxbow Books: 134–45. https://doi.org/10.2307/j.ctv13gvg6k.22

Rijnsdorp, A.D. and Millner, R.S. (1996) 'Trends in population dynamics and exploitation of North Sea plaice (*Pleuronectes platessa* L.) since the late 1800s', *ICES Journal of Marine Science*, Oxford University Press, vol. 53, no. 6: 1170–84. https://doi.org/10.1006/jmsc.1996.0142

Rindorf, A., Gislason, H., Burns, F., Ellis, J.R. and Reid, D. (2020) 'Are fish sensitive to trawling recovering in the Northeast Atlantic?', *Journal of Applied Ecology*, vol. 57, no. 10: 1936–47. https://doi.org/10.1111/1365-2664.13693

Risdon, T. (1811) *The Chorographical Description Or Survey of the County of Devon*, Plymouth, Rees & Curtis.

Ritchie, J. (1920) *The Influence of Man on Animal Life in Scotland*, Cambridge, Cambridge University Press.

Roberts, S. (2007) 'Eurasian Woodcock *Scolopax rusticola* Linnaeus', in Forrester, R.W., Andrews, I.J., McInerny, C.J., Murray, R., McGowan, R.Y., Zonfrillo, B., Betts, M.W., Jardine, D.C., Grundy, D., Andrews, I.J. and Scott, H.I. (eds), *The Birds of Scotland*, Aberlady, Scottish Ornithologists' Club: 653–6.

Robson, E.D. (2019) 'Improvement and environmental conflict in the northern fens 1560–1665', University of Cambridge [Online]. https://doi.org/10.17863/CAM.37259

Rogers, S.I. and Ellis, J.R. (2000) 'Changes in the demersal fish assemblages of British coastal waters during the 20th century', *ICES Journal of Marine Science*, vol. 57, no. 4: 866–81. https://doi.org/10.1006/jmsc.2000.0574

Rollie, C. (2007) 'Ring Ouzel *Turdus torquatus* Linnaeus', in Forrester, R.W., Andrews, I.J., McInerny, C.J., Murray, R., McGowan, R.Y., Zonfrillo, B., Betts, M.W., Jardine, D.C., Grundy, D., Andrews, I.J. and Scott, H.I. (eds), *The Birds of Scotland*, Aberlady, Scottish Ornithologists' Club: 1130–4.

Roos, A.M. (2015) *The Correspondence of Dr. Martin Lister (1639–1712). Volume One: 1662–1677*, Leiden, Brill. https://doi.org/10.1163/9789004263321

Roos, A.M. (2018) *Martin Lister and his remarkable daughters: art and science in the seventeenth century*, Oxford, Bodleian Library Press.

Rose, H. and Smith-Jones, C. (2020) 'Red deer', in Crawley, D., Coomber, F., Kubasiewicz, L., Harrower, C., Evans, P., Waggitt, J., Smith, B. and Matthews, F. (eds), *Atlas of the Mammals of Great Britain and Northern Ireland*, Exeter, Pelagic Publishing: 82–3.

Rosell, R., Harrod, C., Griffiths, D. and McCarthy, T.K. (2004) 'Conservation of the Irish populations of the pollan *Coregonus autumnalis*', *Biology and Environment: Proceedings of the Royal Irish Academy*, Royal Irish Academy, vol. 104, no. 3: 67–72. https://doi.org/10.3318/BIOE.2004.104.3.67

Rowlands, null (1721) 'XII. Part of a letter from the Reverend Mr. Rowlands, to the Reverend Mr. Derham, Prebendary of Windsor, and F.R.S. concerning the stocking of the river mene with oysters', *Philosophical Transactions of the Royal Society of London*, vol. 31, no. 369: 250–1. https://doi.org/10.1098/rstl.1720.0061

Russell, C.W. and Prendergast, J.P. (1871) *The Carte Manuscripts in the Bodleian Library Oxford*, London, George E. Eyre and William Spottiswoode.

Rutty, J. (1772) *An essay towards a natural history of the County of Dublin*, Dublin, W. Sleater, vol. 1.

Ryan, C., Calderan, S., Allison, C., Leaper, R. and Risch, D. (2022) 'Historical occurrence of whales in Scottish Waters inferred from whaling records', *Aquatic Conservation: Marine and Freshwater Ecosystems*: 1–18. https://doi.org/10.1002/aqc.3873

Sabin, R., Bendrey, R. and Riddler, I. (1999) 'Twelfth-century porpoise remains from Dover and Canterbury', *Archaeological Journal*, vol. 156, no. 1: 363–70. https://doi.org/10.1080/00665983.1999.11078911

Saborowski, R. and Hünerlage, K. (2022) 'Hatching phenology of the brown shrimp *Crangon crangon* in the southern North Sea: inter-annual temperature variations and climate change effects', *ICES Journal of Marine Science*, Oxford University Press, vol. 79, no. 4: 1302–11. https://doi.org/10.1093/icesjms/fsac054

Sacheverell, W. (1859) *An account of the Isle of Man*, Cumming, J.G. (ed.), Douglas, Manx Society.

Sax, B. (2007) 'How Ravens Came to the Tower of London', *Society & Animals*, vol. 15, no. 3: 269–83. https://doi.org/10.1163/156853007X217203

Scharff, R.F. (1893) 'The Frog a Native of Ireland', *The Irish Naturalist*, vol. 2, no. 1: 1–6.

Scharff, R.F. (1915a) 'On the Irish names of mammals', *The Irish Naturalist*, vol. 24, no. 3: 45–53.

Scharff, R.F. (1915b) 'On the Irish names of birds', *The Irish Naturalist*, vol. 24, no. 7: 109–29.

Scharff, R.F. (1916) 'On the Irish names of invertebrate animals', *The Irish Naturalist*, vol. 25, no. 9: 140–52.

Scharff, R.F. (1918) 'The Irish Red Deer', *The Irish Naturalist*, vol. 27, no. 10/11: 133–9.

Scharff, R.F., Coffey, G., Cole, G.A.J., Ussher, R.J. and Praeger, R.L. (1902) 'The Exploration of the Caves of Kesh, County Sligo', *The Transactions of the Royal Irish Academy*, vol. 32: 171–214.

Scharff, R.F., Ussher, R.J., Cole, G.A.J., Newton, E.T., Dixon, A.F. and Westropp, T.J. (1906) 'The Exploration of the Caves of County Clare. Being the Second Report from the Committee Appointed to Explore Irish Cave', *The Transactions of the Royal Irish Academy*, vol. 33: 1–76.

Scott, D. (2020) 'Red fox', in Crawley, D., Coomber, F., Kubasiewicz, L., Harrower, C., Evans, P., Waggitt, J., Smith, B. and Matthews, F. (eds), *Atlas of the Mammals of Great Britain and Northern Ireland*, Exeter, Pelagic Publishing: 64–5.

Sea Watch Foundation (2020) *Species fact sheets* [Online]. Available at https://www.seawatchfoundation.org.uk/information-and-fact-sheets/ (Accessed 30 August 2022).

Senn, H., Ghazali, M., Kaden, J., Barclay, D., Harrower, B., Campbell, R.D., Macdonald, D.W. and Kitchener, A.C. (2019) 'Distinguishing the victim from the threat: SNP-based methods reveal the extent of introgressive hybridisation between wildcats and domestic cats in Scotland and inform future in-situ and ex-situ management options for species restoration', *Evolutionary Applications*, vol. 12: 399–414. https://doi.org/10.1111/eva.12720

Serjeantson, D. (2001) 'The great auk and the gannet: a prehistoric perspective on the extinction of the great auk', *International Journal of Osteoarchaeology*, vol. 11, no. 1–2: 43–55. https://doi.org/10.1002/oa.545

Serjeantson, D. (2010) 'Extinct Birds', in O'Connor, T. and Sykes, N.J. (eds), *Extinctions and Invasions*, Oxford, Windgather Press: 146–55. https://doi.org/10.2307/j.ctv13gvg6k.23

Serjeantson, D. and Woolgar, C.M. (2006) 'Fish consumption in medieval England', in Woolgar, C.M., Serjeantson, D., and Waldron, T. (eds), *Food in Medieval England: Diet and Nutrition*, Oxford, Oxford University Press: 102–30.

Serjeantson, Dale (2006) 'Birds: food and a mark of status', in Woolgar, C.M., Serjeantson, D. and Waldron, T. (eds), *Food in Medieval England: Diet and Nutrition*, Oxford, Oxford University Press: 131–47.

Service, R. (1891) 'The Old Fur Market of Dumfries', *The Scottish Naturalist*, vol. 33: 97–102.

Sguotti, C., Lynam, C.P., García-Carreras, B., Ellis, J.R. and Engelhard, G.H. (2016) 'Distribution of skates and sharks in the North Sea: 112 years of change', *Global Change Biology*, vol. 22, no. 8: 2729–43 https://doi.org/10.1111/gcb.13316

Sharplin, J. (2020) 'European hare', in Crawley, D., Coomber, F., Kubasiewicz, L., Harrower, C., Evans, P., Waggitt, J., Smith, B. and Matthews, F. (eds), *Atlas of the Mammals of Great Britain and Northern Ireland*, Exeter, Pelagic Publishing: 28–9.

Sharrock, J.T.R. (1974) 'The changing status of breeding birds in Britain and Ireland', in Hawksworth, D.L. (ed.), *The Changing Flora and Fauna of Britain*, London, Systematics Association, Academic Press: 203–20.

Sheehy, E. and Lawton, C. (2014) 'Population crash in an invasive species following the recovery of a native predator: the case of the American grey squirrel and the European pine marten in Ireland', *Biodiversity and Conservation*, vol. 23, no. 3: 753–74. https://doi.org/10.1007/s10531-014-0632-7

Shelley, J. (2004) 'Atlantic salmon *Salmo salar*', in Davies, C.E., Shelley, J., Harding, P.T., McLean, I.F.G., Gardiner, R., Peirson, G. and Quill, V. (eds), *Freshwater Fishes in Britain*, Colchester, Harley Books: 108–10.

Shrubb, M. (2013) *Feasting, Fowling and Feathers: A History of the Exploitation of Wild Birds*, London, Poyser.

Shubin, N. (2019) 'Extinction in deep time: Lessons from the past?', in Dasgupta, P., Raven, P. and McIvor, A.L. (eds), *Biological Extinction: New Perspectives*, Cambridge, Cambridge University Press: 22–33. https://doi.org/10.1017/9781108668675.004

Shuttleworth, C. (2020) 'Red squirrel', in Crawley, D., Coomber, F., Kubasiewicz, L., Harrower, C., Evans, P., Waggitt, J., Smith, B. and Matthews, F. (eds), *Atlas of the Mammals of Great Britain and Northern Ireland*, Exeter, Pelagic Publishing: 32–3.

Sibbald, R. (1692) *Phalainologia Nova*, Edinburgh, Joannis Redi.

Sibbald, R. (1697) *Auctarium Musaei Balfouriani, e Musaeo Sibbaldiano*, Edinburgh, Academic Press.

Sibbald, R. (1699) *Provision for the Poor in Times of Dearth and Scarcity*, Edinburgh, James Watson.

Sibbald, R. (1701) 'An account of the fishes and other aquatick animals taken in the Firth of Forth', in *Caetologia, etc.*, unpublished ms., National Library of Scotland Adv Mss 33.5.16.

Sibbald, R. (1710) *Vindiciae Scotiae Illustratae*, Edinburgh, Andreas Symson.

Sibbald, R. (1803) *The History, Ancient and Modern, of the Sheriffdoms of Fife and Kinross*, Adamson, L. (ed.), London, R. Tullis.

Sibbald, R. (1892) *Sibbald's History & Description of Stirling-shire, Ancient and Modern 1707*, Stirling, R.S. Shearer and Son.

Sibbald, R. (2020) *The Wild Plants of Scotland and The Animals of Scotland*, Raye, L. (ed.), Cardiff, KDP.

Simms, J.G. (1960) 'Co Donegal in 1739', *Bliainiris Thír Chonaill/Donegal Annual*, vol. 4, no. 3: 203–8.

Sinclair, J. (1793) *The Statistical Account of Scotland, vol. 7*, Edinburgh, William Creech.

Sinclair, J. (1796) *The Statistical Account of Scotland, vol 17*, Edinburgh, William Creech.

Skinner, A., Young, M. and Hastie, L. (2003) *Ecology of the Freshwater Pearl Mussel*, Peterborough.

Sleigh, C. (2012) 'Jan Swammerdam's frogs', *Notes and Records: The Royal Society Journal of the History of Science*, vol. 66, no. 4: 373–92. https://doi.org/10.1098/rsnr.2012.0039

Smaal, A.C. (2002) 'European Mussel Cultivation along the Atlantic Coast: Production Status, Problems and Perspectives BT', in Vadstein, O. and Olsen, Y. (eds), *Sustainable Increase of Marine Harvesting: Fundamental Mechanisms and New Concepts: Proceedings of the 1st Maricult Conference held in Trondheim, Norway*, Dordrecht, Springer Netherlands: 89–98. https://doi.org/10.1007/978-94-017-3190-4_8

Smeenk, C. (1997) 'Strandings of sperm whales *Physeter macrocephalus* in the North Sea: history and patterns', *Bulletin de l'Institut Royal des Sciences Naturelles de Belgique. Entomologie & Biologie*, vol. 67: 15–28.

Smith, C. (1746) *The Antient and Present State of the County and City of Waterford*, Dublin, A. Reilly.

Smith, C. (1750a) *The Ancient and Present State of the County and City of Cork*, Dublin, A. Reilly.

Smith, C. (1750b) *The Antient and Present State of the County and City of Cork, vol. ii*, Dublin.

Smith, C. (1756) *The Antient and Present State of the County of Kerry*, Dublin, printed for the author: and sold by Messrs. Ewing, Faulkner, Wilson, and Exshaw.

Smith, D. (2021) 'Ring Ouzel *Turdus torquatus* Mwyalchen y Mynydd', in Pritchard, R., Hughes, J., Spence, I.M., Haycock, B., Brenchley, A., Green, J., Thorpe, R. and Porter, B. (eds), *The Birds of Wales*, Liverpool, Liverpool University Press: 444–5.

Smith, L.T. (1907) *The itinerary of John Leland in or about the years 1535–1543, parts I to III*, London, George Bell & Sons.

Smith, L.T. (1908) *The itinerary of John Leland in or about the years 1535–1543, parts IV and V*, London, George Bell & Sons.

Smith, N.D. (1990) 'The ecology of the slow-worm (*Anguis fragilis* L.) in southern England', University of Southampton.

Smith, W. and Webb, W. (1656) *The Vale-Royall of England or The County Palatine of Chester*, London, Daniel King.

Smout, T.C. (2009) 'The pinewoods and human use 1600–1900', in Smout, T.C. (ed.), *Exploring Environmental History*, Edinburgh, Edinburgh University Press: 71–85. https://doi.org/10.1515/9780748635146-007

Smyth, J. (1885) *A description of the Hundred of Berkeley, vol. iii*, Maclean, J. (ed.), Gloucester, John Bellows.

Snell, C. (2006) 'Status of the Common tree frog in Britain', *British Wildlife*, vol. 17, no. 3: 153–60.

Snell, C.A. (1991) 'Disappearance of Britain's tree frog (*Hyla arborea*) colonies', *Bulletin of the British Herpetological Society*, vol. 38: 40.

Snell, C.A. (2016) 'The Northern Pool Frog', *British Wildlife*, vol. 28, no. 1: 2–11.

Snell, C., Tetteh, J. and Evans, I.H. (2005) 'Phylogeography of the pool frog (*Rana lessonae* Camerano) in Europe: evidence for native status in Great Britain and for an unusual postglacial colonization route', *Biological Journal of the Linnean Society*, vol. 85, no. 1: 41–51. https://doi.org/10.1111/j.1095-8312.2005.00471.x

Solmundsson, J., Jonsson, E. and Bjornsson, H. (2010) 'Phase transition in recruitment and distribution of monkfish (*Lophius piscatorius*) in Icelandic waters', *Marine biology*, vol. 157, no. 2: 295–305. https://doi.org/10.1007/s00227-009-1317-8

Southall, T.D. and Tully, O. (2014) *Solway Cockle Fishery Study: A Review of Management Options for the Solway Firth Cockle Fishery*, Dumfries.

Southward, A.J., Boalch, G.T. and Maddock, L. (1988) 'Fluctuations in the herring and pilchard fisheries of Devon and Cornwall linked to change in climate since the 16th century', *Journal of the Marine Biological Association of the United Kingdom* 2009/05/11, Cambridge University Press, vol. 68, no. 3: 423–45. https://doi.org/10.1017/S0025315400043320

Speed, J. (1611) *The Theatre of the Empire of Great Britaine*, London, John Sudbury & George Humble.

Spence, I.M. (2021) 'Skylark *Alauda arvensis* Ehedydd', in Pritchard, R., Hughes, J., Spence, I.M., Haycock, B., Brenchley, A., Green, J., Thorpe, R. and Porter, B. (eds), *The Birds of Wales*, Liverpool, Liverpool University Press: 388–9.

Staines, B.W., Langbein, J. and Burkitt, T.D. (2008) 'Red deer *Cervus elaphus*', in Harris, S. and Yalden, D. (eds), *Mammals of the British Isles: Handbook*, 4th edn, Southampton, Mammal Society: 397–406.

Stanbury, A. and UK Crane Working Group (2011) 'Changing Status of the Common Crane in the UK', *British Birds*, vol. 104: 432–47.

Stanbury, A., Eaton, M., Aebischer, N., Balmer, D., Brown, A., Douse, A., Patrick, Lindley, McCulloch, N., Noble, D. and Win, I. (2021) 'The status of our bird populations: the fifth Birds of Conservation Concern in the United Kingdom, Channel Islands and Isle of Man and second IUCN Red List assessment of extinction risk for Great Britain', *British Birds*, vol. 114: 723–47.

Stearn, W.T. (1966) *Botanical Latin*, 4th edn, 2005, Newton-Abbot, David & Charles.

Stefani, F., Gentilli, A., Sacchi, R., Razzetti, E., Pellitteri-Rosa, D., Pupin, F. and Galli, P. (2012) 'Refugia within refugia as a key to disentangle the genetic pattern of a highly variable species: the case of *Rana temporaria* Linnaeus 1758 (Anura, Ranidae)', *Molecular Phylogenetics and Evolution*, vol. 65, no. 2: 718–26. https://doi.org/10.1016/j.ympev.2012.07.022

Stelfox, A.W. (1965) 'Notes on the Irish "Wild Cat"', *The Irish Naturalists' Journal*: 57–60.

Stephen, A.C. (1953) 'Scottish turtle records previous to 1953', *Scottish Naturalist*, vol. 65: 108–14.

Stephen, A.G. (1944) 'IX. – The Cephalopoda of Scottish and Adjacent Waters', *Transactions of the Royal Society of Edinburgh* 2012/07/06, vol. 61, no. 1: 247–70. https://doi.org/10.1017/S0080456800018123

Street, K., Stidwell, H., Sturgeon, S., Jenkin, A. and Trundle, C. (2020) *Cornwall IFCA monthly shellfish permit statistics analysis, summary statistics 2016–2019*, Hayle.

Stringer, A. (1714) *The Experienc'd Huntsman*, Belfast, James Blow.

Summers, R. (2007) 'Scottish crossbill *Loxia scotica* Hartert', in Forrester, R.W., Andrews, I.J., McInerny, C.J., Murray, R., McGowan, R.Y., Zonfrillo, B., Betts, M.W., Jardine, D.C., Grundy, D., Andrews, I.J. and Scott, H.I. (eds), *The Birds of Scotland*, Aberlady, Scottish Ornithologists' Club: 1430–2.

Sutcliffe, S. (2021) 'Herring Gull *Larus argentatus* Gwylan Penwaig', in Pritchard, R., Hughes, J., Spence, I.M., Haycock, B., Brenchley, A., Green, J., Thorpe, R. and Porter, B. (eds), *The Birds of Wales*, Liverpool, Liverpool University Press: 227–30.

Sutton, D. (2010) *Hector Boece: Scotorum Historia* (1575 edn) [Online]. Available at http://www.philological.bham.ac.uk/boece/ (Accessed 11 June 2016).

Svanberg, I. and Cios, S. (2014) 'Petrus Magni and the history of fresh-water aquaculture in the later Middle Ages', *Archives of Natural History*, vol. 41, no. 1: 124–30. https://doi.org/10.3366/anh.2014.0215

Swandro-Orkney Coastal Archaeology Trust (2020) *Musings in honour of World 'Save the Frogs' Day* [Online]. Available at https://www.swandro.co.uk/post/musings-in-honour-of-world-save-the-frogs-day (Accessed 11 July 2022).

Sykes, N. (2010) 'European fallow deer', in O'Connor, T. and Sykes, N.J. (eds), *Extinctions and Invasions: a Social History of British Fauna*, Oxford, Oxbow Books: 51–8. https://doi.org/10.2307/j.ctv13gvg6k.13

Sykes, N. and Carden, R.F. (2011) 'Were fallow deer spotted (OE* pohha/* pocca) in Anglo-Saxon England? Reviewing the evidence for *Dama dama dama* in early medieval Europe', *Medieval Archaeology*, vol. 55, no. 1: 139–62. https://doi.org/10.1179/174581711X13103897378483

Sykes, N.J. (2001) *Norman Conquest: A Zooarchaeological Perspective*, Southampton, University of Southampton.

Sykes, N.J. (2006) 'The impact of the Normans on hunting practices in England', in Serjeantson, D. and Waldron, T. (eds), *Food in Medieval England: Diet and Nutrition*, Oxford, Oxford University Press: 162–75.

Symson, A. (1823) *A Large Description of Galloway*, Maitland, T. (ed.), Edinburgh, W. and C. Tait.

Tapper, S. (1999) *A Question of Balance*, Fordingbridge, The Game Conservancy Trust.

Tapper, S. (1992) *Game Heritage*, Fordingbridge, Game Conservancy Ltd.

Taverner, C. (2019) 'Consider the oyster seller: street hawkers and gendered stereotypes in early modern London', *History Workshop Journal*, vol. 88: 1–23. https://doi.org/10.1093/hwj/dbz032

Taylor, J. (1630) *All the Workes of Iohn Taylor the Water Poet*, London, Iames Boler.

Taylor, S. and Dale, S. (1730) *The history and antiquities of Harwich and Dovercourt*, London, C. Davis.

Taylor, W.L. (1948) 'The Distribution of Wild Deer in England and Wales', *Journal of Animal Ecology*, vol. 17, no. 2: 151–4. https://doi.org/10.2307/1477

Teacher, A.G.F., Cunningham, A.A. and Garner, T.W.J. (2010) 'Assessing the long-term impact of ranavirus infection in wild common frog populations', *Animal Conservation*, vol. 13, no. 5: 514–22. https://doi.org/10.1111/j.1469-1795.2010.00373.x

Teacher, A.G.F., Garner, T.W.J. and Nichols, R.A. (2009) 'European phylogeography of the common frog (*Rana temporaria*): routes of postglacial colonization into the British Isles, and evidence for an Irish glacial refugium', *Heredity*, vol. 102, no. 5: 490–6. https://doi.org/10.1038/hdy.2008.133

Thomas, K. (1991) *Man and the Natural World: Changing Attitudes in England 1500–1800*, London, Penguin.

Thompson, D. (2020a) 'Grey seal', in Crawley, D., Coomber, F., Kubasiewicz, L., Harrower, C., Evans, P., Waggitt, J., Smith, B. and Matthews, F. (eds), *Atlas of the Mammals of Great Britain and Northern Ireland*, Exeter, Pelagic Publishing: 130–1.

Thompson, D. (2020b) 'Harbour seal', in Crawley, D., Coomber, F., Kubasiewicz, L., Harrower, C., Evans, P., Waggitt, J., Smith, B. and Matthews, F. (eds), *Atlas of the Mammals of Great Britain and Northern Ireland*, Exeter, Pelagic Publishing: 132–3.

Thompson, D.B.A. (2007a) 'European Golden Plover', in Forrester, R.W., Andrews, I.J., McInerny, C.J., Murray, R., McGowan, R.Y., Zonfrillo, B., Betts, M.W., Jardine, D.C., Grundy, D., Andrews, I.J. and Scott, H.I. (eds), *The Birds of Scotland*, Aberlady, Scottish Ornithologists' Club: 581–4.

Thompson, D.B.A. (2007b) 'Grey Plover *Pluvialis squatarola* (Linnaeus)', in Forrester, R.W., Andrews, I.J., McInerny, C.J., Murray, R., McGowan, R.Y., Zonfrillo, B., Betts, M.W., Jardine, D.C., Grundy, D., Andrews, I.J. and Scott, H.I. (eds), *The Birds of Scotland*, Aberlady, Scottish Ornithologists' Club: 585–7.

Thompson, P.M. (2008) 'Common seal *Phoca vitulina*', in Harris, S. and Yalden, D.W. (eds), *Mammals of the British Isles: Handbook*, 4th edn, Southampton, Mammal Society: 528–37.

Thorpe, L. (1978) *Gerald of Wales: The Journey through Wales and The Description of Wales*, London, Penguin.

Ticehurst, N.F. (1926) 'An historical review of the laws, orders and customs anciently used for the preservation of swans in England', *British Birds*, vol. 19, no. 8: 186–205.

Timoney, M.A. (2013) 'Hints towards a Natural and Topographical History of the County of Sligoe, by the Rev. William Henry 1739', in Timoney, M.A. (ed.), *Dedicated to Sligo: Thirty-four Essays on Sligo's Past*, Ballymote, Publishing Sligo's Past: 129–56.

Topsell, E. (1658) *The History of Four-Footed Beasts and Serpents*, London, E. Cotes. https://doi.org/10.5962/bhl.title.79388

Tregenza, N.J.C. (1992) 'Fifty years of cetacean sightings from the Cornish coast, SW England', *Biological Conservation*, vol. 59, no. 1: 65–70. https://doi.org/10.1016/0006-3207(92)90714-X

Tully, O. and Clarke, S. (2012) *The status and management of oyster (*Ostrea edulis*) in Ireland*, Oranmore, The Marine Institute, Fisheries Ecosystems Advisory Services, vol. Irish Fish.

Urry, A. (2021) 'Alfred Newton's second-hand histories of extinction: hearsay, gossip, misapprehension (William T. Stearn Student Essay Prize 2020)', *Archives of Natural History*, vol. 48, no. 2: 244–62. https://doi.org/10.3366/anh.2021.0720

Ussher, R.J. and Warren, R. (1900) *The Birds of Ireland*, London, Gurney and Jackson.

van den Hurk, Y. (2020) *On the Hunt for Medieval Whales: Zooarchaeological, historical and social perspectives on cetacean exploitation in medieval northern and western Europe*, BAR Publishing. https://doi.org/10.30861/9781407357201

van den Hurk, Y., Rielly, K. and Buckley, M. (2021) 'Cetacean exploitation in Roman and medieval London: Reconstructing whaling activities by applying zooarchaeological, historical, and biomolecular analysis', *Journal of Archaeological Science: Reports*, vol. 36: 102795. https://doi.org/10.1016/j.jasrep.2021.102795

van den Hurk, Y., Spindler, L., McGrath, K. and Speller, C. (2022) 'Medieval Whalers in the Netherlands and Flanders: zooarchaeological analysis of medieval cetacean remains', *Environmental Archaeology*, vol. 27, no. 3: 243–57. https://doi.org/10.1080/14614103.2020.1829296

van Wijngaarden-Bakker, L.H. (1974) 'The Animal Remains from the Beaker Settlement at Newgrange, Co. Meath: First Report', *Proceedings of the Royal Irish Academy. Section C: Archaeology, Celtic Studies, History, Linguistics, Literature*, vol. 74: 313–83.

Veale, E.M. (1957) 'The Rabbit in England', *The Agricultural History Review*, vol. 5, no. 2: 85–90.

Veale, E.M. (2003) *The English Fur Trade in the later Middle Ages*, Loughborough, London Record Society, vol. 38.

Vehkaoja, M. and Nummi, P. (2015) 'Beaver facilitation in the conservation of boreal anuran communities', *Herpetozoa*, vol. 28: 75–87.

Venn, J. (1912) *The Works of John Caius, MD*, Cambridge, Cambridge University Press.

Viana, D.S., Blanco-Garrido, F., Delibes, M. and Clavero, M. (2022) 'A 16th-century biodiversity and crop inventory', John Wiley & Sons, Inc. Hoboken, USA. https://doi.org/10.1002/ecy.3783

Walker-Meikle, K. (2012) *Medieval Pets*, Woodbridge, Boydell Press.

Wallace, J. (1693) *A Description of the Isles of Orkney*, Edinburgh, John Reid.

Wallace, J. (1700) *An account of the Islands of Orkney*, London, Printed for Jacob Tonson.

Wallis, J. (1769) *The natural history and antiquities of Northumberland*, London, W. and W. Strahan.

Waltho, C. (2007) 'Common Eider *Somateria mollissima mollissima* (Linnaeus)', in Forrester, R.W., Andrews, I.J., McInerny, C.J., Murray, R., McGowan, R.Y., Zonfrillo, B., Betts, M.W., Jardine, D.C., Grundy, D., Andrews, I.J. and Scott, H.I. (eds), *The Birds of Scotland*: 244–8.

Ward, A. and Smith-Jones, C. (2020) 'Roe deer', in Crawley, D., Coomber, F., Kubasiewicz, L., Harrower, C., Evans, P., Waggitt, J., Smith, B. and Matthews, F. (eds), *Atlas of the Mammals of Great Britain and Northern Ireland*, Exeter, Pelagic Publishing: 88–9.

Watson, W.J. (1937) *Scottish Verse from the Book of the Dean of Lismore*, Scottish G., Edinburgh, Scottish Gaelic Texts Society.

Weinert, M., Mathis, M., Kröncke, I., Neumann, H., Pohlmann, T. and Reiss, H. (2016) 'Modelling climate change effects on benthos: Distributional shifts in the North Sea from 2001 to 2099', *Estuarine, Coastal and Shelf Science*, vol. 175: 157–68. https://doi.org/10.1016/j.ecss.2016.03.024

Went, A.E.J. (1945) 'The Irish Pilchard Fishery', *Proceedings of the Royal Irish Academy. Section B: Biological, Geological, and Chemical Science*, vol. 51: 81–120.

Went, A.E.J. (1957) 'The pike in Ireland', *The Irish Naturalists' Journal*: 177–82.

Westcote, T. (1845) *A View of Devonshire in MDCXXX*, Exeter, William Roberts.

Wheeler, A. (1974) 'Changes in the freshwater fish fauna of Britain', in Hawksworth, D.L. (ed.), *The Changing Flora and Fauna of Britain*, London, Systematics Association, Academic Press: 157–78.

Wheeler, A.C. (1979) *The Tidal Thames: The History of a River and its Fishes*, London, Routledge & Kegan Paul.

Wheeler, A. and Jones, A.K.G. (1989) *Fishes*, Cambridge, Cambridge University Press.

Wheeler, A. and Maitland, P.S. (1973) 'The scarcer freshwater fishes of the British Isles', *Journal of Fish Biology*, vol. 5, no. 1: 49–68. https://doi.org/10.1111/j.1095-8649.1973.tb04430.x

White, G. (1977) *The Natural History of Selborne*, Mabey, R. (ed.), London, Penguin.

White, T.H. (1954) *The Book of Beasts: being a Translation from a Latin Bestiary of the Twelfth Century*, New York, Dover Publications.

Whitehead, H. (2002) 'Estimates of the current global population size and historical trajectory for sperm whales', *Marine Ecology Progress Series*, vol. 242: 295–304. https://doi.org/10.3354/meps242295

Whitehead, H. (2006) 'Sperm whales in ocean ecosystems', in Estes, J.A., Demaster, D.P., Doak, D.F., Williams, T.M. and Brownell, R.L. (eds), *Whales, Whaling, and Ocean Ecosystems*, Berkeley, University of California Press: 324–33. https://doi.org/10.1525/california/9780520248847.003.0025

Williams, G. (1974) *Welsh Poems, Sixth Century to 1600*, Berkeley, University of California Press. https://doi.org/10.1525/9780520319493

Williams, S., Perkins, S.E., Dennis, R., Byrne, J.P. and Thomas, R.J. (2020) 'An evidence-based assessment of the past distribution of Golden and White-tailed Eagles across Wales', *Conservation Science and Practice*, vol. 2, no. 8: e240. http://dx.doi.org/10.1111/csp2.240

Williamson, K. (1975) 'Birds and climatic change', *Bird Study*, vol. 22, no. 3: 143–64. https://doi.org/10.1080/00063657509476459

Williamson, T. (2013) *An Environmental History of Wildlife in England 1650–1950*, London, Bloomsbury Academic.

Willughby, F. and Ray, J. (1676) *Ornithologiae*, London, John Martyn.

Willughby, F. and Ray, J. (1686) *De historia piscium libri quatuor*, Oxford, Sheldonian Theatre.

Wilson, B. (2008) 'Bottlenose dolphin *Tursiops truncatus*', in Harris, S. and Yalden, D.W. (eds), *Mammals of the British Isles: Handbook*, 4th edn, Southampton, Mammal Society: 709–15.

Wilson, D.R.B. (1995) 'The ecology of bottlenose dolphins in the Moray Firth, Scotland: a population at the northern extreme of the species' range', University of Aberdeen.

Winfield, I.J. (2004a) 'Whitefish *Coregonus lavaretus*', in Davies, C.E., Shelley, J., Harding, P.T., McLean, I.F.G., Gardiner, R., Peirson, G. and Quill, V. (eds), *Freshwater Fishes in Britain*, Colchester, Harley Books: 104–6.

Winfield, I.J. (2004b) 'Arctic charr *Salvelinus alpinus*', in Davies, C.E., Shelley, J., Harding, P.T., McLean, I.F.G., Gardiner, R., Peirson, G. and Quill, V. (eds), *Freshwater Fishes in Britain*, Colchester, Harley Books: 113–15.

Withers, C.W.J. (2001) *Geography, Science and National Identity: Scotland since 1520*, Cambridge, Cambridge University Press.

Wood, C. (2018) *The Diver's Guide to Marine Life of Britain and Ireland*, 2nd edn, Plymouth, Wild Nature Press.

Wood, C.J. (1998) 'Movement of bottlenose dolphins around the south-west coast of Britain', *Journal of Zoology*, Cambridge University Press, vol. 246, no. 2: 155–63. https://doi.org/10.1111/j.1469-7998.1998.tb00144.x

Wood, H. (1930) 'The Public Records of Ireland before and after 1922', *Transactions of the Royal Historical Society*, vol. 13: 17–49. https://doi.org/10.2307/3678487

Woodman, P.C. (2014) 'Ireland's native mammals: a survey of the archaeological record', *The Irish Naturalists' Journal*, vol. 33: 28–43.

Worthington, T., Kemp, P., Osborne, P.E., Howes, C. and Easton, K. (2010) 'Former distribution and decline of the burbot (*Lota lota*) in the UK', *Aquatic Conservation: Marine and Freshwater Ecosystems*, vol. 20, no. 4, pp. 371–7. https://doi.org/10.1002/aqc.1113

Wotton, E. (1552) *De Differentiis Animalium*, Paris. https://doi.org/10.5962/bhl.title.39610

Wycherley, J., Doran, S. and Beebee, T.J.C. (2002) 'Frog calls echo microsatellite phylogeography in the European pool frog (*Rana lessonae*)', *Journal of Zoology*, vol. 258, no. 4: 479–84. https://doi.org/10.1017/S0952836902001632

Yalden, D. (1999) *The History of British Mammals*, Cambridge, T. & A.D. Poyser Natural History.

Yalden, D.W. (1987) 'The natural history of Domesday Cheshire', *Naturalist*, vol. 112: 125–31.

Yalden, D.W. (2007) 'The older history of the White-tailed Eagle in Britain', *British Birds*, vol. 100, no. 8: 471–80.

Yalden, D.W. and Albarella, U. (2009) *The History of British Birds*, Oxford, Oxford University Press. https://doi.org/10.1093/acprof:oso/9780199217519.001.0001

Yale, E. (2016) *Sociable Knowledge: Natural History and the Nation in Early Modern Britain*, Philadelphia, University of Pennsylvania Press. https://doi.org/10.9783/9780812292251

Yapp, W.B. (1981) *Birds in Medieval Manuscripts*, New York, Schocken Books.

Yoo, G. (2018) 'Wars and wonders: the inter-island information networks of Georg Everhard Rumphius', *The British Journal for the History of Science*, vol. 51, no. 4: 559–84. https://doi.org/10.1017/S0007087418000742

Young, A. (2007a) 'Black-billed Magpie *Pica pica* (Linnaeus)', in Forrester, R.W., Andrews, I.J., McInerny, C.J., Murray, R., McGowan, R.Y., Zonfrillo, B., Betts, M.W., Jardine, D.C., Grundy, D., Andrews, I.J. and Scott, H.I. (eds), *The Birds of Scotland*, Aberlady, Scottish Ornithologists' Club: 1341–3.

Young, A. (2007b) 'Hooded crow *Corvus cornix* (Linnaeus)', in Forrester, R.W., Andrews, I.J., McInerny, C.J., Murray, R., McGowan, R.Y., Zonfrillo, B., Betts, M.W., Jardine, D.C., Grundy, D., Andrews, I.J. and Scott, H.I. (eds), *The Birds of Scotland*, Aberlady, Scottish Ornithologists' Club: 1360–3.

Zeiler, J.T. and Kompanje, E.J. (2010) 'A killer whale (*Orcinus orca*) in the castle: first find of the species in a Dutch archaeological context', *Lutra*, vol. 53, no. 2: 101–3.

Zonfrillo, Bernie (2007) 'Northern Gannet', in Forrester, R.W., Andrews, I.J., McInerny, C.J., Murray, R., McGowan, R.Y., Zonfrillo, Bernard, Betts, M.W., Jardine, D.C., Grundy, D., Andrews, I.J. and Scott, H.I. (eds), *The Birds of Scotland*, Aberlady, Scottish Ornithologists' Club: 394–8.

Index and Glossary of Species Names

As well as functioning as an index, the following list glosses the early modern names for the animals described in this atlas. This allows readers to look up early modern terms to identify the species intended. Question marks (?) indicate where the identification of a name is uncertain, and the term 'and others' indicates where names had a general meaning that could refer to multiple species. Note that spelling was not always fixed in the early modern period so variations on these terms were common. Only the most common variants are included here.